Climate Change Management

Series Editor

Walter Leal Filho, International Climate Change Information and Research Programme, Hamburg University of Applied Sciences, Hamburg, Germany

The aim of this book series is to provide an authoritative source of information on climate change management, with an emphasis on projects, case studies and practical initiatives – all of which may help to address a problem with a global scope, but the impacts of which are mostly local. As the world actively seeks ways to cope with the effects of climate change and global warming, such as floods, droughts, rising sea levels and landscape changes, there is a vital need for reliable information and data to support the efforts pursued by local governments, NGOs and other organizations to address the problems associated with climate change. This series welcomes monographs and contributed volumes written for an academic and professional audience, as well as peer-reviewed conference proceedings. Relevant topics include but are not limited to water conservation, disaster prevention and management, and agriculture, as well as regional studies and documentation of trends. Thanks to its interdisciplinary focus, the series aims to concretely contribute to a better understanding of the state-of-the-art of climate change adaptation, and of the tools with which it can be implemented on the ground.

Notes on the quality assurance and peer review of this publication

Prior to publication, the quality of the works published in this series is double blind reviewed by external referees appointed by the editor. The referees are not aware of the author's name when performing the review; the referees' names are not disclosed.

Michael Oyedele Oyenuga · Rubee Singh ·
Stella Bolanle Apata · Shahbaz Khan · Vikas Kumar

Smart Technologies for Climate Change and Net Zero Policies: Practical Approaches Towards Sustainability

Springer

Michael Oyedele Oyenuga
School of Business
Woxsen University
Hyderabad, Telangana, India

Rubee Singh
Institute of Business Management
GLA University
Mathura, Uttar Pradesh, India

Stella Bolanle Apata
Department of Arts Education
Adekunle Ajasin University
Akungba Akoko, Nigeria

Shahbaz Khan
Department of Industrial Engineering
University of Tabuk
Tabuk, Saudi Arabia

Vikas Kumar
University of Portsmouth
Portsmouth, UK

ISSN 1610-2002 ISSN 1610-2010 (electronic)
Climate Change Management
ISBN 978-3-031-92223-7 ISBN 978-3-031-92221-3 (eBook)
https://doi.org/10.1007/978-3-031-92221-3

This Springer imprint is published by the registered company Springer Nature Switzerland AG
The registered company address is: Gewerbestrasse 11, 6330 Cham, Switzerland

Preface

The urgency to combat climate change has led to the emergence of innovative technologies that support net-zero policies and sustainability initiatives. The global push for sustainability and climate action has never been more urgent. As nations, businesses, and individuals strive to mitigate the effects of climate change, emerging technologies are proving to be powerful enablers in the pursuit of net-zero policies. The convergence of digitalization, artificial intelligence, blockchain, and renewable energy innovations presents unprecedented opportunities to accelerate sustainability efforts and drive transformative change. This authored book, *Smart Technologies for Climate Change and Net Zero Policies: Practical Approaches Towards Sustainability*, was born out of the necessity to explore and understand the intricate relationship between technology, policy, and sustainability.

It seeks to provide a comprehensive analysis of how smart technologies can be effectively leveraged to achieve climate neutrality and net-zero emissions while fostering economic growth and social equity. One of the critical challenges in the transition to a sustainable future is the integration of new technologies into existing systems. Policymakers and industry leaders must navigate regulatory barriers, financial constraints, and technological complexities to implement solutions that deliver tangible environmental benefits.

This book addresses these challenges by presenting an interdisciplinary perspective on sustainable innovation, climate change and the effects of net zero in various sectors by leveraging smart technologies that draw insights from researchers from fields such as environmental science, engineering, economics, and policy studies. The book is structured to offer a logical progression from foundational concepts to real-world applications. The initial chapters provide an overview of net-zero policies emphasizing the need for urgent action in light of the ongoing climate crisis. It then explores groundbreaking technologies, such as AI-driven predictive analytics for energy management, blockchain for carbon credit verification, and IoT-based smart infrastructure, smart agriculture, and smart cities. These discussions are supported by case studies that illustrate successful implementations of sustainability-driven innovations across different sectors.

This book's significant focus is on stakeholders' roles in driving sustainability efforts. Governments, businesses, academia, and civil society all play a crucial role in shaping the technological landscape for a greener future. The book underscores the importance of collaborative approaches, such as public-private partnerships and international cooperation, in overcoming barriers to sustainability adoption. Furthermore, this book highlights the ethical considerations and social implications of various emerging technologies. While digital solutions offer remarkable potential, they also raise concerns related to data privacy, cybersecurity, and the digital divide. Addressing these challenges is vital to ensuring that technological advancements benefit all communities equitably without exacerbating existing inequalities.

As we look towards the future, the need for innovation-driven sustainability solutions will continue to grow. This book aims to serve as a valuable resource for researchers, students, policymakers, and industry professionals seeking to understand and harness the power of smart technologies for climate action. By fostering knowledge-sharing and critical dialogue, we hope to inspire bold strategies and informed decision-making that will pave the way for a net-zero, sustainable world. The journey towards sustainability is complex, but with the right technologies, policies, and collaborative efforts, it is possible to create a resilient and prosperous future for generations to come. It is our hope that this book will contribute to this ongoing journey, equipping readers with the knowledge and inspiration needed to drive meaningful change in the fight against climate change.

The book consists of five parts with 28 chapters, where Part I lays the groundwork for the success of net-zero policies by discussing the essential strategies and perspectives that need to be considered for achieving them. Part II focuses on various technological solutions for net-zero emissions. Part III focuses on providing deeper insight into various practical solutions. These policy and governance strategies are necessary for a net-zero transition in accordance with the UN Agenda 21. Part IV provides a broad overview of a variety of factors, obstacles, opportunities, and challenges that are expected to impact the future. Part V presents a number of different approaches to sustainable development and the adoption of technology that can help in achieving net-zero emissions for sustainable climate change.

Hyderabad, India — Michael Oyedele Oyenuga
Mathura, India — Rubee Singh
Akungba Akoko, Nigeria — Stella Bolanle Apata
Tabuk, Saudi Arabia — Shahbaz Khan
Portsmouth, UK — Vikas Kumar

Acknowledgements We would like to express our gratitude to all those who have contributed to the creation of this book. Firstly, we thank the researchers and experts who have shared their knowledge and insights with us, providing valuable information on the current state of Smart Technologies for Climate Change and Net Zero Policies: Practical Approaches towards Sustainability. We also extend our appreciation to the organizations and institutions that have supported our work, including government agencies, non-governmental organizations, and academic institutions.

Finally, we would like to acknowledge the invaluable contributions of our colleagues and editors, who have worked tirelessly to ensure that this book is informative, engaging, and accessible to a wide range of readers. We hope that this book will inspire further research and action towards smart technology, climate-neutral policies and net-zero emissions to promote sustainability as per the United Nations' agenda. We would like to express our sincere gratitude to the experts and stakeholders who contributed their valuable insights and perspectives to this book. Their contributions have been instrumental in shaping our understanding of the challenges and opportunities involved in achieving net-zero emissions, climate change, and addressing global warming with the use of practical approaches towards sustainability at a global level. We hope that this book will serve as a useful resource for policymakers, practitioners, and researchers working towards a zero-carbon future for the continent.

We are immensely thankful to the Book series editors, Prof. Walter Leal Filho, Editorial Director/Publishing Editor, Dr. Christian Witschel, and the entire team at Springer Nature for bringing out this publication in a very professional and timely manner.

Michael Oyedele Oyenuga
Rubee Singh
Stella Bolanle Apata
Shahbaz Khan
Vikas Kumar

Competing Interests The authors have no competing interests to declare that are relevant to the content of this manuscript.

About This Book

Smart technologies are playing a transformative role in shaping net-zero policies and sustainability efforts worldwide. This book explores the intersection of climate neutrality, carbon emissions, innovation, policy, and sustainability, presenting cutting-edge practical solutions that support global climate goals. This book delves into how artificial intelligence, blockchain, smart grids, carbon capture, and renewable energy innovations are revolutionizing industries and driving decarbonization.

A key highlight of this book is its focus on technological solutions. This unpacks how technology can drive sustainability at scale, from renewable energy breakthroughs to digital innovations such as IoT-based climate monitoring and blockchain-enabled emissions tracking. Case studies of successful implementations across different continents illustrate the transformative power of innovation in energy, urban planning, and industrial processes. Focusing on practical applications, the book covers various sectors, including manufacturing, transportation, agriculture, and urban planning. It illustrates how digital twin technology, smart grids, and precision farming can drive sustainability while fostering economic growth. Furthermore, it emphasizes the importance of policy frameworks and regulatory mechanisms that support technological adoption for environmental resilience.

Recognizing that technology alone cannot solve the climate crisis, the book also explores the role of governance and policy in enabling sustainable transitions. It critically evaluates international agreements like the Paris Accord, examines successful national policy frameworks, and investigates financial mechanisms such as carbon pricing and green investment strategies. With detailed insights into public-private partnerships and legal frameworks, this section empowers policymakers, business leaders, and environmental advocates with the knowledge to enact meaningful change. Beyond policies and technology, the book highlights barriers and opportunities in net-zero implementation. Addressing challenges such as political resistance, infrastructure limitations, and financing gaps, it presents strategic pathways for overcoming obstacles and seizing emerging opportunities. Additionally, it envisions a sustainable future where digital sustainability, energy efficiency, and low-carbon economies become the norm rather than the exception.

Through a multidisciplinary approach, this book provides insights into the integration of digital technologies in sustainable finance, sustainable urban planning, and circular economy models. It highlights the role of government policies, corporate initiatives, and public-private partnerships in accelerating the transition to a sustainable future. Targeted at researchers, policymakers, industry leaders, and students, this book serves as a comprehensive guide to understanding the opportunities and challenges of deploying smart technologies for net-zero and sustainability objectives. It provides case studies, policy recommendations, and strategic insights, making it an essential resource for anyone looking to contribute to sustainable development and climate action.

Contents

Part I Understanding Net-Zero Policies

1 **The State of Net Zero Policies: The Urgency of Net Zero Sustainability and Building a Sustainable Future** 3
Introduction 3
The Urgency of Net Zero for Developing Economies 6
Net Zero Policies: Key Challenges Unique to Developing Nations 9
Global Commitments and Expectations for the Future 10
Implications 11
Conclusion, Limitations and Future Scope 12
References 12

2 **The Impact of Climate Change on Developing Countries** 15
Introduction 15
Climate Vulnerabilities in Developing Nations 16
Adaptive Capacities: Regional Variations 18
Case Studies from Africa: Climate Impacts and Responses 19
Case Studies from Asia: Coping with Climate Disasters 20
Case Studies from Latin America: Adapting to Climate Change 22
Implications of the Study 24
Conclusion, Limitations and Future Scope 25
References 25

3 **Economic Growth Versus Digital Sustainability: Striking the Balance** 29
Introduction 29
Economic Growth and Environmental Degradation 30
The Tension Between Development and Sustainability 31
Sustainable Development Goals (SDGs) and Net Zero 32

Balancing Short-Term Growth with Long-Term Climate Goals 35
Balancing Economic Growth with Climate Goals: Overcoming Challenges 37
Implications 38
Conclusion, Limitations and Future Scope 38
References 39

4 Socio-economic Impacts of Net Zero Policies 41
Introduction 41
Poverty and Inequality Exacerbated by Net Zero Policies 42
Displacement and Migration Due to Environmental Factors 44
Impacts on Health and Livelihoods 45
Climate-Induced Food and Water Insecurity 45
Strategies to Reduce the Socioeconomic Burden 46
Strategies to Reduce the Socioeconomic Burden 47
Implications 48
Conclusion, Limitations and Future Scope 49
References 49

5 Climate Resilience and Adaptation Strategies for Developing Nations 53
Introduction 53
Building Climate Resilience in Agriculture 54
Climate-Proofing Infrastructure and Transportation Systems 56
Strengthening Health Systems for Climate Resilience 57
Community-Based Adaptation (CBA) Strategies 59
National-Level Adaptation Frameworks and Planning 60
Implications 61
Conclusion, Limitations, and Future Scope 61
References 62

6 International Aid and Climate Finance for Developing Economies 65
Introduction 65
Global Climate Finance Mechanisms: Overview 66
Green Climate Fund and Its Role in Developing Countries 67
Bilateral and Multilateral Aid for Climate Action 68
The Role of Private Sector Finance in Climate Initiatives 69
Challenges and Opportunities in Accessing Climate Finance 70
Challenges in Accessing Climate Finance 70
Opportunities in Accessing Climate Finance 71
Implications 72
Conclusion, Limitations, and Future Scope 72
References 73

Part II Technological Solutions for Net-Zero Emissions

7 Technological Innovations in Renewable Energy 79
Introduction 79
Solar Energy Innovations for Rural Electrification 80
Wind Energy Projects in Low-Income Countries 82
Hydropower: Small-Scale Solutions for Local Communities 84
Off-Grid Renewable Energy Technologies: A Path to Energy Access 85
Case Studies of Successful Renewable Energy Projects: Transforming Energy Landscapes Globally 87
Implications 89
Conclusion, Limitations and Future Scope 89
References 90

8 Digital Technologies and Climate Action 93
Introduction 93
The Role of AI in Climate Prediction and Adaptation 94
The Role of the Internet of Things (IoT) in Real-Time Environmental Monitoring 95
Data Analytics for Optimizing Energy Use 97
Blockchain Technologies for Tracking Emissions 98
Case Studies of Digital Technology in Climate Initiatives 99
Implications 100
Conclusion, Limitations, and Future Scope 100
References 101

9 Green Infrastructure and Smart Urbanization 105
Introduction 105
Green Infrastructure and Urban Planning: Principles and Applications 107
Smart Cities: Integrating Sustainability into Urbanization 108
Case Studies of Smart City Initiatives in Developing Countries 111
Barriers to Sustainable Urbanization in Developing Economies 112
Implications 113
Conclusion 114
References 114

10 Circular Economy and Waste Reduction Technologies 117
Introduction 117
The Principles of a Circular Economy 118
Innovations in Waste Management: Recycling and Reuse 120
Cleaner Production: Advancing Sustainable Industrial Practices 121

Manufacturing as a Catalyst of the Circular Economy ... 122
Case Study Projects on Circular Economy and Waste Reduction ... 124
Implications ... 127
Conclusion ... 127
References ... 128

11 Energy Efficiency Technologies for Low-Income Countries ... 131
Introduction ... 131
Energy-Saving Technologies for Households ... 133
Industrial Energy Efficiency Solutions ... 135
Smart Grids and Energy Distribution Efficiency ... 136
Case Studies on Implementing Energy Efficiency in Low-Income Settings ... 137
Rwanda: Agro-Processing Industries ... 137
India: Perform, Achieve, and Trade (PAT) Scheme ... 137
China: Eco-Industrial Parks ... 138
Vietnam: Process Optimization in Manufacturing ... 138
Kenya: Pay-As-You-Go Solar Solutions ... 138
South Africa: Industrial Waste-To-Energy Initiatives ... 138
Bangladesh: Textile Sector Efficiency ... 139
Brazil: Sugarcane Ethanol Plants ... 139
Ethiopia: Cement Industry Optimization ... 139
Philippines: Energy Efficiency in SMEs ... 139
Germany: Advanced Heat Recovery in Steel Plants ... 140
Indonesia: Palm Oil Industry Efficiency ... 140
Turkey: Glass Manufacturing Energy Efficiency ... 140
The Role of Government Incentives in Promoting Energy Efficiency ... 140
Implications ... 142
Conclusion ... 143
References ... 143

Part III Policy and Governance for Net-Zero Transitions

12 Global Net-Zero Commitments: International Agreements ... 147
Introduction ... 147
The Paris Agreement: An Overview ... 147
Key Elements of the Paris Agreement ... 148
Mechanisms for Implementation ... 148
Challenges Facing the Paris Agreement ... 148
Net Zero Targets in International Climate Accords ... 149
The Framework of the Paris Agreement ... 149
The Importance of Net-Zero Targets ... 150
Challenges in Achieving Net-Zero Targets ... 150

Strategies for Effective Implementation 150
The Role of the UN and Other Multilateral Organisations 151
Mechanisms for Achieving Net-Zero Targets 151
Challenges Faced by Multilateral Organizations 152
Monitoring and Accountability Mechanisms 153
Key Mechanisms for Monitoring and Accountability 153
Challenges in Monitoring and Accountability 154
Implications for Global Climate Policy 154
Implications of Global Commitments for Developing Nations 155
Economic Implications 155
Social Implications 156
Environmental Implications 156
Conclusion 157
References 157

13 National Policy Approaches to Net Zero in Developing Countries 161
Introduction 161
Nationally Determined Contributions (NDCs) Under the Paris Agreement 161
Analysis of Successful Policy Frameworks 162
Policy Challenges in Developing Economies 163
The Role of Political Leadership in Climate Policy 164
Implications 167
Conclusion 168
References 169

14 Carbon Pricing and Financial Incentives 173
Introduction 173
Advantages and Disadvantages of Carbon Taxes 174
Advantages of Carbon Taxes 174
Disadvantages of Carbon Taxes 174
Comparative Analysis 176
Financial Incentives for Renewable Energy Adoption 176
Effectiveness of Financial Incentives 177
Challenges and Considerations 177
The Role of Carbon Credits in Developing Economies 178
Key Benefits of Carbon Credits 178
Challenges in Implementing Carbon Credit Systems 178
Case Studies on the Success and Failure of Carbon Pricing 179
Barriers to Implementing Carbon Pricing in Low-Income Regions 180
Implications and Conclusion 181
References 182

15 Public–Private Partnerships for Net Zero Emissions and Climate Action 185
Introduction 185
The Importance of Private Sector Investment 186
Examples of Successful Public–Private Climate Partnerships 187
The Canada Line Project in British Columbia 187
The Green Climate Fund (GCF) 187
The Global Infrastructure Facility (GIF) 188
The Renewable Energy and Energy Efficiency Partnership (REEEP) 188
The Water Infrastructure Public–Private Partnerships 188
The Climate Smart Agriculture Initiative 188
The Urban Climate Change Resilience Initiative (UCCRI) 189
Encouraging Corporate Social Responsibility for Net Zero 189
The Role of Corporate Social Responsibility in Achieving Net Zero 189
Case Studies from Developing Nations 190
Implications 193
Conclusion 194
References 194

16 Strengthening Legal and Institutional Frameworks for Net Zero Economy 199
Introduction 199
Building Legal Frameworks for Climate Action 200
Key Elements of Effective Legal Frameworks 200
The Role of Judiciary Systems in Climate Governance 201
Strengthening Institutions for Climate Monitoring and Enforcement 202
Legal Challenges in Holding Industries Accountable 203
International Legal Frameworks and Their Applicability to Developing Nations 204
Barriers to Effective Implementation 205
Implications 206
Conclusion 207
References 207

Part IV Barriers and Opportunities in Net-Zero Implementation

17 Barriers to Technology Adoption in Developing Countries 213
Introduction 213
Infrastructure Challenges: Energy Grids and Internet Access 214
Energy Grids 214
Internet Access 215
High Upfront Costs of Green Technologies 216
Financial Barriers to Green Technology Adoption 216
Renewable Energy Technologies 216

Addressing the Financial Barrier ... 217
Technical Skills Gap and Capacity Limitations ... 217
Capacity Limitations in Organizations ... 218
Political and Bureaucratic Roadblocks ... 218
Nature of Political Roadblocks ... 218
Bureaucratic Challenges ... 219
Overcoming Political and Bureaucratic Roadblocks ... 219
Overcoming Barriers Through Innovation and Policy ... 220
The Role of Innovation ... 220
Policy Interventions ... 220
Addressing the Skills Gap ... 221
Implications ... 221
Conclusion ... 222
References ... 222

18 Political and Social Challenges to Net Zero Policies Action ... 227
Introduction ... 227
Lack of Political Will in Certain Regions ... 228
Competing Political Interests ... 228
Governance and Institutional Challenges ... 228
Socio-Economic Factors ... 229
Community Acceptance of Green Initiatives ... 229
Awareness and Education ... 229
Participation and Engagement ... 230
Socio-Economic Considerations ... 230
Governance and Policy Frameworks ... 230
Social Resistance to Changing Consumption Habits ... 231
Cultural Norms and Social Identity ... 231
Economic Considerations ... 232
Communication and Awareness ... 232
Governance Issues: Corruption and Inefficiency ... 232
Inefficiency in Governance ... 233
Opportunities for Overcoming Political and Social Challenges ... 233
Implications ... 235
Conclusion ... 236
References ... 237

19 Financing the Transition to Net Zero ... 241
Introduction ... 241
International Climate Finance: Opportunities and Limitations ... 242
International Climate Finance Opportunities ... 242
Limitations of International Climate Finance ... 243

Green Bonds and Sustainable Investment Strategies 243
The Rise of Green Bonds 244
Sustainable Investment Strategies 244
The Role of Private Capital in Climate Financing 245
Opportunities for Private Capital in Climate Financing 245
Challenges in Mobilizing Private Capital 246
Public Funding: National Budgets and Climate Priorities 246
Importance of Public Funding for Climate Action 246
Integrating Climate Priorities into National Budgets 247
Barriers to Accessing and Utilizing Climate Finance 248
Implications 248
Conclusion 250
References 250

20 Energy Transition: Fossil Fuels Versus Renewables 253
Introduction 253
The Case for Renewables 254
The Economics of Fossil Fuel Dependence 254
Economic Benefits of Fossil Fuel Dependence 255
Economic Drawbacks of Fossil Fuel Dependence 255
Environmental Effects of Fossil Fuel Dependence 256
The Energy Security Debate in Transitioning to Renewables 256
The Need for Transition 256
Reliability of Renewable Energy Sources 257
Phasing Out Fossil Fuels: Political and Economic Challenges 257
Political Challenges 257
Economic Challenges 258
Balancing the Transition 259
The Role of International Organizations in Supporting the Transition 259
Case Studies from Countries Transitioning from Fossil Fuels 260
Bulgaria 261
China 261
Indonesia 261
The Western Balkans 262
The European Union 262
Implications 262
Conclusion 263
References 264

21 Capacity Building for Green Technology 267
Introduction 267
Developing Technical Skills in Renewable Energy and Digital Technologies 268
Vocational Training and Education for Climate Jobs 269
The Role of Universities and Research Institutions 270
International Knowledge Exchange Programs 271
Building Local Capacity for Sustainable Development 272
Implications 273
Conclusion, Limitations and Future Scope 274
References 274

Part V Sectoral Approaches and Future Outlook

22 Sustainable Agriculture and Land Use for Emission Reduction 279
Introduction 279
Sustainable Farming Practices 280
Agroforestry and Carbon Sequestration in Agricultural Lands 281
Climate-Smart Agriculture Technologies 282
Reducing Methane and Nitrogen Emissions in Agriculture 283
Case Studies of Successful Land Use Practices 284
Implications 286
Conclusion, Limitations and Future Scope 286
References 287

23 Low-Carbon Transport and Mobility Solutions 291
Introduction 291
Electrification of Transport Systems 292
Public Transportation: Reducing Emissions Through Mass Transit 293
Non-Motorized Transport: Bicycles and Pedestrian Systems 294
Innovations in Green Aviation and Shipping 295
Case Studies of Low-Carbon Mobility in Developing Regions 296
Implications 298
Conclusion, Limitations and Future Scope 299
References 299

24 Decarbonizing Industrial Processes 303
Introduction 303
Reducing Emissions in Manufacturing 304
Innovations in Clean Energy for Heavy Industries 305
Circular Economy in Industrial Processes 306
Case Studies of Decarbonizing Industries in Developing Countries 307
Challenges to Decarbonizing the Industrial Sector 308
Implications 310
Conclusion, Limitations and Future Scope 310
References 311

25 Green Technologies for Water Management 315
Introduction 315
Sustainable Water Usage in Agriculture 316
Water Conservation Technologies 317
Reducing Emissions in the Water Supply Chain 318
Case Studies of Green Water Management Technology 320
The Role of Policy in Ensuring Sustainable Water Practices 321
Implications 322
Conclusion, Limitations and Future Scope 323
References 323

26 The Future of Net Zero and Digital Sustainability in Developing Regions 327
Introduction 327
Scaling up Green Technologies 328
Opportunities for International Collaboration 329
Emerging Technologies for Net Zero 330
The Road Ahead for Sustainable Development 332
Implications 333
Conclusion, Limitations and Future Scope 334
References 334

27 Net Zero Technology and Smart Agriculture 339
Introduction 339
Understanding Net Zero Technology in Agriculture 340
Smart Agriculture: Innovations and Applications 341
Sustainable Practices for Net Zero Agriculture 342
Digital and Technological Interventions for Net Zero Agriculture 343
Challenges and Future Prospects 344
Implications 345
Conclusions, Limitations and Future Scope 346
References 346

28 Smart City and Urban Planning for Climate Change Innovation 351
Introduction 351
The Nexus Between Smart Cities and Climate Change 352
Technological Innovations for Smart and Sustainable Cities 353
Sustainable Urban Planning Strategies 354
Climate Adaptation and Mitigation Policies 355
Case Studies of Smart Cities Driving Climate Innovation 356
San Diego, California 356
Vienna, Austria 357
Stockholm and Malmö, Sweden 357
Yokohama, Japan 357
Dubai, UAE 358

Implications 358
Conclusion, Limitations, and Future Scope 359
References 359

About the Authors

Dr. Michael Oyedele Oyenuga (Ph.D.) is an Associate Professor and Assistant Dean of International Relations for the School of Business at Woxsen University, Hyderabad, India, he has established himself as a thought leader in consumer behavior, digital sustainability, and climate-conscious economic policies. His academic journey, spanning multiple institutions across Nigeria, the USA, India, and Malaysia has equipped him with a multidisciplinary perspective that uniquely positions him to drive conversations on sustainability, technology, and net-zero policies. A distinguished researcher with a robust publication record in high-impact, Scopus-indexed journals, Dr. Oyenuga has contributed significantly to discussions on green marketing, circular economy models, and corporate social responsibility in emerging economies. His research delves into the intersection of economic development and environmental sustainability, exploring pathways for integrating low-carbon technologies into business strategies. As a reviewer for leading journals, including *Heliyon*, *Journal of Consumer Behaviour*, *Discover Sustainability* and *Asia-Pacific Journal of Regional Science*, among others, Dr. Oyenuga has demonstrated an exceptional ability to critically assess and shape high-quality academic discourse. His contributions to academic administration are equally remarkable. At Veritas University, Abuja, Nigeria, he served in various leadership roles, including Head of the Department of Marketing, Coordinator of Faculty Research and Publications, and Senate Committee Member.

Dr. Oyenuga is a Carbon Literacy Certificate recipient (Nottingham Business School, UK, and PRME), he completed the Climate Governance, Diplomacy and Negotiations Leadership Program organized by the African Group of Negotiators Experts Support (AGNES) and honored with awards, including the Manaaki Scholarship (New Zealand Government).

Dr. Rubee Singh (Ph.D., D.Litt.) is an Assistant Professor of HR at Institute of Business Management, GLA University India. She is a Postdoctoral (D.Litt.) Researcher at Kumaun University, Nainital, India. She holds a Ph.D. in Management from Noida International University, India, an MBA in HR from Dr. A. P. J. Abdul

Kalam Technical University, Lucknow, India. She has more than 11 years of experience in research and teaching. Apart from teaching and research, she has handled many administrative positions in premier institutes affiliated with A.K.T.U. She has conducted several FDPs, Conferences, Workshops, and HR Conclaves in the area of Sustainability, CSR, AI, HR, Human Factors, Women Empowerment and Human rights. The George Washington University of Peace, USA, awarded her as Chartered Global Peace Building Professional (CGPP) for her outstanding hard work in social work. She is a United Nations Volunteer, and a contributing author for Voices of Youth (UNICEF).

She is an honorary life member of Women Economic Forum and National Secretary of IATM (Youth Wing)-International Anti-Terrorism Movement India. Dr. Singh is a Managing Editor of IJARSH India. She is serving on the Editorial Board of more than 30 international journals. She serves as the Advisory Board Member of the *Heliyon Journal*, Elsevier, Research in Globalization (Elsevier), and reviewer of many reputed journals like *Technological Society*, *Social Responsibility Journal*, *European Journal of Innovation Management* (EJIM), *British Food Journal*, and *Sustainability*. She is an associate Editor for Circular Economy and Sustainability, Springer. She is a life member of International Association of Engineers, Thailand, IMRTC, USA, ISRD, UK, etc. She is a bestselling author with significant experience as an excellent Editor, Eminent researcher, great academician, Global Ambassador (Children's), and Social Activist. She has presented many papers at national and international conferences, published 25+ academic books including Springer, Elsevier, De Gruyter, Emerald Publishing, and more than 100+ research papers in refereed/UGC index including SCI, ABDC-A, Scopus. She has an excellent research track record with 1100+ citations on Google Scholar and H-index-21. Her areas of research are Sustainability, Digital HRM, Sustainable HRM, Circular Economy, Digital CE, Climate Change, Net Zero, Industry 4.0, Artificial Intelligence, Blockchain Technology, Sustainable Innovation, Eco-Innovation/Digital Innovation, and Corporate Social Responsibility. Dr. Singh has been honored with many prestigious awards, which include consecutive Best Researchers Award of the Years 2022–2023 and 2023–2024 by GLA University, India, Child Ambassador Award of the Year 2022, Iconic Faculty Award-2019, International Young Researchers Award-2021, and Young Scientist Award-2020. The list of awards and appreciation is long, but Dr. Singh has always prioritized her work over recognition, making her unique and commendable in her field.

Dr. Stella Bolanle Apata (Ph.D.) is a Distinguished Academic, Researcher, and Language Education Expert dedicated to fostering knowledge at the intersection of literacy, technology, and sustainable development. As a Lecturer at Adekunle Ajasin University, Nigeria, she has made groundbreaking contributions to digital literacy, artificial intelligence in education, and eco-linguistics—areas vital to shaping global sustainability narratives. With a Ph.D. in Language Education, her research extends beyond traditional boundaries, exploring innovative solutions to enhance climate-conscious education and policy discourse. She is an ardent advocate for interdisciplinary research, integrating emerging technologies into pedagogy to drive transformative learning and environmental consciousness. Her scholarly publications in high-impact journals reflect a deep commitment to equipping future generations with the skills and knowledge necessary for a Net Zero economy. She has presented at global conferences, influencing policy dialogues on digital sustainability, climate literacy, and green technologies. A recipient of multiple academic honors and a mentor to aspiring researchers, Dr. Apata's vision is anchored in leveraging language and technology as tools for societal change.

Dr. Shahbaz Khan (Ph.D.) is an Assistant Professor at the Department of Industrial Engineering, University of Tabuk, Kingdom of Saudi Arabia. He has published more than 100+ research articles in many reputed journals including Scopus, SCI, and ABDC-A categories such as Business Strategy and Environment, Annals of Operation Research, and Operations Management Research. He has an excellent research track record with 5500 citations on Google Scholar and an H-index of 38. Dr. Khan was among the Top 2% of Highly Cited Researchers in 2024 acknowledged by Stanford University and Elsevier scientists globally. He has obtained his Ph.D. and M.Tech. degrees with specialization in Industrial and Production Engineering from Jamia Millia Islamia, India. He is a gold medalist in M.Tech.

He is a member of the International Association of Engineers, Thailand. His research interests are Supply Chain Management, Digital Supply chain management, Circular Economy and Sustainability, Industry 4.0, Blockchain, Artificial Intelligence, benchmarking, and decision modeling for sustainable business. He has published five books, including Springer, Elsevier, and De Gruyter. He is serving as the Advisory Board Member of the *Heliyon Journal*, Elsevier and Associate Editor in *Green Technologies and Sustainability*. He is the reviewer of many reputed journals such as the *Journal of Business Research*, *Trends in Food Science and Technology*, *International Journal of Fuzzy Systems*, *Engineering Applications of Artificial Intelligence*, *Soft Computing*, and *Sustainability*.

Prof. Vikas Kumar (Ph.D.) is a Deputy Vice-Chancellor Research and Innovation at University of Portsmouth, UK. Previously, he was an Associate Dean for Research Innovation and Enterprise at the Faculty of Business, Law and Social Sciences, Birmingham City University, UK. He is also a Professor of Operations and Supply Chain Management. Professor Kumar has over a decade of teaching and research experience. Professor Kumar earned his Ph.D. in Management from the University of Exeter, UK. He did his graduation (B.Tech.) from the National Institute of Advanced Manufacturing Technology Ranchi (Formerly NIFFT), India. He has published over 350 journal articles and conference papers. He has an excellent research track record with 20,000+ citations on Google Scholar and an H-index of 77. Professor Kumar was among the Top 1% of Highly Cited Researchers in 2024. For the third consecutive year in 2024, Stanford University and Elsevier acknowledged him as one of the Top 2% of scientists globally.

He serves on the editorial board of several international journals and has guest-edited special issues in high-impact journals. He is also a reviewer of more than 15 international journals, including ABS 4/4* and ABS 3* journals. Professor Kumar works very closely with industries and has generated research funding of over £1 million from various research agencies such as Innovate UK, EPSRC, British Council, British Academy, Newton Fund, and Science Foundation of Ireland. He has worked on several international collaborative projects with researchers from UK universities and Brazil, Vietnam, Thailand, Indonesia, India, Mexico, Taiwan, Columbia, Peru, Costa Rica, and Turkey. His research interests are Net-Zero Emissions, Circular Economy, Sustainability, Industry 4.0, Digital Supply Chains, Sustainable Supply Chains, and Operational Excellence.

Part I
Understanding Net-Zero Policies

Chapter 1
The State of Net Zero Policies: The Urgency of Net Zero Sustainability and Building a Sustainable Future

Abstract This chapter explores net zero policies and aspirations, emphasizing the evaluation of best practices for adequate growth. Net zero refers to the state where greenhouse gas emissions, after removal, equal those produced, preventing the onset of global climate change. Achieving net zero aligns with the Paris Agreement goal of limiting global temperature rise to 1.5 °C above pre-industrial levels. The discussion highlights the challenges and opportunities for affluent and developing nations. Developed nations often lead in implementing net zero policies, while underdeveloped countries face unique vulnerabilities to climate change impacts. Despite their limited current emissions, emerging nations exhibit significant potential for future greenhouse gas emissions due to urbanization and development, complicating their net zero pathways. The chapter stresses integrating developing nations into global initiatives, as these countries require substantial financial support and access to advanced technology. Net zero efforts intersect with sustainability concepts but face horizontal political challenges, as achieving net zero is more complex for developing nations than for affluent ones. The chapter evaluates global initiatives and the implications of net zero policies for short- and long-term sustainable growth, and underscores the necessity for equitable collaborations to ensure developing nations receive the resources to contribute effectively to climate goals.

Introduction

Net zero is an essential component of the global ecosystem. The objective of climate change is to achieve "Net Zero," a goal that all of humanity has united to pursue. The equation for a net zero-sum game of emissions is greenhouse gas emissions and removal. If humanity can achieve equilibrium, the formalization of things should not lead to global warming and the deterioration of the environment. The 2015 Paris Agreement is a significant factor in the humanitarian intent to prevent the increase in global temperatures beyond 1.5 °C from pre-industrial levels. However, this objective necessitates socioeconomic equity, technological advancements, political policy changes, and unprecedented global collaboration. Developed nations have

M. 0. Oyenuga et al., *Smart Technologies for Climate Change and Net Zero Policies: Practical Approaches Towards Sustainability*, Climate Change Management,
https://doi.org/10.1007/978-3-031-92221-3_1

been industrialised for centuries; they have emitted greenhouse gases at a rate that exceeds that of any other nation in many centuries. Additionally, they possess the financial and technological resources to swiftly surpass developing and third-world nations in pursuing Net Zero and have developed many ambitious decarbonization strategies, including implementing renewable energy sources, enhanced industrial efficiency, and even large-scale regulatory strategies. These nations are expected to help the developing nations to achieve their net zero by 2050.

Overview of net zero goals

Net Zero refers to the condition where the quantity of greenhouse gases (GHGs) emitted into the atmosphere is equal to the quantity of GHGs removed from it. To attain net zero carbon emissions in Africa, a balance must be established across the complex sectors of transportation, heating, aviation, marine, and industrial activity (Nwokolo et al. 2023a, b, c, d, e, f, g, h, i). Subsequently, initiatives like afforestation, reforestation, and artificial interventions like carbon capture and storage must be implemented to achieve a net zero condition. Only then can an overabundance of greenhouse gases not infiltrate a reality that exacerbates global warming. According to the Intergovernmental Panel on Climate Change (IPCC), to avert the most severe consequences of climate change, humanity must achieve Net Zero CO_2 emissions and limit global temperature rise to 1.5 °C above preindustrial levels. This necessitates a comprehensive transformation across all sectors—energy, transportation, manufacturing, and land use along with the innovation and deployment of carbon capture technologies.

The global climate agenda and the role of developing nations

Historically, trends indicate that developed nations have minimal involvement in the escalating impacts of climate change; however, they stand to benefit significantly from international climate initiatives and obligations while also facing substantial risks, if they are not encouraged to engage in the climate agenda. By the late twentieth century, initiatives to globalise climate change policy had already taken place, including establishing the Intergovernmental Panel on Climate Change (IPCC) in 1988 and the United Nations Framework Convention on Climate Change (UNFCCC) in 1992. However, it was not until the adoption of the Kyoto Protocol in 1997 that the first global initiative recognizing climate change as a collective challenge emerged, with industrialized nations formally committing to reducing greenhouse gas emissions. However, unlike developed countries, developing nations such as China and India were not legally obligated to emission reduction targets due to the UNFCCC's principle of "common but differentiated responsibilities" (UNFCCC 1997), which evaluated the historical emissions causality between developing and developed nations.

The latest international climate pact is the 2015 Paris Agreement, which aims to restrict temperature increases to below 2 °C over preindustrial levels, intending to limit the rise to 1.5 °C. The Kyoto Protocol established specific per-country quotas with stricter requirements for developed nations. In contrast, the Paris Agreement is a quasi-inclusive framework that encourages both developed and developing nations to submit their intentions through Nationally Determined Contributions (NDCs) based on situational assessments (UNFCCC 2015).

The agreement recognizes the commitment of developing nations to mitigate and adapt while also highlighting the obstacles and necessity for international finance and support to achieve these objectives.

The Global South is experiencing the impacts of climate change, including droughts and increasing sea levels, which necessitate adaptation measures. The Sahel area of Africa is experiencing heightened droughts and famines, although the coasts of South Asia, encompassing India, Bangladesh, Indonesia, and several island countries, are progressively inundated (Nwokolo et al. 2023a, b, c, d, e, f, g, h, i). Nwokolo et al. (2023a, b, c, d, e, f, g, h, i) highlight that numerous African nations persist in relying on fossil fuels for their development, industrialisation, and energy needs. As a result, upcoming efforts, focused on urbanisation or meeting energy demands could intensify obstacles to sustainable development. Developing nations face a significant challenge; while they are making strides, they still lack sufficient infrastructure and access to energy technology (Oyenuga and Omale 2024).

The Green Climate Fund (GCF), created internationally under the UNFCCC, provides developing nations with financial resources to implement mitigation or adaptation programs (GCF 2022). This is highly beneficial, as without financial possibilities, developing nations may never attain renewable energy and sustainable energy efficiency or acquire the resilience needed to adapt to climate change, as they consistently run at a deficit. Assistance is seldom provided promptly; transfers may require months or years, and many are accompanied by limitations that limit their applicability to specific projects. Many developing nations are anticipated to complete their respective international climate obligations, nevertheless, this is more readily articulated than accomplished. One of the most pressing challenges is fund. Numerous developing nations are so heavily indebted that they cannot dedicate resources to climate initiatives. Nwokolo et al. (2023a, b, c, d, e, f, g, h, i) assert that Africa requires billions of dollars to revamp its infrastructure and transition to sustainable energy. Nonetheless, it possesses the least capacity to gain finance for climate initiatives since stated goals are consistently subordinated to the imperatives of poverty alleviation, healthcare, and education.

Kumar et al. (2024) discovered that several developing nations rely on antiquated, high-emission energy systems due to the absence of affordable alternatives; hence, without the provision of low-cost solutions by developed nations, no alternatives will be utilised.

Moreover, susceptibility to climate response evolves throughout time due to institutions and governance. The susceptibility to climate response escalates as improved governance and transparency enable these nations to address their climate requirements in the future better (Nwokolo et al. 2023a, b, c, d, e, f, g, h, i). Sub-Saharan Africa possesses abundant solar resources, enabling solar energy equity to furnish millions with clean electricity and reduce reliance on coal and oil power plants (Nwokolo et al. 2023a, b, c, d, e, f, g, h, i). Nevertheless, achieving energy access necessitates both local and foreign expenditures for infrastructure development and technology progress.

However, any positive outcomes from a global climate program will be rendered futile without international collaboration. Article 9 of the Paris Agreement mandates the provision of sufficient fund, technological transfer, and capacity building to assist developing nations (Nwokolo et al. 2023a, b, c, d, e, f, g, h, i). Furthermore, bilateral agreements should reinforce these initiatives, while public–private-NGO partnerships should facilitate resource access to mitigate barriers to climate change action in all countries.

The Urgency of Net Zero for Developing Economies

Due to their socioeconomic condition, countries that are vulnerable to climate change are countries with agriculture dependence, and limited adaptation and with emerging economies. Many developing regions, such as sub-Saharan Africa and Southeast Asia, are geographically vulnerable to extreme weather, rising seas, and storms (Nwokolo et al. 2023a, b, c, d, e, f, g, h, i). For instance, droughts and floods increase in the Sahel and desert parts of Africa and island nations in Asia due to sea level rise and flooding, threat the nation's food supply, water, and quality of life. Food and water shortages worsen poverty and inequality, preventing vulnerable communities from recovering (Nwokolo et al. 2023a, b, c, d, e, f, g, h, i). Thus, Net Zero is more important for these expanding regions since they cannot afford to lose people or face a climate disaster. Affected developing nations suffer severe socioeconomic effects from climate change. The World Bank (2022) estimates that 100 million individuals would have an increased risk of poverty, with the several challenges being experienced by them due to climate change in struggling developing nations.

Therefore, the importance of Net Zero to developing nations is not just due to their vulnerability to climate change, but also because of their potential for low-carbon growth. Many developing nations have abundant renewable energy resources, such as solar, wind, and hydropower, yet a significant portion of this potential remains untapped (Oyetunde et al. 2023). Utilizing these resources can reduce greenhouse gas emissions, create job opportunities, and promote low-carbon sustainable development (IRENA 2021). Sub-Saharan Africa holds significant potential for solar energy generation on an international scale. If effectively harnessed, this potential could reduce reliance on fossil fuels while improving electricity access and industrialisation prospects (Nwokolo et al. 2023a, b, c, d, e, f, g, h, i).

Furthermore, investing in renewable energy aligns with the Sustainable Development Goals mandate, SDG 7, which aims to ensure access to affordable, reliable, sustainable, and modern energy for all. Increased access to renewable energy promotes sustainable socioeconomic development through equitable rural electrification, improved access to education and healthcare, and enhanced efficiency for small businesses. IRENA (2021) supports this, stating that if investment in renewables in developing countries doubles by 2050, there could be 42 million job opportunities in the global market, with the majority in developing nations. A low-carbon economy by

2050 has the potential to transform the development trajectory of emerging nations. The key is to make it happen.

Net Zero is not just a commendable objective; but both urgent and opportunistic; unfortunately, developing nations have significant obstacles that hinder their ability to achieve it. Initially, there were no funds to assists countries being bedeviled by it. This is because many developing nations are already indebted. Most lack the budgetary capacity to enhance their economies, much less to establish clean energy infrastructure and implement climate change mitigation strategies (Barrett 2022). Numerous initiatives remain unaccomplished in enabling developing nations to obtain financing for climate initiatives. The Green Climate Fund (2022) states that wealthy countries are obligated to establish a $100 billion yearly initiative since 2020 to support climate initiatives in underdeveloped nations; nevertheless, this commitment remains unfulfilled. Consequently, vulnerable developing nations at risk lack enough climatic infrastructure to effectively address climate change. A further impediment is the accessibility of decarbonization technology.

Carbon capture and sequestration technologies, industrial energy efficiency, and climate-resilient agricultural practices exist; however, they are expensive and not readily available for implementation in developing nations. Kumar et al. (2024) assert that, regarding intellectual assets and its expensive maintenance, several developing nations continue to depend on high-emission practices and are unable to alter their carbon legacies without the requisite technology. Also, Kumar et al. (2024) opine that South Asia depends on inexpensive, readily available nonrenewable resources such as coal, which dictate their energy requirements; hence, their achieving a Net Zero transition is complex. Moreover, institutional and administrative failures intensify the Net Zero challenges that emerging nations face.

Moreover, political instability, corruption, and poor bureaucracy hinder adequate climate change initiatives and reduce the efficacy of international assistance. Simultaneously, local politicians may lack the regulatory competence necessary to initiate action—securing suitable fund but lacking the requisite rules or experience to foster sustainable climate change initiatives can hinder efforts to combat climate change (Nwokolo et al. 2023a, b, c, d, e, f, g, h, i). As a result, a comprehensive regulatory framework with enduring institutional ability, would facilitate emerging nations to swiftly pursue a more sustainable growth model and achieve Net Zero status (Oyedele 2025).

It is worthy of note that the Global Climate Fund (GCF) and World Bank are the only two of several entities that offer help or, at the very least, access to clean technology for poor countries (Green Climate Fund 2022). Nwokolo et al. (2023a, b, c, d, e, f, g, h, i) report that the potential for global linkages and knowledge, is essential for addressing the technological and financial challenges, that developing nations encounter in meeting their climate obligations.

Financing is a critical barrier for developing nations facing climate change. Nations must fund climate change initiatives such as renewable energy projects, sustainable agriculture, disaster preparedness, and climate-resilient infrastructure but many developing countries, burdened by significant debt and constrained cash reserves, cannot allocate additional revenue to these projects. (Nwokolo et al. 2023a, b, c, d, e, f, g, h, i).

The Green Climate Fund (2022) claims that affluent nations attempt to provide $100 billion yearly to disadvantaged populations worldwide are often unfulfilled or lacking supplementary support, indicating that developing countries require equitable financing initiatives. In vulnerable climates such as sub-Saharan Africa and South Asia, where climate-induced natural disasters occur, anticipated financing for sustainable goals, whether for infrastructural modifications or disaster relief efforts, is limited (World Bank 2022). According to Nwokolo et al. (2023a, b, c, d, e, f, g, h, i), although Africa has emerged as one of the most quickly developing regions globally in recent years, characterized by enhanced economic autonomy and improved quality of life due to foreign donor resources, which have facilitated advancements in healthcare, education, and building engineering, yet, Africa requires additional international resources and financing to develop sustainable energy systems and infrastructure that align with international net zero objectives, energy-dependent growth strategies, regulatory frameworks, and the growing needs of its population.

Clean energy technologies encompass solar panel and wind turbine generation, as well as energy-efficient solutions for manufacturing systems that are both cost-effective and nearly obligatory (IRENA 2021). Moreover, the implementation of such technologies can inhibit the realization of latent human resource potential; for instance, carbon capture systems, climate-responsive agricultural systems, and early warning detection systems necessitate systematic modifications informed by community feedback to function effectively and attain optimal efficiency (Kumar et al. 2024). Nwokolo et al. (2023a, b, c, d, e, f, g, h, i) reiterate that Africa's net-zero ambitions stem from regional capabilities, political constraints, and the potential to utilize resources; consequently, solutions lie in renewable energy sources solar and wind as well as energy-efficient systems for carbon capture and hydrogen energy. The execution of these ideas necessitates governmental commitment, stakeholder support, and sustained funding.

Furthermore, clean technology is frequently inaccessible due to rights restrictions and costly licensing, hindering these nations from implementing clean innovations that may reduce emissions and substantially enhance their already diminished capacities (Kumar et al. 2024).

Numerous nations exhibit political instability and corruption in governance, hindering the efficient execution of climate policy (Nwokolo et al. 2023a, b, c, d, e, f, g, h, i). When fund designated for climate adaptation is misappropriated for political advantage or redirected owing to poor governance, the integration leads to difficulties in achieving the UN Sustainable Development Goals (SDGs). Numerous mitigation and adaptation initiatives should rely on both micro-community and macro-community activities; hence, openness in resource allocation is essential (World Bank 2022). Nwokolo et al. (2023a, b, c, d, e, f, g, h, i) reiterate that Africa's sustainable energy transition depends on policy, technology, and financing solutions, necessitating judicious governance and regulation of external capital for renewable energy to maintain a delicate equilibrium among economic growth, environmental sustainability, and social development. Agriculture is a significant socioeconomic activity in several regions of Africa and Asia, yet it is a primary source of greenhouse gas emissions and one of the industries most impacted by climate change. Sustainability requires stable climatic and weather conditions (Nwokolo et al. 2023a, b, c, d, e, f, g, h, i).

Net Zero Policies: Key Challenges Unique to Developing Nations

Social inequality ranks as the second most significant global issue for emerging nations, behind climate change. The most impoverished—women, children, and the destitute—are disproportionately affected by climatic disasters and possess diminished prospects for rehabilitation or prevention. Coastlines in South Asia, ravaged by cyclones and inhabited by low-income families, and drought-stricken villages in South Africa, where transient populations reside, suffer adversely due to the absence of social safety nets, inadequate healthcare, deteriorating transportation, and insufficient telecommunications (World Bank 2022). This imbalance exacerbates risk factors for vulnerable emerging nations striving to adapt and establish resilient communities. climatic change vulnerabilities are exacerbated by gender, with women—particularly in rural developing regions—being more susceptible to climatic disasters due to their reliance on and the resources allocated to water, food, and caring. Nwokolo et al. (2023a, b, c, d, e, f, g, h, i) assert that women in sub-Saharan Africa exist beyond the confines of gendered expectations. Consequently, they are intensified by the climate catastrophe and limited access to financial resources, land opportunities, or collaborative initiatives that enhance their participation in mitigation or adaption strategies.

Developing nations, especially in Africa, lack the financial wherewithal to forgo low carbon initiatives or to integrate renewable energy into their manufacturing processes. Africa's primary obstacles to achieving net zero emissions are inadequate legislative frameworks and budgetary constraints (Chusi et al. 2024). Moreover, emerging nations require fossil fuel energy to facilitate significant urbanization. These nations will ensure net-positive growth; developed countries provide financial resources for the execution and ongoing support of sustainable initiatives. These nations perceive the initiatives of the developed world as imperialistic endeavours that hinder the industrialisation efforts of the developing world (Shakya et al. 2023). Third, there exists relative net-zero regulatory governance that precludes conversion and sustained net-zero functionality. In contrast to building, driven by political initiatives and socio-environmental progress, poor governance arises from political instability and corrupt ethical practices. Low-income countries, in comparison, lack access to renewable technologies that attract developed nations. For instance, low-income countries are unable to achieve net-zero building projects due to the absence of the same construction materials and advancements seen in affluent ones (Li and

Gou 2024a, b). Fifth, a significant portion of the low-income global population is susceptible to energy instability.

Samanamud and Boone (2024) evaluate the significance of governance with technical and budgetary sustainability. In this regard, climate change complicates environmental and land use issues for developing economies. For example, the Andean temperatures fluctuate so swiftly that inhabitants experience internal climate and socioeconomic transformations; rather than adjusting to climate change, it would be more advantageous to integrate intersectional realities. Gupta (2024) notes that the climates in emerging economies necessitate institutional and socio-cultural legacies to convert equity-based constraints into entrepreneurial traits; access to funding and policy modifications are essential pathways for enhanced economic sustainability. Achieving net-zero in the developing world is, therefore, contingent upon the global community, supporting legislative reforms, investing in renewable energy, ensuring socioeconomic and human rights justice, and advancing technologies that will need long-term financial commitment. Unless these substantial imbalances are addressed, the transition to a viable and sustainable global economic system will remain fragmented, catastrophic and a mirage.

Global Commitments and Expectations for the Future

Regardless of the medium, whether maritime or terrestrial, there exists a worldwide imperative for a commitment to a Net Zero norm in the context of climate-related activities and promises. The Paris Agreement aims to restrict global temperature rise to 2 °C above pre-industrial levels, with an optimal target of 1.5 °C. Many nations across various concerned sectors seek a long-term decarbonisation strategy, with many committing to achieve Net Zero by 2050. The marine sector aims to achieve this through enhanced efficiency, biofuels, and innovative ammonia-based green fuels (Kramel et al. 2024). The construction sector aims to achieve a projected 96% decrease in demand-side emissions for residential heating and cooling by 2050 through demand-side and technical modifications (Mastrucci et al. 2024). However, despite these objectives, there remains a practical and attainable aspect to contest. Some scholars challenge the conclusions and objectives of some studies, asserting that achieving Net Zero is huge and almost unattainable; Therefore, they propose that climate stabilisation may be attained with an only 40% reduction by the 2070s, rather than the 2050 target of total reductions (Spencer 2024).

A debate is underway, as some remain ambivalent about the efficacy of actual carbon sequestration and the meaningfulness of global mitigation efforts to date. Furthermore, countries that will not achieve carbon neutrality particularly many developing nations still possess the potential to prevent global temperature increase. BECCS (bioenergy with carbon capture and storage) is a method of climate change mitigation that encompasses the creation of bioenergy followed by carbon capture. According to Li and Gou (2024a, b), BECCS is classified as a negative emission technology since it concurrently generates energy and reduces atmospheric carbon levels.

The technology's yield derives from cultivating biomass that sequesters atmospheric carbon via photosynthesis, followed by the combustion or utilisation of the bioenergy to generate power, during which excess carbon capture is pursued and either sequestered underground or repurposed for alternative applications, as noted by Li and Gou (2024a, b). Nonetheless, BECCS can significantly contribute to the reduction of surplus emissions in nations that have not yet embraced carbon neutrality. Countries that continue to generate billions of greenhouse gases annually can use BECCS in the short term to prevent negative emissions, while also utilising this system to advance long-term climate change mitigation objectives. Countries in need can implement BECCS in the short term if they own suitable carbon capture locations and significant biomass concentrations. Nonetheless, the expenses related to installation, alterations in land use, and assured dependable availability to sustainable biomass pose obstacles to widespread adoption.

Therefore, Net Zero initiatives depend on technology, financing, and political actions that promote fairness within a sustainable development framework. Some experts contend that even if the world achieves Net Zero, a temperature change will occur at some time. This escalation signifies the necessity for emissions monitoring and the stabilisation of climate change until at least 2050. Consequently, these initiatives must be advocated by governments and their investors to enable firms to realistically and fairly achieve a low-carbon future.

Implications

Net Zero objectives are interconnected with several Sustainable Development Goals (SDGs). They are directly associated with SDG 13 (Climate Action) and indirectly with SDG 7 (Affordable and Clean Energy) through investments in renewable energy, as well as with SDG 9 (Industry, Innovation, and Infrastructure) by emphasizing sustainable activities within industries. Moreover, low-carbon projects may be associated with SDG 2 (Zero Hunger) through climate-sensitive agriculture practices and SDG 11 (Sustainable Cities and Communities) via climate-sensitive urban development strategies. Consequently, advancing towards Net Zero facilitates the achievement of climate change-related Sustainable Development Goals (SDGs) and broader socioeconomic development initiatives.

Moreover, subsequent studies might assess the feasibility of various carbon capture methods, the socioeconomic implications of transitioning to Net Zero economies, and additional microregional prospects for emissions reductions. Exploring financial alternatives such as green bonds and climate funds may address the funding deficit in developing nations. Evaluating effective adaption strategies about local requirements with fair governance frameworks may identify the most effective solutions.

Conclusion, Limitations and Future Scope

The necessity for a Net Zero policy arises from the imminent threat of a worsening crisis due to existing activities that adversely affect greenhouse gas emission levels; Net Zero policies mitigate emissions and establish removal efforts equitably. A Net Zero strategy is associated with the objectives of the Paris Agreement, which seeks to limit temperature rises to no more than 1.5 degrees Celsius. To attain Net Zero by 2050, a radical and comprehensive policy strategy for the Energy, Transportation, Agricultural, and Industrial sectors is essential, which requires renewable energy resources, sustainable policy initiatives, and international cooperation. Developing nations contribute insignificantly to global climate change; however, these nations face imminent crises financial instability, reliance on technology, and political and socio-economic governance therefore, only altruistic initiatives will ascertain whether conditional efforts yield beneficial or detrimental outcomes in the effective and efficient execution of a Net Zero policy approach. Sustainable effectiveness arises from recognized facilitative elements, equal resource availability, and opportunities for engagement for all individuals. Ultimately, Net Zero aims to benefit the world and its people through transparent and facilitative initiatives to expedite global contact.

References

Chusi TN, Bouraima MB, Su Q, Badi I, Oloketuyi EA, Qiu Y (2024) Evaluating the barriers to the transition to net-zero emissions in developing countries: a multi-criteria decision-making approach [Community Development]. https://doi.org/10.59543/comdem.v1i.10067

Green Climate Fund (GCF) (2022) Annual report on climate action financing. https://www.greenclimate.fund

Gupta R (2024) Exploring the impact of socio-cultural factors on entrepreneurship development in emerging markets. Emerg Mark J 15(3):98–120. https://www.researchsquare.com/article/rs-3938479/v1

International Renewable Energy Agency (IRENA) (2021) Renewable energy benefits: leveraging local industries. IRENA, Abu Dhabi. https://www.irena.org/publications/2021/Jun/Renewable-energy-benefits-Leveraging-local-industries

Joshi A, Singh R, Kumar V, Rani S (2024) Net zero strategy: exploring opportunities, confronting challenges for developing countries. In: Transition towards a sustainable future: net zero policies and environmental sustainability, pp 233–252. Springer Nature Singapore, Singapore. https://doi.org/10.1007/978-981-97-5756-5_11

Kramel D, Krey V, Fricko O, Maczek F, Muri H, Strømman AH (2024) Maritime sector transition pathways towards net-zero within global energy scenarios. Res Square. https://doi.org/10.21203/rs.3.rs-4471628/v1

Kumar K, Singh R, Sharma A, Lehri BPS (2024) Climate change adaptation strategies for achieving net-zero economy. In Transition towards a sustainable future: net zero policies and environmental sustainability, pp 163–183. Springer Nature Singapore, Singapore. https://doi.org/10.1007/978-981-97-5756-5_8

Li J, Gou Z (2024a) Addressing the development gap in net-zero energy buildings: a comparative study of China, India, and the United States. Energy Sustain Dev 76:101418. https://doi.org/10.1016/j.esd.2024.101418

Li Y, Gou X (2024b) Addressing the development gap in net-zero energy buildings. J Environ Eng 28(4):198–215. https://www.sciencedirect.com/science/article/abs/pii/S0973082624000449

Mastrucci A, Boza-Kiss B, van Ruijven B (2024) Towards net-zero emissions in global residential heating and cooling: a global scenario analysis. Environ Res Lett 19(1). https://doi.org/10.1088/1748-9326/aca0c1

Nwokolo SC, Singh R, Khan S, Kumar A (2024) Threats to the rapidity of sustainability transitions posed by technological changes. In: Technological pathways for Africa' s net-zero economy. Elsevier. https://doi.org/10.1016/B978-0-443-31486-5.00002-1

Nwokolo SC, Singh R, Khan S, Kumar A, Luthra S (2023a) Africa's awakening to climate action. In: Africa's path to net-zero: exploring scenarios for a sustainable energy transition, pp 299–310. Springer Nature Switzerland, Cham. https://doi.org/10.1007/978-3-031-28293-5_15

Nwokolo SC, Singh R, Khan S, Kumar A, Luthra S (2023b) Technological pathways to net-zero goals in Africa. In: Africa's path to net-zero: exploring scenarios for a sustainable energy transition, pp 93–210. Springer Nature Switzerland, Cham. https://doi.org/10.1007/978-3-031-28293-5_5

Nwokolo SC, Singh R, Khan S, Kumar A, Luthra S (2023c) Decarbonizing hard-to-abate sectors in Africa. In: Africa's path to net-zero: exploring scenarios for a sustainable energy transition, pp 211–236. Springer Nature Switzerland, Cham. https://doi.org/10.1007/978-3-031-28293-5_6

Nwokolo SC, Singh R, Khan S, Kumar A, Luthra S (2023d) Global investment and development in Africa. In: Africa's path to net-zero: exploring scenarios for a sustainable energy transition, pp 15–58. Springer Nature Switzerland, Cham. https://doi.org/10.1007/978-3-031-28293-5_2

Nwokolo SC, Singh R, Khan S, Kumar A, Luthra S (2023e) Africa's awakening to climate action. In: Africa's path to net-zero: exploring scenarios for a sustainable energy transition, pp 299–310. Springer Nature Switzerland, Cham. https://doi.org/10.1007/978-3-031-44514-9_15

Nwokolo SC, Singh R, Khan S, Kumar A, Luthra S (2023f) Technological pathways to net-zero goals in Africa. In: Africa's path to net-zero: exploring scenarios for a sustainable energy transition, pp 93–210. Springer Nature Switzerland, Cham. https://doi.org/10.1007/978-3-031-44514-9_5

Nwokolo SC, Singh R, Khan S, Kumar A, Luthra S (2023g) Decarbonizing hard-to-abate sectors in Africa. In: Africa's path to net-zero: exploring scenarios for a sustainable energy transition, pp 211–236. Springer Nature Switzerland, Cham. https://doi.org/10.1007/978-3-031-44514-9_6

Nwokolo SC, Singh R, Khan S, Kumar A, Luthra S (2023h) Global investment and development in Africa. In: Africa's path to net-zero: exploring scenarios for a sustainable energy transition, pp 15–58. Springer Nature Switzerland, Cham. https://doi.org/10.1007/978-3-031-44514-9_2

Nwokolo SC, Singh R, Khan S, Kumar A, Luthra S (2023i) Introduction: Africa's net zero transition. In: Africa's path to net-zero. CSR, sustainability, ethics & governance. Springer, Cham. https://doi.org/10.1007/978-3-031-44514-9_1

Nwokolo SC, Singh R, Khan S, Kumar A, Luthra S (2023) Impacts of Climate Change in Africa. In: Africa's Path to Net-Zero. CSR, Sustainability, Ethics & Governance. Springer, Cham. https://doi.org/10.1007/978-3-031-44514-9_7

Oyedele OM (2025). The cobra effect: exploring the dark side of innovation and environmental sustainability in developing nations. In: Singh R, Joshi A, Filho WL, Khan S (eds) Zero carbon industry, eco-innovation and environmental sustainability. World sustainability series. Springer, Cham. https://doi.org/10.1007/978-3-031-80220-1_15

Oyenuga MO, Omale SA (2024) Is Africa Jinxed? Exploring the challenges of technology access and adoption in Africa. Afr J Econ Sustain Dev 7(4):142–161. https://doi.org/10.52589/AJESDULN1LRNF

Oyetunde TO, Oyenuga MO, Adoga GJ (2023) Green management practices and organisational efficiency in higher institution. Budapest Int Res Crit Linguist Educ (BirLE) J 6(3):276–290. https://bircu-journal.com/index.php/birle/article/view/7716

Samanamud MZ, Boone RB (2024) Potential ecological and land management impacts of climate change on Andean ecosystems and agropastoral communities. Environ Sci Rev 21(1):34–57. https://essopenarchive.org/doi/full/10.22541/essoar.171690754.41564822

Shakya SR, Nakarmi AM, Prajapati A, Pradhan BB, Rajbhandari US, Rupakheti M, Lawrence M (2023) Environmental, energy security, and energy equity (3E) benefits of net-zero emission strategy in a developing country: a case study of Nepal. Energy Rep 9:1–17. https://doi.org/10.1016/j.egyr.2023.01.055

Spencer RF (2024) Is Net Zero necessary? Meeting the Paris Agreement temperature target with 39% global emissions reductions by the 2070s. Clim Policy 24(2):123–135. https://doi.org/10.1080/14693062.2023.2165678

United Nations Framework Convention on Climate Change (UNFCCC) (1997) Kyoto protocol to the United Nations framework convention on climate change. https://unfccc.int

United Nations Framework Convention on Climate Change (UNFCCC) (2015) Paris Agreement. https://unfccc.int/process-and-meetings/the-paris-agreement/the-paris-agreement

World Bank (2022). Poverty and climate impacts: vulnerability and adaptation. World Bank, Washington, DC. https://documents.worldbank.org/en/publication/documents-reports/documentdetail/534871468155709473/poverty-and-climate-change-reducing-the-vulnerability-of-the-poor-through-adaptation

Chapter 2
The Impact of Climate Change on Developing Countries

Abstract Developing nations exhibit low greenhouse gas emissions; however, they encounter climate change impacts that exceed their forecasts. Although the chapter indicates that individuals with less exposure may experience relative ease of vulnerability, developing nations face heightened risks from extreme weather patterns, hurricanes, rising ocean levels, and declines in agricultural production which often results in food insecurity, inadequate water supply, social unrest, and diminished quality of life. Indigenous worldviews inform the development of novel adaptation strategies while developing nations assess their endurance through the analysis of geographical and socioeconomic factors. Developing nations' drought- and desertification-prone zones share similarities in migration, reforestation, and the establishment of climate-resilient regions. Case studies from Asia demonstrate resilience in the face of typhoons, heat waves, and floods, which highlight the ability of individuals to adapt to congested urban environments while Latin America is addressing deforestation and water shortages resulting from climate-induced migration. This chapter highlights the necessity for developing nations to adjust to climate change and global efforts to establish resilient, sustainable communities.

Introduction

Climate change may not represent the most critical moral dilemma of the twenty-first century; however, it is certainly one of the most significant issues. The nations most impacted by the global effects of climate change are those least prepared to tackle the problem, specifically the developing world (Oyenuga and Omale 2024). Developing nations demonstrate negligible increases in greenhouse gas emissions while facing significant impacts from rising sea levels, extreme storms, and challenges to agricultural sustainability. Ironically, nations with minimal contributions are now the primary victims. The situation presents an ethical dilemma concerning the concepts of time and space. Changes in precipitation patterns and increased temperatures can alter growing seasons and threaten food availability. Inadequate health systems and socio-political dynamics render these nations vulnerable, lacking the financial

M. O. Oyenuga et al., *Smart Technologies for Climate Change and Net Zero Policies: Practical Approaches Towards Sustainability*, Climate Change Management,
https://doi.org/10.1007/978-3-031-92221-3_2

resources, political will, or health infrastructure necessary to address the injustices intensified by climate change. The adverse effects of this exposure are disheartening. Individuals face relocation, experience starvation, and lose employment due to this significant disruption of the natural order. Increasing sea levels threaten coastal communities, with the potential to inundate small island nations. Agricultural failures resulting from droughts impede crop cultivation, whereas unforeseen floods cause significant damage to harvests, disproportionately impacting women and children.

Improvements derived from ancient infrastructure established over millennia underscore the importance of Indigenous knowledge systems for effective, sustainable, and culturally relevant solutions though these regional initiatives are insufficient. Climate adaptation requires substantial financial and technological resources, which are often unavailable to many impoverished nations. The global community acknowledges the need for adaptation, and efforts have been made to assist through initiatives like the Paris Agreement and the Green Climate Fund. However, a significant gap remains between the commitments made and the support delivered. Proposals have reached up to hundreds of billions, including a suggested $100 billion annually from wealthy nations; however, numerous climate-vulnerable states have yet to receive the funding necessary for effective adaptation initiatives.

Climate Vulnerabilities in Developing Nations

Given the increasing prevalence of climate change worldwide, it is clear that emerging nations are the most susceptible. Singh et al. (2025) assert that sustainability should be prioritized in the contemporary corporate landscape since it originates from the circular economy to promote more sustainable business practices within the ESG framework. However, Joshi et al. (2024) caution that the Paris Agreement facilitates nations in achieving net-zero emissions through various initiatives to keep global temperatures at or below 2° C; sustainable infrastructures and collaborative efforts support low-carbon economies. These nations, which statistically contribute the least to greenhouse gas emissions, face the most severe climatic challenges, including climate disasters, food and water access and availability limitations, health concerns, and financial difficulties. Hence, circular economies are the solution for greenhouse emissions. Singh et al. (2025) emphasize the growing awareness among industries, governments, and consumers regarding the detrimental environmental impacts of data consumption. Consequently, Singh et al. (2023) observe an increasing preference for a net-zero economy as companies recognize the undeniable reality of climate change. Detrimental climate practices adversely impact businesses both in the short term through costs and extreme weather events and in the long term by disrupting valued activities within the value chain. Consequently, by structural equation modelling, including 241 respondents from Indian manufacturing organizations, the authors elucidate the elements of climate change adaptation, namely company awareness and the extent to which corporate behaviours and financial resources influence sustainability initiatives or their absence.

Coastal developed nations, the Small Island Developing States (SIDS), face increased vulnerability from increasing sea levels. The IPCC indicates that low-lying areas are subjected to sea level rise, resulting in coastal erosion and salt-water intrusion, which causes population displacement (IPCC 2021). Natural calamities, including droughts and floods, have recently escalated across Africa. Natural catastrophes that impair resource production and food security arise from a sector very susceptible to the fluctuations of the global population since several nations depend on agriculture for jobs and revenue generation. Nonetheless, as highlighted by Nwokolo et al. (2023a, b, c, d, e, f), diminished precipitation dependability has adversely affected agricultural production and food security, placing millions in danger throughout the continent.

Many sectors in emerging nations depend on agriculture, natural resources, and tourism (Oyenuga and Akwen 2025). Climatic change impacts agriculture and natural resources by reducing agricultural output and depleting resources; in tourism, resources (such as reefs and forests) and climatic factors (including storms and flooding) have significant influences. Moreover, poorer nations are less inclined to possess varied industrial alternatives; hence, when adversely affected by climate change, they possess lower resources for recovery, rendering them increasingly susceptible to persistent recession. Moreover, many nations possess inadequate levels and access to resources that hinder efficient adaptation. Nwokolo et al. (2023a, b, c, d, e, f) assert that funding for improved infrastructure and sustainable energy alternatives exacerbate climate vulnerabilities in most African states. Therefore, without increased financial remuneration from global entities or developed nations, they cannot acquire the resilience necessary to address and maintain recovery from climate-induced calamities. As a result, resource deficiencies indicate that underdeveloped countries are increasingly susceptible to climate change.

Furthermore, the marginalized global population endures the most severe consequences of climate change; the vulnerable, including women, children, and nations with the lowest socioeconomic status, have the most significant effects. For example, during a natural catastrophe, poor socioeconomic groups are least likely to own enough roofing resources to be accessed by disaster recovery initiatives or healthcare operations; their rehabilitation is further prolonged, exacerbating an already complex scenario. Nonetheless, a further critical domain of risk is to health. The World Health Organisation (WHO) estimates that climate change will result in an extra 250,000 fatalities worldwide annually from 2030 to 2050, attributable to hunger, malaria, diarrhoea, and heat stress (WHO 2020). These health challenges predominantly arise in poor nations with constrained and already burdened budgets and institutions.

Adaptive Capacities: Regional Variations

Many poor countries lack the requisite technical expertise and institutional aptitude to implement suitable modifications (Oyedele 2025). Consequently, the vulnerability levels experienced by industrialized nations may not uniformly apply to all socioeconomic strata; thus, modified findings are included (Nwokolo et al. 2023a, b, c, d, e, f). The failure of the Paris Agreement to deliver the expected $100 billion annually to poor nations by 2020 substantiates that claim. The concept of foreign aid to developing nations for establishing a low-carbon society is also reiterated by Nwokolo et al. (2023a, b, c, d, e, f). Initiatives are also executed by international institutions, such as the Green Climate Fund and the World Bank, to provide financial and logistical assistance to alleviate the impacts of climate vulnerability. Furthermore, it will substantiate the necessity of a worldwide policy for climate adaptation. Socioeconomic status is the primary determinant of a region's adaptability. The wealthier regions, Europe and North America are more adept at adaptation due to their availability of financial resources, technical infrastructures, and sociocultural support systems, facilitating the integration of anticipated demands into existing practices or innovations.

The European Commission's evaluation of Environmental Sustainability about environmental change dictates Europe's capacity for climate change adaptation, underscoring the necessity for sustainability in its agriculture, water, and energy sectors (European Commission 2021). A significant portion of the developing world possesses climate-sensitive businesses, such as agriculture, rendering them susceptible to vulnerabilities. These areas, which have little financial resources, depend on these industries for everyday operations and lack the potential for long-term modifications. Africa's financial deficiencies diminish the likelihood of sustainability (Nwokolo et al. 2023a, b, c, d, e, f).

Challenges persist in societies where socioeconomic disparities impede adaptable capacity; rural inhabitants often lack the same access and inclusion as urban counterparts, exacerbating the split between urban adaptive capability and rural acceptance (Roy et al. 2020). Coastal locations and arid regions are particularly vulnerable to significant weather events and have greater adaptation challenges. The Small Island Developing States (SIDS) in the Caribbean and the Pacific Islands are severely challenged by rising sea levels and storm surges. These places have limited territory and few resources, making them more vulnerable and difficult to adapt. On the other hand, locations with seemingly better geographical conditions, such as milder climates or more water, have an inherent advantage in terms of climate resilience. It can be argued that the Sahel is not better at handling floods and heat waves, given its hot desert climate with limited freshwater, than Northern Europe, which enjoys temperate climate conditions with plentiful freshwater. According to Nwokolo et al. (2023a, b, c, d, e, f), geographical vulnerabilities persist, including extended droughts that severely diminish the agricultural potential of the Sahel, where farming contends with insufficient precipitation and irrigation. Based on the most susceptible climates, therefore, regional adaptation is possible but only under given circumstances. Governance and institutional set-ups already present at the regional level are pivotal in adapting to climate change. Regions with sounder structures demonstrate greater adaptability than those regions without such institutional frameworks since such regions have sufficient resources for carrying out researches, implementation of changes, and long-term alterations. Pilli-Sihvola et al. (2018) show that Germany and Sweden have national and municipal governance frameworks that integrate adaptation strategies relevant to the nanoscale into their climate change programs. Nwokolo et al. (2023a, b, c, d, e, f) note that although several African countries have various national adaptation strategies and plans, their implementation has been constrained by dysfunctional institutions and corrupt political leadership.

Case Studies from Africa: Climate Impacts and Responses

Africa is one of the most vulnerable continents in the world. Despite making a minimal contribution to the total greenhouse gases worldwide, this region faces extreme temperatures, variability in precipitation, protracted droughts, and destructive flooding that often lead to food insecurity and water shortages. Some of these effects notably deteriorate the quality of life. For instance, a few African nations have initiated activities aimed at adaptation and mitigation efforts. According to the United Nations Economic Commission for Africa, the development of renewable energy in Africa calls for investment needs, progressive policies, international collaboration, and technology innovation (Nwokolo et al. 2023a, b, c, d, e, f). The Horn of Africa, that is, Ethiopia, Somalia, and Kenya, experienced drought for consecutive years from 2011 to 2022, thus leading to increased hunger and displacements. Ethiopia joined forces with the UN organizations to enhance drought vulnerability through the building of new schemes and dams on water (World Bank 2020). The Productive Safety Net Program (PSNP) provides cash transfers and food assistance for vulnerable populations when droughts hit. According to climate change projections, droughts will intensify in these areas, meaning that more enhanced adaptation measures should be put in place.

The devastating impact of Cyclone Idai in 2019 was a manifestation of climate vulnerability in Mozambique. The catastrophe flooded nearly three million people in Mozambique, Zimbabwe, and Malawi. Damages to the infrastructure led to the displacement of 400,000 people with losses of more than $2 billion. Mozambique established a climate-resilient reconstruction program with help from the World Bank and several international organizations (World Bank 2021). There has been the development of infrastructure through the construction of elevated roads, reinforced bridges, and drainage improvement systems. Furthermore, early warning systems and disaster risk management policies have been enacted to reduce fatalities in subsequent storms.

The Sahel region, from Senegal to Sudan, is a vast area of significant desertification because of increasing temperatures and decreasing precipitation. Agricultural land loss is one of the significant threats to millions of livelihoods (Nwokolo et al. 2023a, b, c, d, e, f). In 2007, the African Union started the Great Green Wall initiative to restore 100 million hectares of degraded areas by 2030. Niger, Burkina Faso, and Senegal have undertaken agroforestry projects and sustainable land management practices. FMNR strategy implemented by Niger has restored five million hectares of the area while improving soil fertility and providing an economic opportunity. Rising sea levels and coastal erosion pose a great threat to densely populated regions in West Africa, including Lagos, Nigeria, and Accra, Ghana. It is estimated by the World Bank that 40 per cent of the coastal population in West Africa are at risk of displacement. In Ghana, government-funded coastal defence systems, such as seawalls and mangrove restoration projects, are meant to safeguard the most vulnerable areas. Nigeria is embracing community-based adaptation approaches, including relocation of at-risk communities inland and provision of alternative livelihoods for persons affected by environmental degradation (Nwokolo et al. 2023a, b, c, d, e, f).

Water scarcity is one of the significant challenges in Southern Africa, especially in South Africa, Namibia, and Botswana. In 2018, Cape Town was threatened by "Day Zero," where the city's water supply was almost depleted. The government introduced severe water conservation measures, increased wastewater recycling, and allocated funds for desalination plants. Namibia is now embarking on massive desalination projects that will reduce dependency on variable rainfall patterns. It's the continent of significant climate change issues, where desertification plagues the Sahel, drought haunts the Horn of Africa, and erosion on the coastal shores of West Africa. Efforts like the Great Green Wall, cyclone-resistant infrastructure in Mozambique, and water management projects in Southern Africa provide very useful lessons on how to advance global climate resilience. Improving these efforts calls for more resources, international collaboration, and increasing consciousness to expedite Africa's adaptation and mitigation to climate change.

Case Studies from Asia: Coping with Climate Disasters

Asia ranks as one of the most climate-vulnerable regions worldwide. Climatic disasters, such as floods, cyclones, droughts, and heat waves, are widespread across diverse geographical and socioeconomic contexts, affecting a population exceeding 4.5 billion. As a result, Asian nations were required to adopt various strategies for climate change adaptation and disaster management.

The monsoon season annually causes riverine flooding, and melting from the Himalayas further complicates the challenges encountered by Bangladesh. Recent studies indicate that climate change has intensified conditions through increased rainfall and projected rising sea levels, potentially flooding substantial areas of low-lying Bangladesh (Rahman and Salehin 2013). The World Bank (2021) indicates that 80% of Bangladesh is vulnerable to flooding, with climate projections suggesting an increase in this percentage in the coming decades. Bangladesh employs a comprehensive flood management strategy that includes both structural and non-structural measures. This strategy features the construction of new embankments, flood retention basins, early warning systems, and community-based flood preparedness and forecasting initiatives led by local politicians and NGOs/Raiders.

The government of Bangladesh partners with the World Bank on the Bangladesh Coastal Embankment Improvement Project to protect approximately 1.5 million people from tidal floods and storm surges (World Bank 2020). Also, communities collaborate with local governments and NGOs to implement climate adaptation measures, such as flood early warning systems and cyclone shelters (Huq and Reid 2007). The Community-Based Flood Risk Management (CBFRM) project, which includes an NGO component, enhances community understanding of floods and preparedness through sensitisation workshops, simulated disasters, and the establishment of flood shelters (Islam et al. 2020). Consequently, various initiatives are being implemented in Bangladesh to mitigate floods and reduce fatalities and disasters; however, consistent success is not attained.

The Philippines experiences over 20 typhoons annually, with many leading to fatalities and an increasing frequency of storms attributed to climate change, resulting in economic devastation. Typhoon Haiyan (Yolanda), classified as a Category 5 storm in November 2013, was one of the most powerful storms documented in the Philippine area of responsibility, causing over 6,000 deaths and displacing millions globally (Cruz et al. 2019). The Philippine response team has achieved notable advancements in disaster management and infrastructure post-Typhoon Haiyan. The National Disaster Risk Reduction and Management Plan (NDRRMP) outlines objectives related to risk mitigation, preparedness, and recovery. The national government allows local government entities to develop disaster risk reduction strategies that address the unique vulnerabilities of their regions (Yap et al. 2020). Tacloban City, the region most impacted by Typhoon Haiyan, subsequently established storm-resistant evacuation centres and an early warning system through its local government to alert citizens of approaching storms (ADB 2021).

India has experienced more intense heat waves in the last two decades, a trend expected to worsen as a result of global warming. A severe heat wave in 2015 resulted in over 2,000 fatalities, primarily in Andhra Pradesh and Telangana (Imam 2020). As India anticipates more severe and frequent heat waves, urban areas are adopting comprehensive Heat Action Plans (HAPs) instead of concentrating solely on the most vulnerable populations affected by extreme heat, such as the elderly, manual labourers, and slum residents. The Heat Action Plan (HAP) implemented in Ahmedabad, Gujarat, serves as a prominent example of effective acclimatisation to heat waves. Established in 2013 by the Natural Resources Defence Council (NRDC) and the Indian Institute of Public Health. This HAP includes an early warning system, awareness and education initiatives, and, since 2019, organises training for physicians to identify and manage heat-related ailments and diseases (NRDC 2019). Cooling centres have been established, water distribution is implemented during heat waves when individuals exhibit increased sensitivity, and work shifts for outdoor labourers have been modified to reduce exposure. Following Ahmedabad's Heat Action Plan, Nagpur and Surat developed their versions of a heat action plan customised to their specific requirements. The union government aims to regulate the HAP due to its acknowledgement that heat waves represent a considerable public health and socioeconomic risk at the national level (Imam 2020).

Vietnam exhibits significant potential for economic growth; however, it remains vulnerable to drought due to its intrinsic meteorological traits, especially in the central and southern regions. The Mekong Delta, often referred to as Vietnam's "rice bowl," serves as a vital region for food production. The ongoing drought and salinity intrusion have recently affected this region, impacting the nation's self-sufficiency (World Bank 2020). Moreover, influenced by climate change marked by irregular precipitation and extended natural disasters—it is essential to assess drought resistance to execute these adjustments. As a result, various programs have encouraged the Vietnamese government to assess drought resistance and undertake enhancements. Regions identified as vulnerable to drought are shifting from rice cultivation to more sustainable or drought-resistant crops, including maize and cassava (Nguyen et al. 2020). Irrigation systems are being modified to improve application efficiency.

The government has promoted integrated water resource management (IWRM) to enhance interregional and intersectoral utilisation. IWRM acknowledges the specific needs of domestic consumption in contrast to industrial or agricultural demands. International organisations, such as the ADB, have performed assessments and consultations that have supported the funding and management of various projects, allowing Vietnam to prevent or alleviate drought-related issues.

By 2035, regions in these countries are expected to encounter unsustainable freshwater resources as a result of climate change and the rapid rate of glacial melt (Immerzeel et al. 2020). Another disadvantage that the people living at low altitudes suffer is glacial lake outburst floods due to the melting of glaciers. Nepal is now engaged in monitoring glaciers and addressing GLOF issues. In 2016, a major achievement was the lowering of Imja Lake—one of the most hazardous glacial lakes in Nepal by three meters through artificial drainage to mitigate significant glacial lake outburst floods. Facilitating water discharge created room for less amount of out flow and stabilized sediments and ICIMOD 2018. Partnership with UNDP and ICIMOD executed this project. For instance, a model has set for other states to counter effect of melting the glaciers. As well as policymaker of Bhutan developed entire adaptation plan as a part to strengthen rural mountain communities. This includes improved early warning systems, optimized land-use planning, and additional non-agricultural, non-climate-related livelihood options (Chhetri et al. 2020). All these are vital because climate change impacts the long-term water sustainability of the region gradually and risks catastrophe.

Case Studies from Latin America: Adapting to Climate Change

Latin America is especially vulnerable to climate change because of its geographical and socioeconomic conditions. Geographic vulnerabilities come from a continent characterized by varied landscapes and ecosystems with considerable dependence on climatic-sensitivity-sensitive economic activities such as agriculture and tourism, along with pronounced social inequalities. Regions are thus sensitive to natural calamities, unpredictable precipitation patterns, and temperature fluctuations. Millions in the continent have lost quality life as a result of industrial hindrances, agricultural challenges, environmental problems, and structural decline. Most Latin American countries use different measures to deal with climate vulnerability, ranging from natural resource management, food and water safety programs, and development policies.

The Amazon rainforest in Brazil is referred to be the "lungs of the planet," serving as a carbon sink and a reservoir of global biodiversity. Nonetheless, climate change, deforestation, and issues related to illegal land acquisition adversely affect the region, resulting in rapid biome alterations that jeopardize fragile species and humans reliant on the area for everyday activities and subsistence. Consequently, a coordinated initiative to restore the biome via agroforestry and reforestation serves as a responsive measure to climatic challenges and an endeavour to mitigate greenhouse gas emissions as temperatures increase and precipitation patterns alter. A notable program is the Amazon Sustainable Landscapes Program, which advocates for reforestation and sustainable land management to save biodiversity and reduce carbon emissions. Furthermore, Brazil's "Cultivando Água Boa" program exemplifies a successful forestry and water management project.

Communities cooperate to rehabilitate indigenous flora around water bodies and within watersheds, restoring ecosystems and enhancing water retention—essential for agricultural and potable water requirements in a region experiencing more regular droughts. Another method of adaptation arises from the expansion of agroforestry in Brazil.

Brazil is implementing a program to promote a good transition towards sustainable farming practices and reforestation to address its socioeconomic demands. Nevertheless, several difficulties persist around unlawful deforestation and land disputes. Furthermore, Colombia, a leading global coffee grower, sustains an agrarian economy with coffee as its primary export. Similar to other agricultural commodities, coffee is susceptible to fluctuations in temperature and alterations in precipitation patterns linked to global climate change. Fluctuating temperatures and precipitation have resulted in diminished yields and heightened disease prevalence specifically, coffee rust disease which devastates fields extensively and pushes farmers to the verge of destitution following a single ravaged crop. Consequently, after the intervention of the government and NGOs, Colombian coffee producers began modifying their cultivation practices for climate resilience. The Colombian Coffee Growers Federation has been promoting the cultivation of a more resilient coffee plant that is less vulnerable to disease and climate-related growth variations.

Additionally, Colombia participates in a UNDP collaboration with the "Adaptation Fund," which allocates financial resources for resilience initiatives in agriculture. Consequently, farmers participate in programs on climate-smart agriculture that instruct them on soil retention, pest control, and irrigation to enhance resistance to fluctuations. These initiatives enhance climate change adaptation, strengthening the resilience of the Colombian coffee sector, yet disparities and accessibility continue to pose challenges.

The Andean glaciers of Peru fulfil the nation's drinking, irrigation, and hydroelectric needs; nevertheless, increasing temperatures are accelerating glacier melt, resulting in diminished drinkable water and the occurrence of glacial lake outburst floods. Excessive glacier melt presents significant socioeconomic challenges for residents of the Andean area, both urban and rural. Peru has achieved notable advancements in water management and catastrophe risk mitigation for adaptation. The "Glacier Management Project" is the national initiative to evaluate glacial lakes and develop early warning systems to avert catastrophic floods. The Peruvian government has initiated watershed management projects that enhance water retention capacity by creating artificial lakes and reservoirs to collect water during wet seasons for use in dry seasons. Moreover, Peru has implemented its Ecosystem-based Adaptation (EbA) initiatives, encompassing reforestation of the Andean highlands and wetland restoration, which is encouraging.

Mexico City is the preeminent metropolis in Latin America. Rising heat waves and temperatures adversely affect poor socio-economic groups, as well as the young and elderly populations. This is why Mexico City has sought to enhance urban resilience and reduce heat exposure. The "Green Roof Program" allows for greater vegetation on rooftops, which decreases ambient temperatures, enhances air quality, and lowers air conditioning expenses, functioning as a kind of natural climate control. Moreover, Mexico City has developed additional parks and green corridors across the metropolitan landscape to serve as shaded zones and cooling stations during extreme heat. Additionally, regarding international collaboration, Mexico City participates in the Rockefeller Foundation's effort "100 Resilient Cities," which seeks to develop resilience strategies globally. The instances of Brazil, Colombia, Peru, and Mexico illustrate the particular deficiencies in climate change adaptation within Latin America, as well as the innovative adaptations that arise in response to these challenges. Each of these nations aims to tackle climate change resilient development through agroforestry, water conservation, climate-smart agriculture, and urban resilience, despite limited project finance and sporadic infrastructural solutions.

Implications of the Study

Key stakeholders, including educators, must advocate to national governmental agencies for a development process centred on climate change mitigation concerning agricultural needs, infrastructure, and health accessibility to improve sustainable viability initiatives (Omale et al. 2024). Furthermore, supporting policies that advance renewable resources, sustainable construction, and environmentally friendly maintenance (Oyetunde et al. 2023) can reduce vulnerabilities to climate change while promoting economic development in the regions. Modifications to the Sustainable Development Goals via international climate agreements require a strategic approach to ensure that developed countries meet their financial and advisory assistance obligations.

The Sustainable Development Goals strengthen the relationship between vulnerability and engagement in climate-related development issues. SDG 13 (Climate

Action) primarily intersects with adaptation, as the most vulnerable regions necessitate increased resilience to endure the substantial impacts of climate change. Additionally, SDG 2 (Zero Hunger) is linked to climate-sensitive agriculture, especially in areas where climatic variability significantly impacts agricultural practices and jeopardises food security. Additionally, SDG 7 (Affordable and Clean Energy) and SDG 9 (Industry, Innovation, and Infrastructure) emphasise that adaptation and mitigation necessitate the development of clean, renewable energy sources and resilient infrastructure to reduce climate vulnerabilities. Thus, a larger intersection of SDGs with adaptation reveals increased vulnerability and developmental necessity. In conclusion, future research should compare previous adaptation efforts to determine their historical effectiveness on a large scale and assess the feasibility of regional adaptations.

Conclusion, Limitations and Future Scope

While climatically vulnerable developing nations confront numerous challenges from climate impacts, which alter lifestyles, economic stability, and geopolitical relations complicating the situation internally, external geopolitical factors exacerbating the complexity include geographical locations that increase disaster susceptibility, reliance on climate-dependent economies for fiscal viability, and transportation and infrastructure corridors that hinder effective facilitation. Numerous initiatives are underway to address vulnerabilities, including climate-resilient agriculture practices, sustainable building advancements, and informed policy adaptations across all sectors. However, adaption necessitates that International Financial Institutions serve as the financiers of technical and fiscal progress through the Green Climate Fund and Adaption Fund.

References

Asian Development Bank (ADB) (2021) Building resilience to climate change in the Philippines. ADB, Manila. https://www.adb.org/publications/building-resilience-climate-change-philippines

Chhetri P, Sharma KR, Shrestha RB (2020) Building resilience in the Himalayan region: addressing glacier melting and disaster risks. ICIMOD, Kathmandu. https://www.icimod.org/resource/31292

Cruz RV, De Guzman RG, Landicho LD (2019) Typhoon Haiyan recovery and rehabilitation: lessons learned in building disaster resilience. Nat Hazards 97(3):781–800. https://link.springer.com/article/10.1007/s11069-019-03586-x

European Commission (2021) EU adaptation strategy. https://climate.ec.europa.eu/eu-action/adaptation-climate-change/eu-adaptation-strategy_en

Huq S, Reid H (2007) Community-based adaptation: a vital approach to the threat climate change poses to the poor. Clim Dev. https://www.tandfonline.com/doi/abs/10.1080/17565529.2008.10829003

ICIMOD (2018) Lowering the risk of GLOFs in Nepal. https://www.icimod.org/initiative/reducing-glof-risk-nepal
Imam M (2020) Heat action plans in India: a policy analysis of climate adaptation in urban centers. J Clim Policy 11(1):115–129. https://www.tandfonline.com/doi/abs/10.1080/14693062.2020.1816368
Immerzeel WW, van Beek LPH, Bierkens MFP (2020) Climate change will affect the Asian water towers. Nature 588(7837):82–85. https://www.nature.com/articles/s41586-020-2964-5
IPCC (2021) Sixth assessment report: climate change 2021. https://www.ipcc.ch/report/ar6/wg1/
Islam MR, Uddin AM, Rahman M (2020) Community-based flood risk management in Bangladesh: lessons learned and the way forward. Environ Manag 55(2):254–268. https://link.springer.com/article/10.1007/s00267-019-01224-8
Joshi A, Singh R, Kumar V, Rani S (2024) Net zero strategy: exploring opportunities, confronting challenges for developing countries. In: Transition towards a sustainable future, pp 233–252. Springer. https://link.springer.com/chapter/10.1007/978-981-97-5756-5_11, ISBN 9789819757558
Natural Resources Defense Council (NRDC) (2019) Ahmedabad's heat action plan: Reducing deaths from extreme heat. https://www.nrdc.org
Nguyen VT, Pham HN, Le TA (2020) Coping with drought in the Mekong delta: transforming agriculture in Vietnam. Environ Sustain 3(4):151–167. https://link.springer.com/article/10.1007/s42398-020-00099-8
Nwokolo SC, Singh R, Khan S, Kumar A, Luthra S (2023a) Africa's awakening to climate action. In: Africa's path to net-zero. CSR, sustainability, ethics & governance. Springer, Cham. https://doi.org/10.1007/978-3-031-44514-9_9
Nwokolo SC, Singh R, Khan S, Kumar A, Luthra S (2023b) Remedies to the challenges of renewable energy deployment in Africa. In: Africa's path to net-zero. CSR, sustainability, ethics & governance. Springer, Cham. https://doi.org/10.1007/978-3-031-44514-9_3
Nwokolo SC, Singh R, Khan S, Kumar A, Luthra S (2023c) Decarbonizing hard-to-abate sectors in Africa. In: Africa's path to net-zero. CSR, sustainability, ethics & governance. Springer, Cham. https://doi.org/10.1007/978-3-031-44514-9_6
Nwokolo SC, Singh R, Khan S, Kumar A, Luthra S (2023d) Global investment and development in Africa. In: Africa's path to net-zero. CSR, sustainability, ethics & governance. Springer, Cham. https://doi.org/10.1007/978-3-031-44514-9_2
Nwokolo SC, Singh R, Khan S, Kumar A, Luthra S (2023e) Technological pathways to net-zero goals in Africa. In: Africa's path to net-zero. CSR, sustainability, ethics & governance. Springer, Cham. https://doi.org/10.1007/978-3-031-44514-9_5
Nwokolo SC, Singh R, Khan S, Kumar A, Luthra S (2023f) Influencing the scale of Africa's energy transition. In: Africa's path to net-zero. CSR, sustainability, ethics & governance. Springer, Cham. https://doi.org/10.1007/978-3-031-44514-9_4
Omale SA, Yusuf SO, Oyenuga MO, Ikemefuna M, Ojo SS, Momodu ID (2024) Organizational learning in the post-covid 19 era: a prerequisite for stakeholder satisfactions. Twist 19(3):265–272 https://twistjournal.net/twist/article/view/339
Oyedele OM (2025) The cobra effect: exploring the dark side of innovation and environmental sustainability in developing nations. In: Singh R, Joshi A, Filho WL, Khan S (eds) Zero carbon industry, eco-innovation and environmental sustainability. World sustainability series. Springer, Cham. https://doi.org/10.1007/978-3-031-80220-1_15
Oyenuga MO, Akwen PI (2025) Tourism and intercultural understanding in developing nations. Afr J Cult Hist Relig Tradit 8(1):1–14. https://doi.org/10.52589/ajchrt-okuxnmsq
Oyenuga MO, Omale SA (2024) Is Africa Jinxed? Exploring the challenges of technology access and adoption in Africa. Afr J Econ Sustain Dev 7(4):142–161. https://doi.org/10.52589/AJESDULN1LRNF
Oyetunde TO, Oyenuga MO, Adoga GJ (2023) Green management practices and organisational efficiency in higher institution. Budapest Int Res Crit Linguist Educ (BirLE) J 6(3):276–290 https://bircu-journal.com/index.php/birle/article/view/7716

Pilli-Sihvola K, Harjanne A, Haavisto R (2018) Adaptation by the least vulnerable: managing climate and disaster risks in Finland. Int J Disaster Risk Reduct 31:1266–1275. https://research.aalto.fi/en/publications/adaptation-by-the-least-vulnerable-managing-climate-and-disaster

Rahman R, Salehin M (2013) Flood risks and reduction approaches in Bangladesh. In: Shaw R, Mallick F, Islam A (eds) Disaster risk reduction approaches in Bangladesh, pp 65–90. Springer. https://link.springer.com/chapter/10.1007/978-4-431-54252-0_4

Roy J et al (2020). Rural–urban disparities in spatiotemporal pattern of vulnerability to climate change: a study of Madhya Pradesh, India. Environ Earth Sci. 82, Article 588. https://link.springer.com/article/10.1007/s12665-023-11274-7

Singh R, Crowther D, Ravesangar K (2023) Harnessing human capital. In: Transition towards a sustainable future: net zero policies and environmental sustainability, pp 27–46. Springer. https://link.springer.com/chapter/10.1007/978-981-97-5756-5_2

Singh R, Gupta A, Kandpal V (2025). ESG Excellence in the digital world: circular economy E-business solutions. In: Fintech for ESG and the circular economy, pp 193–204. https://doi.org/10.1002/9781394234035.ch10

World Bank (2020) Vietnam: managing drought and water scarcity in a changing climate. World Bank, Washington, DC. https://www.worldbank.org/en/country/vietnam/publication/vietnam-toward-a-safe-clean-and-resilient-water-system

World Bank (2021) Bangladesh coastal embankment improvement project. World Bank, Washington, DC. https://projects.worldbank.org/en/projects-operations/project-detail/P128276

World Health Organization (WHO) (2020) Climate change and health. https://www.who.int/news-room/fact-sheets/detail/climate-change-and-health

Yap DP, Cruz R, Zubair MS (2020) Resilience building in Tacloban after Typhoon Haiyan: a case study of adaptive capacity in a vulnerable urban community. Clim Dev 12(5):413–424. https://doi.org/10.1080/17565529.2019.1613212

Chapter 3
Economic Growth Versus Digital Sustainability: Striking the Balance

Abstract Digital technologies are rapidly reshaping industries and societies, providing both a new set of opportunities and challenges for sustainable growth. In this chapter, we posit that economic development can no longer be modelled separately from our stewardship of the environment, making economic growth and digital sustainability inextricably linked. The chapter examines how digital applications can make resource use more efficient, lower carbon footprints and enable sustainable businesses. It cites cases of organizations, which have successfully executed a strategy of digital sustainability and how they drive innovation and competitiveness alongside other pressing environmental priorities. The chapter shows the different forms of compromises that may likely happen between radical growth in terms of economics versus sustainable practices it points out critical barriers that organizations have to overcome when their growth strategy is compatible with the requirements of sustainability, such as regulatory hurdles, technological constraints, and cultural changes. The rationale of this chapter is to provide actionable insights and recommendations to policymakers, business leaders, and researchers to face the evident challenges posed by economic growth and digital sustainability.

Introduction

Unprecedented economic growth calls for new benchmarks for economic and industrial growth, which in turn call for higher benchmarks for environmental sustainability. Economic growth is essential to understanding and carrying out business strategies and achieving personal success. From a global perspective, it emphasises the need for change, but politically, it may sometimes be a one-party issue. However, like technological breakthroughs, economic growth requires attention to digital sustainability. When economic growth recognises digital transformation as the means to increase productivity and efficiency, it becomes morally acceptable to set conditions that purposefully impede the advancement of non-digital participation. Relying on industrialisation and expanding urban populations to supply socioeconomic possibilities, the relationship between economic growth and the environment

M. 0. Oyenuga et al., *Smart Technologies for Climate Change and Net Zero Policies: Practical Approaches Towards Sustainability*, Climate Change Management,
https://doi.org/10.1007/978-3-031-92221-3_3

remained unstable for decades. Increased output from a socioeconomically advanced environment promotes market development by raising demand for goods and services that are no longer readily available, whether they are virtual or material. Virtual progress is important, even if it might not seem like much to the typical person. Just as the era of socioeconomic growth produced a wealth of resources, credit cards, and microprocessing technology, safe spaces facilitate virtual expansion in a way that is similar to resource-intensive operations, producing energy and virtual waste when the new market meets its needs. For the current generation, this dichotomy poses a serious challenge: How can emerging countries grow economically sustainably without worsening environmental deterioration throughout the world (Oyedele 2025)?

Economic Growth and Environmental Degradation

For many years, there has been a heated discussion about the relative merits of environmental preservation and growth. Politicians, academics, and environmentalists have long argued about whether the benefits of economic growth (Fasanmi et al. 2025) justify the possible harm to the environment or if stabilisation is the best course for creating a better world. Increased industry, urban migration, and changing consumer behaviour (Oyenuga et al. 2021) are all components of economic expansion. These activities have socio-environmental disadvantages even if they are beneficial for generating cash. Thus, like socio-environmental harms, socio-economic growth includes deforestation, excessive pollution of the air and water, and the loss of natural resources (Uddin 2021).

The Environmental Kuznets Curve (EKC) is a topic of significant debate since it implies a link between socio-environmental degradation and economic advancement. A theoretical inverted U-shape is depicted by the Environmental Kuznets Curve (EKC), which suggests that socio-environmental effects increase with national growth but decrease beyond a certain economic threshold (Grossman and Krueger 1995). The Environmental Kuznets Curve (EKC) hasn't, however, been widely assessed with positive findings about its impacts.

Developed countries have government policies and technology to lower carbon emissions (Uddin 2021). On the other hand, wealthy countries nevertheless have socio-environmental problems even if their inhabitants are less affected by socio-economic factors. Environmental deterioration and economic growth are linked to several important aspects. Given that many developing countries still rely on fossil fuels, Gyamfi and Evans contend that energy consumption plays a critical role in both economic growth and decline. Hickel and Kallis (2020) argue that degrowth is preferable to growth since growth is illogical and that sufficiency and lower consumption are better ways to improve the environment. According to Hofstetter, the global energy initiative is uneven, permitting various sectors to continue using energy-saving methods; yet, the absence of strict monitoring over worldwide programs may have unfavourable effects. According to Kumar et al. (2024), who cited the OECD

(2011), international energy programs that target specific businesses are inconsistent, whereas emission laws and pollution fees that encourage avoidance behaviour are comprehensive (United Nations 2015).

The Tension Between Development and Sustainability

Sustainability and development are interconnected with each other. Through the establishment and facilitation of additional production, consumption, infrastructure, and other components, development aims to improve human living, especially economically. But when production and consumption rise, pollution levels rise, waste management problems worsen, natural resource depletion increases and biodiversity concerns for species worsen (Nwokolo et al. 2024).

Through a quasi-state neutral approach, sustainability seeks to address these problems and guarantee that people may meet their basic needs without endangering those of future generations (Joshi et al. 2024; Singh et al. 2025). The idea of development and the existing sustainable goals and increments are central to this argument. While mining and deforestation capitalise on biodiversity while ignoring long-term ecological sustainability, development creates resources for exploitation. Furthermore, attempts to speed up industrial growth which is exacerbated by urbanization lead to increased pollution and greenhouse gas emissions, making sustainable development more difficult in both developed and developing nations. This is how the Amazon is owned by the developed world. Being a developed country, it is driven by commercial interests and looks for short-term financial gains from deforestation, but in the long run, it causes problems for biodiversity and increases carbon emissions (Joshi et al. 2024). However, there is significant air and water pollution as a result of this success. As a result of policy interventions, stricter environmental rules, greater green investments, and improved public participation in clean energy projects have all been implemented (Zhang et al. 2017).

The Sustainable Development Goals, on the other hand, encourage environmental sustainability (Goal 13) and development inclusiveness (Goal 8), allowing countries to create policies that address social and economic justice in addition to environmental issues (United Nations 2015). Consequently, green architecture in poor nations makes it easier to separate negative environmental effects from positive economic effects and to use resources efficiently (UNEP 2011). Furthermore, energy-efficient construction and international corporate social responsibility provide new possibilities for less environmental duty while acknowledging the possibility of financial gain (Kumar et al. 2024).

When socioeconomic non-environmental problems clash with environmental ones, it might be difficult to maintain policy coherence (Singh et al. 2023). Additionally, because they have less access to financial and technological resources, low-income areas are less likely to participate in sustainable development (World Bank 2022; Oyenuga and Omale 2024). On the other hand, innovations in renewable energy and green technology in low-income areas will guarantee long-term economic sustainability and provide new job possibilities. Public health, physical capital, labour productivity, and energy investment are all hampered by carbon emissions, especially those from the use of fossil fuels. While climate change is expected to reduce agricultural output and drive up costs, increased carbon emissions cause temperatures to rise, worsening public health problems (Metcalf and Stock 2020). Furthermore, it makes energy projects more difficult since rising temperatures require more energy to cool (Auffhammer and Mansur 2014).

Sustainable Development Goals (SDGs) and Net Zero

The Sustainable Development Goals (SDGs) and the Net Zero emissions target are essential objectives that international policymakers must embrace to limit human influence on the Earth more effectively. The SDGs signify a global objective for comprehensive development across several sectors. In contrast, the net zero emissions target is the minimum worldwide goal for reducing and offsetting greenhouse gas emissions. By harmonizing objectives through comparison and contrast, policy may be formulated for advancement without compromising ecological integrity.

In 2015, the United Nations introduced seventeen Sustainable Development Goals (SDGs) as part of the 2030 Agenda for Sustainable Development, aimed at eradicating poverty, protecting the planet, and fostering sustainable living for everyone by 2030. The Sustainable Development Goals (SDGs) are comparative initiatives to environmental efforts per socioeconomic and developmental frameworks. Among all the suggested objectives, the thirteenth Climate Action must be prioritized to alleviate climate change and its repercussions. (United Nations 2015). Net Zero is a condition where the total greenhouse gas emissions emitted into the atmosphere are equal to the total greenhouse gas emissions withdrawn, resulting in a net balance of zero. Consequently, it is not neutral. It aligns closely with the Paris Agreement's objectives, which aim to regulate variables to prevent the Earth's temperature from rising more than 2 °C beyond pre-industrial levels, 1.5 °C if substantial efforts are made (IPCC 2021).

The relationship between Net Zero and the Sustainable Development Goals is evident. The shift from reliance on fossil fuels to renewable energy sources necessitates diminished carbon footprints and fair access to sustainable energy. Implementation of solar and wind energy is essential for energy accessibility and climate objectives, particularly in developing nations such as those in Africa (Nwokolo et al. 2023a, b). Moreover, SDG 9 (Industry, Innovation, and Infrastructure) funding will provide sustainable infrastructures and creative developments to decarbonize industrial operations. This pertains to SDG 11 (Sustainable Cities and Communities) as urban areas are the primary focus for climate change interventions, given that over 50% of the global population now inhabits these locations. Urban development initiatives and low-carbon mass transportation networks substantially mitigate potential emissions from a densely concentrated population in a single geographic area (IRENA 2021). This relates to SDG 12 (Responsible Consumption and Production) as climate-positive outcomes arise from sustainable production methods that minimize waste and optimize resource allocation.

Initiatives in the circular economy throughout the industrial and building sectors have demonstrably reduced carbon footprints (Joshi et al. 2024). Moreover, cohesive policy mechanisms incorporating inclusive and equitable concerns are essential to achieving Net Zero and advancing the Sustainable Development Goals (SDGs). Support must be derived from all government, industry, and civil society tiers. International cooperation in renewable energy and technology transfers promotes fairness, enabling resource-poor nations to join with bigger counterparts to achieve Net Zero objectives more effectively than while alone (Krishnan et al. 2024). Moreover, the stipulations of the European Union Climate Law establish the EU as a leader in global climate change efforts and a model for sustainable transformation; consequently, African nations ought to embrace this Law, engage with international partners for guidance and support, and participate in renewable energy training and capacity-building initiatives to ensure the longevity of these actions and innovations for climate sustainability (Nwokolo et al. 2023a, b).

Cross-sectoral policy integration ensures that the objective of achieving Net Zero effectively supports the attainment of development goals. Funding mechanisms are crucial for attaining Net Zero and the corresponding socioeconomic benefits. The ability to access financial avenues via carbon pricing and international climate finance for Net Zero also facilitates access to subsequent funding sources. Funding channels linked to renewable energy technology projects reduce emissions and fulfil SDG 7 (Affordable and Clean Energy) by enhancing technology and addressing employment creation possibilities for SDG 8 (Decent Work and Economic Growth). (Krishnan et al. 2024). Regionals may pursue sustainable development by enhancing institutional capacity and enabling low-carbon technology transfers. Mitigating technical limitations necessitates the education of prospective recipients and collaborative efforts for technology transfers. The transfer of sustainable energy technologies to sub-Saharan Africa mitigates emissions in this vulnerable region while enhancing electricity availability, advancing both Sustainable Development Goals and Net Zero objectives (Nwokolo et al. 2023a, b). These activities illustrate that the Sustainable Development Goals are compatible with Net Zero objectives.

The European Green Deal enables Europe to achieve sustainable development via cross-sectoral policies that offer rapid, creative, and cost-effective solutions for long-term climate change transformation (European Commission 2021). Therefore, aligning the Sustainable Development Goals with Net Zero objectives is complex, while not insurmountably tricky. Through policy integration, enhanced finance, and improved access to technology and innovations, countries benefit significantly from concurrent advancement in planned growth and unanticipated responses to environmental initiatives. International engagement, blended finance, and capacity-building initiatives are essential to ensure that the pursuit of Net Zero does not overshadow the broader objectives of the Sustainable Development Goals (SDGs). When all elements are evaluated about one another rather than as independent variables, the necessary adjustments for climate change will be alleviated.

The worldwide interrelation between SDGs and Net Zero underscores a collective chance to work on a climate-resilient approach and optimize operational efficiency. Globally, the most substantial environmental, social, and economic sustainable solution may be achievable. However, it may originate from renewable energy sources in Africa; Sustainable Development Goal 7, through solar and wind investments, offers accessible, clean, and affordable electricity. The Kenya Lake Turkana Wind Power Project mitigates carbon emissions while enhancing electricity availability in previously unserved rural areas, achieving a genuine net zero outcome (IRENA 2021).

Transitioning to renewable energy meets Net Zero objectives, mitigates, mitigates energy poverty, and promotes energy fairness. Nonetheless, equity is present in financial and energy exchanges that need global consensus. The network focuses on the Sustainable Development Goals (SDGs) and Net Zero to foster sustainable capabilities within European communities. Copenhagen possesses a comprehensive green infrastructure, whilst Stockholm depends on public transit; primarily, both cities evaluate their socioeconomic requirements by SDG 11 (Sustainable Cities and Communities), minimizing carbon footprints despite being located in huge, heavily populated regions. The European Green Deal (European Commission 2021) is Europe's response to integrating the advantages of urbanization with initiatives to combat climate change, promoting sustainable development and economic viability. These findings indicate that legislation and finance at the municipal level facilitate the viability of transitions to sustainable urban development. Community solar networks aim to foster grassroots engagement for both emission reduction and enhanced resilience methods against climate disasters (Connell 2016).

Recently, desertification and climate change have yielded promising advancements. The Great Green Wall extends across the Sahel in Sub-Saharan Africa, aiming to combat desertification and socioeconomic issues by reforesting deforested areas and generating employment with low greenhouse gas emissions. This initiative aligns with the objectives of SDG 15 and acknowledges the interconnections essential for effective greening and Net Zero initiatives. In addition, China exemplifies industrialisation, demonstrating actions that align with both the Sustainable Development Goals and Net Zero objectives. The Chinese government has modified its incorporation of low carbon technologies in the steel and cement sectors, resulting in a reduction in emissions by hundreds of millions of tonnes, without adversely impacting industrial production or hindering economic growth. China's inclination towards domestic technical endeavours rather than worldwide collaborative endeavours demonstrates efficacy in the pursuits of industrialisation and decarbonisation, as advocated by SDG 9 (Industry, Innovation and Infrastructure) (Zhang et al. 2017). The community-focused energy awareness enables Costa Rica's technically viable renewable energy initiative to achieve over 98% renewable energy utilisation, thereby promoting SDG 7 and conforming to the aspirations for a Net Zero future. The completion of this project illustrates the significance of community involvement for the effective sustainability of services after implementation (IRENA 2019).

The link between the Sustainable Development Goals (SDGs) and Net Zero is complicated by many case studies across industries and locations. Research indicates that comparable legal frameworks, financial opportunities, involvement in climate-related activities, and developmental intent/financing are present in renewable initiatives throughout Africa and the circular economy in Asia throughout the initial design and integration phases. As more nations identify this relationship or intersectionality, the potential for substantial advancements towards a sustainable, lasting, and equitable reality will increase.

Balancing Short-Term Growth with Long-Term Climate Goals

A political nuance arises wherein international relations, corporations, and individuals foster a dependence on immediate financial gains that are so minimal to the extent that they undermine environmental objectives and good governance. This evaluates the political intents of climate change climate organisations, the dynamics of political environments, and the objectives at both national and state levels aimed at facilitating climate policy and achievements in environmental matters.

Economic development is periodically evaluated by Gross Domestic Product (GDP), which reflects increased output, consumption, and investment. While economic expansion enhances the standard of life and reduces poverty, it also introduces adverse aspects, such as increased greenhouse gas emissions, accelerated depletion of natural resources, and environmental volatility and deterioration (Stern 2007). Consequently, a fruitful interchange occurs. In the near term, despite prospects for economic expansion, more exploitative practices threaten long-term sustainability initiatives for climate change. The Environmental Kuznets Curve (EKC) posits that as nations progress and income increases, they experience greater environmental degradation and climate change; however, beyond a certain income threshold, affluent countries can afford cleaner technologies and regulatory measures (Grossman and Krueger 1995). However, the issue with the EKC is its relative nature. Applying the Environmental Kuznets Curve (EKC) in a socioeconomic environment may presuppose that historical trends are normalised concerning economic growth and stabilisation; nevertheless, empirical evidence of oversaturation indicates a tendency towards greater deterioration in the long term. The static nature of the EKC is problematic because developing countries may struggle to afford superior technology; impoverished nations with limited financial resources, ineffective governance, and deteriorating infrastructure frequently experience similar exploitative patterns out of necessity (Nwokolo et al. 2023a, b).

In regions with international policy agreements and robust regulatory frameworks, future climate change expectations are associated with present economic advancements. The 2015 Paris Agreement seeks to restrict the increase in global temperature to below 2 degrees Celsius relative to pre-industrial levels (UNFCCC 2015). This requires evaluation based on global greenhouse gas reduction impacts. Another element of the pact is the NDC's Nationally Determined Contributions which mandate each nation to submit its projected reductions in a tailored plan. National intentions and the execution of NDCs differ according to income, political dynamics, and phases of development. Wealthy nations may incorporate climate change into their long-term economic forecasts, but poor nations need immediate resources from economic development and can defer climate measures (Rogelj et al. 2016). In Sub-Saharan Africa, reliance on fossil fuels and energy security is prioritised over less pressing benefits (Nwokolo et al. 2023a, b).

Kenya's wind farms and India's solar systems demonstrate that access to investment in energy policy may yield energy advantages and long-term prosperity (IRENA 2021). Europe's low-emission vehicles and eco-friendly urban transport indicate that it can achieve economic growth while simultaneously minimising its carbon footprint; these initiatives require substantial initial investments that are essential for future sustainability (Creutzig et al. 2015).

The Agriculture Sector is the primary contributor to emissions, particularly through methane and nitrous oxide, hence exacerbating the sustainability challenge. Nevertheless, sustainable alternatives are available. Practices such as regenerative agriculture and precision farming may sacrifice immediate yields for long-term sustainability and resilience, resulting in less negative environmental impact in the future. Furthermore, Africa is in the forefront with two sustainable alternatives: digital agriculture and carbon sequestration (Krishnan et al. 2024).

The EU's Green Deal supports blended policies through decarbonization plans in tandem with economic sustainability (European Commission 2021). Green bonds and climate adjustment funds support sustainable, long-term investment opportunities over even undercapitalized regions (Nwokolo et al. 2023a, b). Publicly available investments in clean processes, such as China's industrial decarbonization efforts, show that clean, innovative pathways to achievement have economic and climate benefits (Zhang et al. 2017). Capacity building through technology transfer supports a just transition to Net Zero (IRENA 2019) while climate creation efforts ensure social feasibility (Connell 2016). The only variable that did not positively correlate was the pandemic. The decision per capita by Germany in 2021 (European Commission 2021) to decrease emissions showed how to sustain a chapter as a global leader without international reliance on patents and per capita reductions. Multicultural research shows the need for Japan to transition to hydrogen (Krishnan et al. 2024) and the International Solar Alliance in India as public–private partnerships geared toward enhancing renewability (Nwokolo et al. 2023a, b). When separation from growth in greenhouse gas emissions occurs alongside stable policy policies that integrate and support from international regions, growth occurs sustainably without decrease. Similarly, nations have undertaken connected climate and economic efforts from the European Union's (EU) Green Deal for a climate-neutral Europe by 2050 using renewable energies, circular economies, and carbon reductions through the EU Emissions Trading System (European Commission 2019, 2021) to Germany's more regulatory approach on one end and China's, the largest renewable resource investor, expansion of its solar and wind capacity and its sustainable Belt and Road Initiative on the other (Zhang et al. 2017; Krishnan et al. 2024).

Also in the collaboration is the United States' Inflation Reduction Act of 2022, which provides tax credits for green energy, electric vehicles, and carbon capture (White House 2022). Africa is not excluded from energy transformation, with geothermal projects in East Africa led by Kenya and Noor Ouarzazate Solar Complex by Morocco (IRENA 2021; Nwokolo et al. 2023a, b) and Japan's extensive investment in hydrogen fuel technology—a public–private partnership for sustainable development (Krishnan et al. 2024). Therefore, no matter where such efforts are made, with supporters and obstacles along the way, these collaborative efforts make sustainability possible for economic and environmental benefits.

Balancing Economic Growth with Climate Goals: Overcoming Challenges

Even though it seems that adequate measures have been enacted, potentially jeopardising foreign financial benefits or access; continuous incremental improvements, both minor and substantial, in the medium and long term would be beneficial, yet appear unachievable without compromising socioeconomic advancement. Nonetheless, a persistent issue remains—the disparity between industrialised and poor nations. Developing nations face severe conditions marked by poverty, inflation, and unemployment, compelling them to seek urgent economic solutions for social issues that necessitate employment, energy, and food while grappling with significant debt, which relegates climate change solutions to a secondary priority relative to these urgent concerns (Sachs 2015, Oyedele et al. 2023).

Many African nations depend on fossil fuel employment and energy due to its current affordability and accessibility, despite its contribution to global warming in two decades (Nwokolo et al. 2023a, b). This imbalance can only be mitigated with substantial international reparative aid, funding, and access to technology. The Green Climate Fund is essential since rich nations possess the means to support poorer nations in addressing their financial and energy educational requirements and solutions throughout time (Green Climate Fund 2022).

Furthermore, long-term climate policies require political motivation, although there is a national focus on immediate political concerns. Such socio-political contexts compel politicians to adopt measures they anticipate will yield returns through socio-political avenues and electoral support. Consequently, climate initiatives are ephemeral and ineffective (Giddens 2009).

Moreover, the industrial sectors in poor nations lack the technological expertise and resources to include more energy-efficient designs and processes, resulting in reliance on carbon emissions (Nwokolo et al. 2023a, b). These limitations can be addressed by international inventions. Technology transfer agreements and public–private partnerships can facilitate a broader range of financing alternatives (European Commission 2019).

The resolution to the dilemma of aligning immediate developmental requirements with enduring climate change consequences lies in amalgamation. Initially, carbon pricing, levies, and cap-and-trade initiatives create a financial motivation for enterprises to diminish emissions, while concurrently generating additional income for future sustainability development projects (Stavins 2008). Secondly, the emergence of green jobs suggests that increased financial investments in renewable resources will diminish carbon footprints while simultaneously generating employment in expanding sectors that balance short-term and long-term viability, proving to be both feasible and profitable (Krishnan et al. 2024). Third, the Paris Agreement promotes enhanced global collaboration, ensuring mutual accountability among nations for thresholds and resource sharing, thus facilitating the achievement of ambitious climate change objectives (UNFCCC 2015). Such solutions necessitate inclusive politics, socio-political frameworks, advanced technology, and global cooperation for universal advantage.

Implications

Harmonising the actions of many nations would require significant international engagement to provide equal finance and access to copyrighted technologies for developing countries.

Net Zero activities greatly support the SDGs. SDG 13 (Climate Action) addresses emissions via mitigation and adaptation, SDG 7 (Affordable and Clean Energy) invests in renewable energy, and SDG 9 (Industry, Innovation, and Infrastructure) promotes sustainable industrialisation. Improved understanding of sustainable agriculture supports SDG 2 (Zero Hunger), while urban climate-adaptive activities support SDG 11. The confluence of these goals shows that a multimodal, holistic approach to sustainability is necessary since one goal depends on the others. Numerous study options exist.

Conclusion, Limitations and Future Scope

As more and more evolution takes place on the digital terrain, there would be both opportunities and threats that would demand careful navigation to achieve desired positive outcomes. In the continued debates on economic growth and digital sustainability, finding an equilibrium between these two critical objectives fosters a sustainable and equitable future. Economic growth, hitherto perceived as a most important measure of national prosperity, must be reborn in light of the integration of sustainability as a key ingredient in practices and policies. It should be noted that long-term growth can no longer be achieved at the expense of ecological integrity or social well-being.

References

Auffhammer M, Mansur ET (2014) Measuring climatic impacts on energy consumption: a review of the empirical literature. Energy Econ 46:522–530. https://doi.org/10.1016/j.eneco.2014.08.017

Connell J (2016) Renewable energy in Pacific Island countries: policy, opportunities, and challenges. Energy Policy 98:676–685. https://doi.org/10.1016/j.enpol.2016.08.004

Creutzig F et al (2015) Global typology of urban energy use and potentials for low-carbon cities. Nat Clim Chang 5(9):819–825. https://doi.org/10.1038/nclimate2776

European Commission (2019) The European green deal. https://cc.europa.eu

European Commission (2021) Emissions Trading System (ETS): a cornerstone of the EU's climate policy. https://ec.europa.eu

Giddens A (2009) The politics of climate change. Polity Press. https://www.wiley.com/en-us/The+Politics+of+Climate+Change-p-9780745646930

Fasanmi OO, Comfort AB, Oyedele OM (2025) Exploring the role of innovation on economic performance of manufacturing firm in Nigeria. J Econ Financ Manag Stud 08(01). https://doi.org/10.47191/jefms/v8-i1-67

Green Climate Fund (2022) Annual report: mobilizing climate finance. https://www.greenclimate.fund

Grossman GM, Krueger AB (1995) Economic growth and the environment. Q J Econ 110(2):353–377. https://doi.org/10.2307/2118443

Hickel J, Kallis G (2020) Is green growth possible? New Polit Econ 25(4):469–486. https://doi.org/10.1080/13563467.2019.1598964

Intergovernmental Panel on Climate Change (IPCC) (2021) Climate change 2021: the physical science basis. Sixth assessment report. https://www.ipcc.ch

International Renewable Energy Agency (IRENA) (2019) Renewable energy: a key to achieving SDGs. https://www.irena.org

International Renewable Energy Agency (IRENA) (2021) Renewable energy and jobs: annual review. https://www.irena.org

Joshi A, Singh R, Kumar V, Rani S (2024) Net zero strategy: exploring opportunities, confronting challenges for developing countries. In: Singh R, Crowther D (eds) Transition towards a sustainable future. Approaches to global sustainability, markets, and governance. Springer, Singapore. https://doi.org/10.1007/978-981-97-5756-5_11

Krishnan R et al (2024) Policy coherence for sustainable development: aligning SDGs and net zero goals. Sustain Policy Rev 10(1):102–120. https://doi.org/10.1016/j.spr.2023.05.009

Kumar K, Singh R, Sharma A, Lehri BPS (2024) Climate change adaptation strategies for achieving net-zero economy. In: Singh R, Crowther D (eds) Transition towards a sustainable future. approaches to global sustainability, markets, and governance. Springer, Singapore. https://doi.org/10.1007/978-981-97-5756-5_8

Metcalf GE, Stock JH (2020) The role of carbon pricing in global climate policy. J Econ Perspect 34(4):87–114. https://doi.org/10.1257/jep.34.4.87

Nwokolo SC, Singh R, Khan S, Kumar A, Luthra S (2023) Technological pathways to net-zero goals in Africa. In: Africa's path to net-zero: exploring scenarios for a sustainable energy transition, pp 93–210. Springer Nature Switzerland, Cham. https://link.springer.com/chapter/10.1007/978-3-031-44514-9_5

Nwokolo SC, Singh R, Khan S, Kumar A, Luthra S (2023b) Influencing the scale of Africa's energy transition. In: Africa's path to net-zero. CSR, sustainability, ethics & governance. Springer, Cham. https://doi.org/10.1007/978-3-031-44514-9_4

Nwokolo SC, Singh R, Khan S, Kumar A, Luthra S (2024) Africa's Awakening to Climate Action. In: Africa's Path to Net-Zero. CSR, Sustainability, Ethics & Governance. Springer, Cham. https://doi.org/10.1007/978-3-031-44514-9_9

Organisation for Economic Co-operation and Development (OECD) (2011) Towards green growth: a summary for policymakers. OECD Publishing, Paris. https://www.oecd.org/greengrowth/towards-green-growth-9789264111318-en.htm

Oyedele OM (2025) The cobra effect: exploring the dark side of innovation and environmental sustainability in developing nations. In: Singh R, Joshi A, Filho WL, Khan S (eds) Zero carbon industry, eco-innovation and environmental sustainability. World sustainability series. Springer, Cham. https://doi.org/10.1007/978-3-031-80220-1_15
Oyedele MO, Marcus GO, Ahungwa AI (2023) Do consumers care about green marketing practices? Insight from a developing nation. Budapest Int Res Crit Inst J (BIRCI J) 6(3):1424–1436. https://bircu-journal.com/index.php/birci/article/view/7665
Oyenuga MO, Ahungwa AI, Onoja E (2021) Effect of brand equity on consumer behaviour among students of Veritas University, Nigeria: a study of apple smartphones. Mark Brand Res 8:48–64. https://doi.org/10.33844/mbr.2021.60329
Oyenuga MO, Omale SA (2024) Is Africa Jinxed? Exploring the challenges of technology access and adoption in Africa. Afr J Econ Sustain Dev 7(4):142–161. https://doi.org/10.52589/AJESDULN1LRNF
Rogelj J et al (2016) Paris Agreement climate proposals need a boost to keep warming well below 2°C. Nature 534(7609):631–639. https://doi.org/10.1038/nature18307
Sachs J (2015). The age of sustainable development. Columbia University Press. https://cup.columbia.edu/book/the-age-of-sustainable-development/9780231173148
Singh R, Crowther D, Ravesangar K (2023) Harnessing human capital. In: Transition towards a sustainable future: net zero policies and environmental sustainability, pp 27–46. Springer. https://link.springer.com/chapter/10.1007/978-981-19-5756-5_2
Singh R, Ravesangar K, Kumar V (2025) Human Capital for Eco-Innovation Eco-System: Assessing the Impact of Sustainable HR Practices on Achieving Environmental Sustainability. In: Singh R, Joshi A, Filho WL, Khan S (eds). Zero Carbon Industry, Eco-Innovation and Environmental Sustainability. World Sustainability Series. Springer, Cham. https://doi.org/10.1007/978-3-031-80220-1_7
Stavins RN (2008) A meaningful U.S. cap-and-trade system to address climate change. Harv Environ Law Rev 32:293–372. https://harvardelr.com/2008/01/01/a-meaningful-u-s-cap-and-trade-system-to-address-climate-change/
Stern, N (2007) The economics of climate change: the stern review. Cambridge University Press. https://www.cambridge.org/core/books/economics-of-climate-change/
Uddin M (2021) Revisiting the environmental kuznets curve: evidence from emerging economies. Environ Econ Policy Stud 23(4):789–810. https://doi.org/10.1007/s10018-021-00305-7
United Nations Environment Programme (UNEP) (2011) Decoupling natural resource use and environmental impacts from economic growth. https://www.unep.org
United Nations Environment Programme (UNEP) (2022) Adapting to climate change: global assessment and strategies. https://www.unep.org
United Nations Framework Convention on Climate Change (UNFCCC) (2015) Paris Agreement. https://unfccc.int
White House (2022) Inflation reduction act of 2022: summary of key climate provisions. https://www.whitehouse.gov
World Bank (2022) Climate change and economic vulnerabilities: Global report. https://www.worldbank.org
Zhang W et al (2017) China's renewable energy strategy: achievements and challenges. Energy Policy 101:512–518. https://doi.org/10.1016/j.enpol.2016.11.023

Chapter 4
Socio-economic Impacts of Net Zero Policies

Abstract Although the goal of Net Zero is to avert climate change by reducing greenhouse gas emissions, its implications on inequality, poverty, and human well-being are complex. This chapter makes the case that Net Zero initiatives exacerbate inequality and poverty. Communities with abundant resources are better equipped to transition to a more resilient and sustainable environment, even in the face of job restructuring, unemployment, rising energy costs, and rising commodity prices. Sea level rise and human displacement from storms are taken into consideration while evaluating ecological resources. People who are already struggling are dispersed by this migration, and host communities are burdened with having to welcome newcomers despite their obligation to assist. However, the evidence for socioeconomic variables impacting short-term livelihoods and those who rely on fossil fuels is uncontroversial. Food and water insecurity is the third indicator of climate change. To provide fair policies and socioeconomic safety nets for underprivileged areas during just transition, this chapter recommends Net Zero-related socioeconomic catastrophe prevention strategies and policy requirements. This means that to implement a Net Zero policy without sacrificing equity and social stability for an ecologically sustainable goal, socioeconomic variables may be purposefully employed and balanced in the planning process.

Introduction

Net Zero, the equilibrium of greenhouse gas emissions produced and eliminated from the atmosphere constitutes a primary objective of global climate change mitigation initiatives. Although much of the literature on Net Zero suggests scientifically and environmentally viable exploration, there is a paucity of focus on socioeconomic feasibility; however, without comprehending this secondary potential, intervention efforts and their outcomes become complex and detrimental. Consequently, attaining Net Zero is not only a technical and engineering accomplishment but a venture rooted in sociology and economics, informed by principles of equality, ethics, inclusion, and justice for everyone. Household effects are tangible consequences of climate change,

M. 0. Oyenuga et al., *Smart Technologies for Climate Change and Net Zero Policies: Practical Approaches Towards Sustainability*, Climate Change Management,
https://doi.org/10.1007/978-3-031-92221-3_4

particularly in areas inhabited by the most vulnerable populations facing poverty, injustice, and resource scarcity. Net Zero efforts aimed at decarbonising diverse businesses and transitioning to renewable energy provide benefits to alleviate imminent climate change. However, they complicate socioeconomic transformation. The change in the employment sector marginalises specific vulnerable groups, including individuals engaged in high carbon-emitting businesses and residents of low-income regions unable to bear rising energy bills, ultimately leading to job losses. These issues complicate the noble objectives of transition and elucidate the implications of social safety nets, fair access to resources and opportunities, and the imperative for transparency and interactive feedback, highlighting the demand for just transitions.

Poverty and Inequality Exacerbated by Net Zero Policies

The net-zero economy has become a prominent issue due to climate change being one of organizations' most urgent concerns today. The problems are apparent, including explicit repair costs for damage, operational interruptions caused by natural catastrophes, and indirect risks associated with value chain disruptions and infrastructure deterioration that complicate corporate operations (Kumar et al. 2024). The decarbonization plan envisions a low-carbon economy characterized by sustainable infrastructure and interagency coordination, with a heightened commitment from emerging nations to decarbonization. Consequently, the developed world is largely positioned to prioritize capital investments in alternative energy, implementation of energy conservation strategies, and education of employees and stakeholders through sustainability initiatives aimed at enhancing environmental consciousness (Nwokolo et al. 2023a, b, c, d, e, f). Consequently, inaction would result in catastrophic consequences, including heightened flooding, increasing sea levels, increased yearly disasters, and severe droughts and famines. Industries are under significant strain.

Industries face primary and secondary vulnerabilities; businesses that are destroyed and closed due to severe hurricanes or other disasters encounter one type of vulnerability, while those suffering from shortages because their satellite facilities are submerged or public infrastructure is devastated face another (Kumar et al. 2024). However, the execution of such changes inadvertently exacerbates poverty and inequality, rendering vulnerable communities increasingly susceptible. Nwokolo et al. (2023a, b, c, d, e, f) assert that the regulation of foreign investments in green technology and exploitation efforts should be constrained. Furthermore, empowered agencies must be held accountable for their energy objectives related to climate change action while aligning developing nations with their expected economic growth and historically supported social development initiatives. Consequently, as the developing world endeavours to collaborate with others to mitigate climate change, it is reasonable to seek investment in technology that extracts carbon dioxide from the atmosphere; however, an agency or facility can only effectively execute this transition if it understands how to manage its most valuable resources optimally (Singh et al. 2024a, b). Net Zero policies aim to diminish greenhouse gas emissions to net zero, ensuring that global temperature increases do not exceed 1.5 degrees Celsius. These policies indicate a shift towards renewable energy, carbon pricing, enhanced energy efficiency, sustainable land use (IPCC 2018), sustainable ecology, and a circular economy.

Nevertheless, despite the laudable objective of these policies to avert calamities and promote an environmentally sustainable future, Net Zero regulations adversely affect socioeconomic position (Oyedele 2025). In contexts where funding is limited, regulatory obstacles hinder the initiation of renewable projects, and inadequate infrastructure impedes the adoption and utilization of renewable technologies, ultimately undermining the economic viability of the region (Nwokolo et al. 2023a, b, c, d, e, f; Oyenuga and Omale 2024). Consequently, this underscores the influence of Net Zero regulations on the socioeconomic conditions of poverty and inequality by recommending measures to mitigate such adverse effects.

Lower-income areas that rely on jobs in carbon-intensive industries suffer the most from the loss of jobs caused by moving away from fossil fuels as a source of energy. Without any equitable mechanisms to help the poorer families that would struggle to pay for rising costs, carbon pricing will also worsen energy poverty (Zhao et al. 2022). Furthermore, low-income groups are discriminated against by carbon taxes (Dorband et al. 2019), and the high cost of green technology prevents them from being accessible (Büchs et al. 2011). If carbon tax funds are not dispersed, the danger of world poverty might rise by 19% by 2030 (Zhao et al. 2022). Moreover, the transition to renewable energy sources results in employment losses within the traditional industry, adversely affecting low-skilled individuals who find themselves in a more precarious position. Popp et al. (2020) assert that the shift to Net Zero generates more green jobs, yet these positions need reskilling, resulting in transient unemployment and income disparity. Hence, the following strategies should be adopted:

1. **Just Transition Frameworks**: Measures must be implemented to safeguard workers from high- to low-carbon job possibilities. These frameworks, include essential retraining requirements, social safety nets, and initiatives, to revitalize local communities (ILO 2015), and offer a beacon of hope, in preventing detrimental economic impacts.
2. **Comprehensive Carbon Pricing**: This initiative must be comprehensive. Carbon pricing strategies must incorporate revenue recycling to benefit low-income groups and prevent regressive effects. A lump-sum refund or a targeted reparative initiative will alleviate the financial burden on disadvantaged groups (Dorband et al. 2019).
3. **Supplementary Policies That Integrate Them**: The inclusion of vulnerable groups in policy discussions on energy transition ensures that their needs are addressed from a policy perspective, promoting a more equitable approach (Newell and Mulvaney 2013).

4. **Enhanced Access to Green Technologies**: Low-income families are more likely to acquire green technology and the associated advantages of energy transition through grants, lending possibilities, and community initiatives, which they may more readily access (Büchs et al. 2011; Nwokolo et al. 2023a, b, c, d, e, f).
5. **Evaluation and Evaluation Procedures**: These refer to the methodologies employed to examine the socioeconomic ramifications of Net Zero, allow for modifications to socioeconomic repercussions, and implement corrective measures for any arising imbalances.

Ultimately, it is evident that policymaking for Net Zero is essential for the innovations required to combat climate change, but the socioeconomic ramifications and demands of such policymaking might exacerbate poverty and inequality.

Displacement and Migration Due to Environmental Factors

Several factors are responsible for people's migration. Environmental migration refers to the movement of individuals across borders prompted by various environmental changes, including natural events (such as floods and hurricanes), progressive alterations (such as desertification and sea-level rise), or, more recently, conflict. The IOM defines "environmental migrants" as individuals or groups compelled to vacate their usual residence, either temporarily or permanently, due to environmental alterations and/or adverse living conditions that jeopardize their livelihood and well-being over an extended duration, as they traverse intra-or international borders (IOM 2019). Environmental deterioration is a significant urban element correlated with migration from rural regions (Nawrotzki and DeWaard 2018).

Environmental factors are increasingly influencing migration patterns worldwide. Understanding the drivers, patterns, and implications of environmental migration is crucial. This understanding is the foundation for developing effective policies and interventions. As climate change continues to impact ecosystems and human livelihoods, proactive measures are essential to address the challenges and opportunities presented by environmental migration. Environmental deterioration presents significant obstacles to health and welfare; exposure to air pollution, natural catastrophes, and resultant relocation causes immediate and long-term suffering. Consequently, it is essential to comprehend the detrimental effects on human existence to avert, adjust to, or mitigate such circumstances (Gemenne 2011; Black et al. 2011; McLeman 2014).

Impacts on Health and Livelihoods

Negative health consequences resulting from environmental degradation include:

1. **Air Pollution**: Exposure to particulate matter (PM2.5), nitrogen dioxide (NO_2), and sulfur dioxide (SO_2) is significantly associated with respiratory and cardiovascular diseases; Cohen et al. (2017) indicate that ambient air pollution was responsible for over 4.2 million premature deaths worldwide in 2015.
2. **Water Quality Concerns**: Contaminated water bodies resulting from industrial effluents and agricultural pesticide and fertilizer runoff are significant problems. The WHO reports that 485,000 individuals perish annually due to cholera and other water quality issues linked to inadequate access to clean water.
3. **Soil Quality Concerns**: Deteriorated quality and contamination of soil impede food accessibility and quality. Lal (2009) contends that soil degradation diminishes agricultural output, resulting in hunger and increased vulnerability to illness.
4. **Climate Change**: Increasing temperatures and unpredictable weather patterns render existence detrimental to health. The Lancet Countdown on Health and Climate Change indicates that climate change contributes to the proliferation of communicable diseases, temperature-associated ailments, and mental health disorders (Watts et al. 2018).

Climate-Induced Food and Water Insecurity

Climate change concerning food and water security is a paramount worldwide issue that adversely affects human health, socioeconomic conditions, and sociopolitical growth. International storms are increasingly frequent or severe, resulting in heightened droughts and hurricanes, while precipitation patterns may shift, causing prolonged rainfall in one region followed by abrupt changes. Consequently, individuals, particularly those in developing nations, cannot depend on stable growing seasons or consistent access to potable water. Agriculture has consistently been one of the most vulnerable human systems to climatic shocks. Variations in temperature and precipitation influence the timing of growth and harvest, as well as the health and mortality of animals and fish. Elevated temperatures, for instance, shorten growth cycles for vital crops such as wheat and maize, hence heightening the probability of diminished yields.

Moreover, elevated CO_2 enhances photosynthetic potential, yet diminishes quality at reduced concentrations, resulting in decreased zinc and iron content in crops. Food supplies are jeopardized by the recent surges of natural calamities, including droughts and floods. In the 2010 Russian heatwave, wheat output diminished, leading to worldwide price surges and food shortages in other countries. This complicates food supply and escalates prices, so placing low-income communities at greater risk of inadequate access to nutritious meals. Moreover, climate change impacts water supply and quality by disrupting the normal hydrological cycle. California's prolonged drought from 2012 to 2016, driven by climate change, severely restricted fresh water access in 2015, resulting in billions in agricultural revenue losses owing to state restrictions and diminishing sources of potable water (Diffenbaugh et al. 2015).

Furthermore, the retreat of glaciers and diminished snowpack result in reduced freshwater supply for melt-dependent people in the Himalayas, hence constraining the activities of millions downstream (Immerzeel et al. 2010). Increased sea levels facilitate the intrusion of saltwater into freshwater aquifers, hence polluting potable water sources and agricultural outputs for populations in susceptible, low-lying coastal regions (Werner and Simmons 2009). Ultimately, the patterns of climate-induced food and water shortages are not spatially distributed over the globe. Developing countries have a significant disadvantage. Sub-Saharan Africa and South Asia are emerging regions characterized by limited adaptive capacity and significant dependence on climate-sensitive sectors. Sub-Saharan Africa's dependence on climate for food production is exacerbated by predicted yield losses of up to 50% for several crops by 2020 (Schlenker and Lobell 2010). South Asia's dependence is linked to the nature of its monsoon seasons; insufficient rainfall or excessive and intense precipitation during the monsoon can result in crop failures and water wastage due to climatological unpredictability in start and intensity (Kumar et al. 2010). Malnutrition resulting from inadequate food supply and limited dietary diversity leads to stunted growth, weakened immune systems, and increased vulnerability to sickness and disease, especially in children. Insufficient access to clean water leads to heightened occurrences of cholera and dysentery, hence hampering future health initiatives (Levy et al. 2016).

Strategies to Reduce the Socioeconomic Burden

The necessity for both adaptation and mitigation strategies to address food and water insecurity resulting from climate change is evident:

1. **Climate-Smart Agriculture**: Lipper et al. assert the need for agricultural strategies that enhance yield and yield potential for climate change adaptation while concurrently diminishing greenhouse gas emissions.
2. **Water Management**: Molden asserts the necessity for technology and practices that enhance water usage efficiency or provide more sustainable water access. This includes drip irrigation, wastewater treatment and recycling initiatives, and rainfall gathering.
3. **Policy Interventions**: It is imperative to formulate and execute policies that bolster sustainable agriculture operations, mitigate water impact thresholds, and assist vulnerable people. This includes policy measures in infrastructure, research, and extension services necessary to improve adaptive capability (FAO 2016).

4. **International Collaboration**: Climate change is a worldwide concern necessitating collective attention. The Paris Agreement seeks to regulate global temperature rises through a framework of adaption and mitigation measures, including help for developing countries. Thus, efforts to mitigate and adapt to climate change benefit us all by promoting sustainability and enhancing access to food and water.

Strategies to Reduce the Socioeconomic Burden

Possible policies and interventions include social policy measures aimed at mitigating the socioeconomic impacts of environmental concerns and climate change. Africa's population is projected to double by 2050, exacerbating existing resource constraints; therefore, development initiatives must align with foreign investors' socially responsible strategies for sustainable energy development in Africa to avoid increasing the carbon footprint. Sustainable development initiatives must provide suitable energy utilization and generation to prevent exacerbating climate change (Nwokolo et al. 2023a, b, c, d, e, f). Moreover, suitable policies and actions would reduce the probability of adverse effects on vulnerable people while fostering chances for sustainable development. These include:

1. **Interventions in Social Policy**: Socio-political policy (a set of rules, regulations, and laws that govern social and political interactions and structures) establishes precarious safeguards. Numerous recent modifications to socio-political policy, as well as prior actions, have aimed to safeguard vulnerable groups, including newcomers and those requiring socio-political reforms. Recent modifications to socio-political policy that have proven enduring for immigrants encompass initiatives for social justice that support disadvantaged groups in need of socio-political fairness. Policies aimed at poverty alleviation that guarantee equitable resource distribution will mitigate socioeconomic issues. SDG 10 aims to reduce inequality both within and between nations; equitable access to resources, promoted throughout the years, is expected to guarantee that by 2030, the income growth of the lowest 40% of individuals in every country will surpass the per capita average growth of that country. By 2030, nations ought to focus on enhancing fairness via taxation, remuneration, and social safety frameworks (Martin 2021). Addressing poverty and its relationship with health is crucial for mitigating the impact of socioeconomic factors on health. The statistics on socioeconomic determinants of health indicate the relationship and resultant health impacts of alleviating poverty. Moreover, employment opportunities generated in high-demand sectors will mitigate poverty and alleviate wealth stagnation at a certain socioeconomic tier (Oh et al. 2021).
2. **Universal Basic Income (UBI)**: A Universal Basic Income (UBI) is a guaranteed socio-economic measure designed to prevent poverty and inequality. Intervention options to mitigate poverty equity should be socio-political rather than dependent on economic reasons for the enduring socio-engagement of human equity.

A micro intervention based on market rationale will not enable impoverished persons to participate in sustainable development; a Universal Basic Income is inherently universal (Lakavath 2021).

3. **Behavioral Interventions**: Behavioural Science contributes to alleviating poverty and socioeconomic disparity by researching and using scientific principles in socially significant initiatives. The behavioural phenomena explaining why individuals excessively respond emotionally and with goal-oriented actions to chronic stressors may eventually mitigate poverty and socioeconomic disparity by reducing the prevalence of such chronic stressors, which have lasting health implications. Community resilience is essential for mitigating obstacles to economic development and the impacts of environmental degradation. The greater a community's resilience to economic (or other) shocks, the less likely it is to disintegrate. Consequently, community resilience is crucial to future sustainable socioeconomic solutions.
4. Modifying socioeconomic welfare policies to enhance the sustainability of poverty eradication efforts. Poverty alleviation policies are distinctive to each country. Nations exist within countries, which are at varying levels of development. Consequently, measures that promote sustained responsible self-sufficiency would benefit individuals and alleviate socioeconomic burdens (Lakavath 2021).
5. **Diet Quality**: Socioeconomic Disparities in Nutritional Quality An inadequate diet is a major factor in illness prevalence, exhibiting strong socioeconomic and demographic disparities. Consequently, enhancing diet quality through socioeconomic disparities in food availability via micro-level interventions would simply intensify health equity disparities (Blake 2016).

Implications

Socio-economic implications of net zero policies are quite profound and far-reaching beyond mere environmental benefits. Commitment to greenhouse gas emissions reduction for net zero by countries and organizations leads towards a network of economic transformation, social equity, and technological innovation complexities.

Net zero can spur significant economic transformation by promoting growth in green industries and new job opportunities. According to Valero (2024), it means that a net zero economy requires new skills and competencies for green jobs. It would support employment in new emerging sectors, such as renewable energy, energy efficiency, and sustainable agriculture, but also the retraining and upskilling of the workforce. This would also mean that the transition process would not displace workers in such traditional industries while achieving justice and inclusiveness during the transition process. Socio-economic impacts related to net zero policies are mainly tied to the issues of equity and accessibility. A socially inclusive way to reach net zero must account for the needs of the most vulnerable populations who are likely affected by climate policies. Benefits of the transition, such as improved public health and reduced energy costs, should be equitably distributed across all segments of society; otherwise, policymakers will fail in the achievement of that goal (Middlemiss 2024). This requires targeted support measures and community engagement to empower those most affected by the transition.

In addition to these environmental aspects, net-zero policies have an economic implication related to fiscal aspects. Investing in sustainable infrastructure and technologies can indeed come at some short-term cost, but such investments will benefit the economy through improved resilience against climate impacts and health cost reductions in the long run, as proposed by various studies such as those of An et al. (2023). Policymakers, therefore, need to have a long-term perspective when considering the net zero initiative in terms of their economic feasibility. Apart from economic feasibility challenges, the net zero transition involves significant technological innovation and infrastructure development challenges. As Mushafiq (2024) asserts, notable investment in low-carbon technologies must be accomplished for net zero policies to gain full implementation. This calls for collaboration between the public and private sectors to spur innovation and ensure that these technologies are available and affordable for all.

Conclusion, Limitations and Future Scope

This chapter underscores the critical role of international collaboration in achieving climate objectives. It emphasizes that, without financial, technological, and educational resources, emerging nations will struggle to attain net zero by 2050. Additionally, the chapter recognizes the potential for grassroots and localized initiatives that include indigenous knowledge and address the specific needs of certain regions. Consequently, the Paris Accord is recognized as one of several worldwide initiatives, alongside research outcomes that contribute to the integration of Sustainable Development Goals (SDGs). Ultimately, Chapter 4 demonstrates that achieving net zero in poor nations is equally sustainable for social fairness and climate change.

References

An K, Zhang S, Zhou J, Wang C (2023) How can computable general equilibrium models serve low-carbon policy? a systematic review. Environ Res Lett 18(3):033002. https://doi.org/10.1088/1748-9326/acbbe2

Black R, Bennett SR, Thomas SM, Beddington JR (2011) Migration as adaptation. Nature 478(7370):447–449. https://doi.org/10.1038/478477a

Blake MK (2016) Socioeconomic inequalities in diet quality: from identifying the problem to implementing solutions. Curr Nutr Reports 5(3):178–186. https://doi.org/10.1007/s13668-016-0167-5

Büchs M, Bardsley N, Duwe S (2011) Who bears the brunt? Distributional effects of climate change mitigation policies. Crit Soc Policy 31(2):285–307. https://doi.org/10.1177/0261018310396036

Cohen AJ, Brauer M, Burnett R, Anderson HR, Frostad J, Estep K, Balakrishnan K, Brunekreef B, Dandona L, Dandona R, Feigin V, Freedman G, Hubbell B, Jobling A, Kan H, Knibbs L, Liu Y, Martin R, Morawska L, Pope CA, Shin H, Straif K, Shaddick G, Thomas M, van Dingenen R, van Donkelaar A, Vos T, Murray CJL, Forouzanfar MH (2017) Estimates and 25-year trends of the global burden of disease attributable to ambient air pollution: an analysis of data from the Global Burden of Diseases Study 2015. The Lancet 389(10082):1907–1918. https://doi.org/10.1016/S0140-6736(17)30505-6

Diffenbaugh NS, Swain DL, Touma D (2015) Anthropogenic warming has increased drought risk in California. Proc Natl Acad Sci 112(13):3931–3936. https://doi.org/10.1073/pnas.1422385112

Dorband II, Jakob M, Kalkuhl M, Steckel JC (2019) Poverty and distributional effects of carbon pricing in low- and middle-income countries – a global comparative analysis. World Dev 115:246–257. https://doi.org/10.1016/j.worlddev.2018.11.015

FAO (2016) The state of food and agriculture: climate change, agriculture and food security. Food and Agriculture Organization of the United Nations. http://www.fao.org/3/a-i6030e.pdf

Gemenne F (2011) Why the numbers don't add up: a review of estimates and predictions of people displaced by environmental changes. Glob Environ Chang 21(Supp. 1):S41–S49. https://doi.org/10.1016/j.gloenvcha.2011.09.005

Immerzeel WW, van Beek LPH, Bierkens MFP (2010) Climate change will affect the Asian water towers. Science 328(5984):1382–1385. https://doi.org/10.1126/science.1183188

Intergovernmental Panel on Climate Change (IPCC) (2018) Global warming of 1.5°C. https://www.ipcc.ch/sr15/

International Labour Organization (ILO) (2015) Guidelines for a just transition towards environmentally sustainable economies and societies for all. https://www.ilo.org/global/topics/green-jobs/publications/WCMS_432859/lang--en/index.htm

International Organization for Migration (IOM) (2019) Environmental migration. https://www.iom.int/environmental-migration

Kumar KK, Rajagopalan B, Hoerling M, Bates G, Cane M (2010) Unraveling the mystery of Indian monsoon failure during El Niño. Science 314(5796):115–119. https://doi.org/10.1126/science.1131152

Kumar K, Singh R, Sharma A, Lehri BPS (2024) Climate change adaptation strategies for achieving net-zero economy. In: Singh R, Crowther D (eds) Transition towards a sustainable future. Approaches to global sustainability, markets, and governance. Springer, Singapore. https://doi.org/10.1007/978-981-97-5756-5_8

Lakavath G (2021) Poverty alleviation strategies: economic, social, and justice perspectives. In: Development challenges and solutions, pp 95–110. Springer. https://doi.org/10.1007/978-3-030-56227-4_6

Lal R (2009) Soil degradation as a reason for inadequate human nutrition. Food Secur 1(1):45–57. https://doi.org/10.1007/s12571-009-0009-z

Levy K, Woster AP, Goldstein RS, Carlton EJ (2016) Untangling the impacts of climate change on waterborne diseases: a systematic review of relationships between diarrheal diseases and temperature, rainfall, flooding, and drought. Environ Sci Technol 50(10):4905–4922. https://doi.org/10.1021/acs.est.5b06186

Martin G (2021) Sustainable development and income inequality: goals and challenges. In: Encyclopedia of sustainability, pp 1–10. Springer. https://doi.org/10.1007/978-3-319-71060-0_37-1

McLeman R (2014) Climate and human migration: past experiences, future challenges. Cambridge University Press. https://doi.org/10.1017/CBO9781139136938

Middlemiss L (2024) Place-based and people-centred: principles for a socially inclusive net zero transition. Geo Geogr Environ 11(2). https://doi.org/10.1002/geo2.157

Mushafiq M (2024) Net-zero policy and forward default risk in the energy sector: evidence of corporate environmentalism using (a)symmetric models. Bus Strateg Environ 34(1):1475–1493. https://doi.org/10.1002/bse.4058

Nawrotzki RJ, DeWaard J (2018) Climate shocks and the timing of migration from Mexico. Popul Environ 39(2):143–164. https://doi.org/10.1007/s11111-017-0284-y
Newell P, Mulvaney D (2013) The political economy of the 'just transition.' Geogr J 179(2):132–140. https://doi.org/10.1111/geoj.12008
Nwokolo SC, Singh R, Khan S, Kumar A, Luthra S (2023a) Technological pathways to net-zero goals in Africa. In: Africa's path to net-zero. CSR, sustainability, ethics & governance. Springer, Cham. https://doi.org/10.1007/978-3-031-44514-9_5
Nwokolo SC, Singh R, Khan S, Kumar A, Luthra S (2023b) Scenarios that could give rise to an African net-zero energy transition. In: Africa's path to net-zero. CSR, sustainability, ethics & governance. Springer, Cham. https://doi.org/10.1007/978-3-031-44514-9_8
Nwokolo SC, Singh R, Khan S, Kumar A, Luthra S (2023c) Decarbonizing hard-to-abate sectors in Africa. In: Africa's path to net-zero. CSR, sustainability, ethics & governance. Springer, Cham. https://doi.org/10.1007/978-3-031-44514-9_6
Nwokolo SC, Singh R, Khan S, Kumar A, Luthra S (2023d) Remedies to the challenges of renewable energy deployment in Africa. In: Africa's path to net-zero. CSR, sustainability, ethics & governance. Springer, Cham. https://doi.org/10.1007/978-3-031-44514-9_3
Nwokolo SC, Singh R, Khan S, Kumar A, Luthra S (2023e) Impacts of climate change in Africa. In: Africa's path to net-zero. CSR, sustainability, ethics & governance. Springer, Cham. https://doi.org/10.1007/978-3-031-44514-9_7
Nwokolo SC, Singh R, Khan S, Kumar A, Luthra S (2023f) Introduction: Africa's net zero transition. In: Africa's path to net-zero. CSR, sustainability, ethics and governance. Springer, Cham. https://doi.org/10.1007/978-3-031-44514-9_1
Oh S, Kim Y, Choi J (2021) Job training and vocational education for poverty alleviation. J Econ Policy Stud 29(3):12–25. https://en.wikipedia.org/wiki/Poverty
Oyedele OM (2025) The cobra effect: exploring the dark side of innovation and environmental sustainability in developing nations. In: Singh R, Joshi A, Filho WL, Khan S (eds) Zero carbon industry, eco-innovation and environmental sustainability. World sustainability series. Springer, Cham. https://doi.org/10.1007/978-3-031-80220-1_15
Oyenuga MO, Omale SA (2024) Is Africa Jinxed? Exploring the challenges of technology access and adoption in Africa. Afr J Econ Sustain Dev 7(4):142–161. https://doi.org/10.52589/AJESDULN1LRNF
Valero A (2024) Net zero and the labour market: evidence from the UK. LSE Public Policy Rev 3(2). https://doi.org/10.31389/lseppr.97
Popp D, Vona F, Marin G, Chen Z (2020) The employment impact of green fiscal push: evidence from the American Recovery Act. Brook Pap Econ Act 2020(1):1–77. https://doi.org/10.1353/eca.2020.0001
Singh R, Crowther D, Ravesangar K (2024a) Harnessing human capital for a sustainable future: HR's pivotal role toward net zero policies. In: Singh R, Crowther D (eds) Transition towards a sustainable future. Approaches to global sustainability, markets, and governance. Springer, Singapore. https://doi.org/10.1007/978-981-97-5756-5_2
Singh R, Gupta A, Kandpal V (2024b) ESG excellence in the digital world. In: Fintech for ESG and the circular economy, pp 193–204. Springer. https://pure.jgu.edu.in/id/eprint/8609/
Watts N, Amann M, Arnell N, Ayeb-Karlsson S, Belesova K, Berry H, Bouley T, Boykoff M, Byass P, Cai W, Campbell-Lendrum D, Chambers J, Daly M, Dasandi N, Davies M, Depoux A, Dominguez-Salas P, Drummond P, Ebi KL et al (2018) The 2018 report of the Lancet Countdown on health and climate change: Shaping the health of nations for centuries to come. The Lancet 392(10163):2479–2514. https://doi.org/10.1016/S0140-6736(18)32594-7
Zhao S, Fujimori S, Takahashi K (2022) Poverty impacts of carbon pricing: Implications for climate policy design. Environ Res Lett 17(3):034027. https://doi.org/10.1088/1748-9326/ac5574

Chapter 5
Climate Resilience and Adaptation Strategies for Developing Nations

Abstract This chapter identifies significant advancements in climate resilience and adaptation from emerging nations susceptible to climate change and unpredictability. It evaluates catastrophe mitigation for sustainable development using innovative, grassroots methodologies. Initially, it notes that underdeveloped nations must promote climate-resilient agriculture. Consequently, the use of climate-resilient agricultural methods is advised. Drought-resistant crops, agroforestry, and effective irrigation techniques enhance food security in precarious situations. This chapter highlights the implementation of climate-resilient infrastructure and transportation. Developing nations may alleviate train disruptions and washed-out roads caused by extreme storms and increasing sea levels through climate-resilient highways, storm-resistant bridges, and multimodal infrastructure systems in urban swamps. The chapter indicates that an effective health systems architecture, illness monitoring and reporting, and skilled medical practitioners are essential to address climate-sensitive health challenges, ranging from high heat to vector- and water-borne infections. Finally, community-oriented adaptation is recommended as a grassroots strategy.

Introduction

The impacts of climate change are increasingly evident, disproportionately affecting developing nations that lack the infrastructure and resources to mitigate its effects. Often grappling with poverty and economic instability (Oyenuga and Omale 2024), these nations face heightened vulnerability to natural disasters, shifting weather patterns, and rising sea levels. As global temperatures rise and environmental unpredictability intensifies, the need for robust climate resilience and adaptation strategies becomes ever more pressing. Central to the discourse is the concept of climate-resilient agriculture, which is critical for ensuring food security and supporting the livelihoods of vulnerable populations. Agriculture, a primary economic activity in many developing nations, is profoundly susceptible to climate variability.

M. 0. Oyenuga et al., *Smart Technologies for Climate Change and Net Zero Policies: Practical Approaches Towards Sustainability*, Climate Change Management,
https://doi.org/10.1007/978-3-031-92221-3_5

Health systems also emerge as a critical component of climate resilience. Rising temperatures and altered precipitation patterns have led to an increase in vector-borne diseases, heat-related illnesses, and malnutrition. Initiatives such as climate-adaptive hospitals, health workforce training, and the integration of telehealth technologies are discussed as effective measures to strengthen public health in the face of climate change. Therefore, this chapter provides a comprehensive overview of climate resilience and adaptation strategies tailored to the unique challenges of developing nations.

Building Climate Resilience in Agriculture

As countries become aware of net-zero objectives intensified by the Paris Agreement's aim to restrict global temperature rises to no more than 2 °C above pre-industrial levels, decarbonization emerges as a universally endorsed standard. Consequently, decarbonizing buildings to achieve a low-carbon economy is the objective of construction policy advancements and inter-agency cooperation (Joshi et al. 2024; Nwokolo et al. 2023a, b, c, d, e, f, g). Climate resilience in agriculture is crucial for food security, vulnerable populations, and ecosystem dynamics. This resilience arises from several advancements necessary to enhance the agricultural system's capacity to predict, withstand, and adjust to climate forecasts and pressures. Agriculture is susceptible to climate and climate change due to the disruption of food production systems, the global economy, and ecological integrity. Climate resilience in agriculture refers to the ability of agricultural systems to predict, adapt to, respond to, and recover from the impacts of climate change.

The use of certain methods enhances the adaptability and sustainability of agricultural systems, ensuring their operation amid shifting climatic patterns. It consists of the following:

1. **Climate-Smart Agriculture (CSA)**: This enables practitioners to sustain food security and agricultural productivity while fostering resilience and mitigating greenhouse gas emissions. These pertain to genetically modified organisms (GMOs) that enhance drought-resistant flora, optimize irrigation and drainage systems, and include activities such as agroforestry. CSA embodies a pragmatic approach to relative sustainability for future agricultural growth against climate change (FAO 2013).
2. Diversifying crops and income reduces dependence on and susceptibility to weather disturbances while enhancing sustainability. For instance, those who engage in polyculture typically through inter-planting or companion planting mitigate their risks and provide an alternate source of income in the event of a crop failure. Moreover, the inclusion of livestock, aquaculture, or agroforestry systems enhances resilience in agricultural systems. Diversity management improves the resilience of agricultural systems (Lin 2011).

3. **Ecologically sound water management techniques**: Resilience is derived from enhanced water management techniques, including rainwater gathering, drip irrigation, and the construction of water retention systems. Such measures mitigate the effects of droughts and variable precipitation. Rockström et al. (2010) assert that inadequate water management will hinder the performance and resilience of rainfed agriculture.
4. **Enhanced Soil Quality**: Soil management practices such as conservation tillage, cover cropping, and organic amendments enhance soil quality, resulting in improved water retention and increased nutrient availability. Superior soils exhibit less susceptibility to erosion and deterioration, hence facilitating more stable and sustainable agricultural opportunities. Soil management influences climate change equality and mitigation (Lal 2009).
5. **Agroforestry Systems**: The tree canopy offers shade, serves as a windbreak and facilitates erosion management. Transitioning from rudimentary agroforestry practices to advanced systems enhances biodiversity and microclimates, protecting susceptible crops from severe ecological stressors. According to Garrity (2004), agroforestry serves to diversify and strengthen vulnerable systems within smallholder farms.
6. **Early Warning Systems and Climate Information Services**: In the absence of climate services, policy formulation becomes arbitrary, and institutional resources remain unfunded for extended periods. This reliance on empirical data provides a stable future characterized by steady climatic patterns, reassuring policymakers and agricultural professionals of the strategy's effectiveness. Persistent climatic changes over the past few decades warrant the establishment of an institutional resource for both small and large-scale participation. A significant portion of this resilience derives from policy and a robust institutional structure. In the presence of governments, there will be access, extension services, and the establishment of conducive settings, among other factors. Climate-resilient agriculture is ineffective without policies (World Bank Group 2023).

Also, farmers require education on sustainable agriculture (Fasanmi et al. 2024), climate change adaptation, and resource management to effectively adopt resilient practices. Agricultural extension services and farmer field schools would serve as suitable mediums for this educational dissemination. Rivera and Alex (2004) and Omale et al,(2024) assert that agricultural extension contributes to resilience and entrepreneurship alongside funding mechanisms and insurance.

The availability of financial tools, such as micro-financing, loans, or weather-indexed insurance. provides a sense of security and support to farmers, enhancing their confidence in the strategy's effectiveness. Hazell et al. (2010) evaluate insurance for agricultural risks, further reinforcing the importance of financial support in our strategy. When communities actively participate in the planning and implementation of adaptations, interventions are more aligned with community needs and cultural contexts. This community-based approach leverages indigenous knowledge and fosters community buy-in, thereby enhancing the sustainability of resilience and adaptation initiatives. Reid et al. propose that community-based adaptation enhances resilience, making the community an integral part of the process. This comprehensive strategy involves climate-smart agriculture, diversification, soil and water conservation, sustainable resource management, and policy intervention to establish an agricultural system adaptable to climate change, ensuring both short- and long-term food security and livelihood sustainability.

Climate-Proofing Infrastructure and Transportation Systems

Climate change endangers global infrastructure and transportation networks. Consequently, climate change necessitates reengineering global infrastructure and transportation networks to prevent further calamity. The ramifications of impacts whether systems are impaired or disruptions arise from severe storms, supply networks, and infrastructure undermined by climate change have hampered global business operations. (Kumar et al. 2024). The first phase involves doing this in Africa. Africa needs a basic framework like the EU Climate Law to support its transition effort and promote sustainable policies, practices, and technology. Secondly, Europe and global stakeholders must collaborate with Africa to secure financial, technical, and human resource investments to facilitate climate change initiatives. Third, African politicians must recognize the significance of educational access and the development of renewable energy skills, with governmental subsidies to stimulate and invest in clean energy technologies to foster creative initiatives. Nwokolo et al. (2023a, b, c, d, e, f, g).

Consequently, climate-resilient infrastructure encompasses planning, design, building, and maintenance modifications aimed at enduring pressures for which resilience potential is present. The strains resulting from climate change include sea-level rise, heightened rainfall, intensified temperature extremes, and increased storms and storm occurrences. The characteristics of location and transit networks can cause varying experiences for residents, resulting in significant alterations to even the most robust infrastructures, delayed operational systems, and unforeseen maintenance expenditures. For instance, intensified storms can undermine the foundation of an above-ground train; heightened rains can saturate a tube station; elevated temperatures can deteriorate the asphalt on surface roads or distort train tracks, all leading to adverse consequences for safety and service (Gössling et al. 2023). As a result, climate resiliency initiatives must extend beyond theoretical applications to mitigate stressors for effective transportation systems (Apata et al. 2024). Planners and policymakers must comprehend the strains imposed by climatic conditions and conduct integrated assessments to evaluate how existing infrastructure will respond to these climate risks, enhancing preparedness for such events (Oyedele 2025).

Vulnerability evaluations identify susceptible assets, whereas climate forecasts can supplement the essential design requirements. This adaptive design recognizes that climate change will inevitably lead to more challenges over time (Chester and Allenby 2019). Furthermore, novel monitoring technologies provide real-time assessment of structural integrity, enabling maintenance before calamities occur.

When a sensor identifies a problem such as an internal defect or an external difficulty like an increased weight on a bridge or ocean acidification destabilising a seawall a network of sensors will evaluate the necessary repairs before further tragedies occur. Repairs throughout time are evaluated based on anticipated maintenance facilitated by climate forecasting (Gössling et al. 2023).

Consequently, climate-resilient infrastructure depends on a legislative and regulatory framework. Chester and Allenby (2019) opine that implementing the revised building regulations and establishing a new assessment standard should be implemented, ensuring all requisite activities adhere to climate resiliency criteria. Furthermore, Chester and Allenby (2019) suggest that there should be provision of financial incentives for individuals engaging in climate-resilient initiatives and government funding for research and development projects related to climate adaptations.

Furthermore, informing stakeholders on the implications of climate-proofing should render activities more feasible (Gössling et al. 2023). It is also essential to evaluate the locations of climate-proofing efforts. The Netherlands employs several flood protection strategies, such as storm surge barriers, and most of its infrastructure is raised to counteract sea level rise. New York City is implementing climate resilience measures through investments in green infrastructure to prevent flooding along the southeastern shore and combat sea-level rise, while also addressing the urban heat island effect (Chester and Allenby 2019). However, these are not the sole factors that diminish the competitively viable attractiveness of renewable energy solutions in Africa.

In addition, the lack of necessary infrastructural systems for the advancement and implementation of sustainable energy technologies is an additional obstacle. Therefore, given the absence of economic advancements in the region, the prospects for growth in sustainable energy solutions are impeded. However, obstacles such as budget limitations, insufficient information, and political apathy hinder the implementation of climate-hardening measures. Mandated climate resilience is a cooperative initiative including the law, the economy, and the people. Ultimately, all critical elements of infrastructure will be renovated and designed with an emphasis on climate resilience, permitting the use of innovative materials and designs, while also contemplating international collaborations for knowledge and resource sharing (Chester and Allenby 2019). Infrastructure and transport networks must be fortified against climatic impacts to guarantee functionality and safety.

Strengthening Health Systems for Climate Resilience

Enhancing international health systems infrastructure to withstand climate change is essential. In other words, health systems need to be bolstered to predict, subsidize, and recover (in a stabilized fashion) climate-induced health issues. It is a predictive approach to health services and public safety in response to environmental changes. Climate change intersects with health through various contexts, complicating the determinants of health. Rising temperatures, altered precipitation patterns, and escalating weather-related disasters exacerbate the incidence of vector-borne diseases, heat-related illnesses, and malnutrition. Research suggests that malaria ascends to higher altitudes due to climate change, necessitating immediate operational measures for adaptation (Githeko et al. 2000). Thus, planners and policymakers must access this information to anticipate future needs if conditions remain unchanged, preventing adverse outcomes. This intersection is already apparent with Health National Adaptation Plans (HNAP), which provides a framework for nations to integrate a climate change responsive approach into their public health strategies.

The HNAP comprises the necessary actions for health systems to attain climate resilience through anticipatory, absorptive, and transformative capacities amid climate change uncertainty (World Health Organisation [WHO] 2021). Furthermore, adequate monitoring is essential for climate-related diseases and environmental health information. For example, early warning systems are developed so that intervention can occur before a public health concern is realized; a similar situation is a weather pattern that's suggestive that, shortly, flu or dengue fever will occur, which is important to know to predict resources and intervention (Hii et al. 2012). In addition, health systems need to be freshly designed or adapted to meet climate-based calamities. Hospitals should be stable during and after catastrophes, and operable for other climate-based difficulties. This covers energy-reliant and water-dependent systems. The notion of climate-adaptive hospitals promotes the idea, that health systems should be able to operate to withstand the pressures of climate change. (Watts et al. 2018).

Health workforce training is a crucial factor in effective climate adaption. Courses on climate-related health consequences, catastrophes, and disaster adaptation will enable training from the most basic levels up to all levels of response. Trained personnel will be more able to aid when climate change brings health-related difficulties. This is needed by the WHO.

In addition, health professionals should work with their people to embrace effective climate-adaptive strategies from the inside out; such integration promotes buy-in and the success of such adaptations. For example, training helps folks better understand what climate forecasts bring to their health and how to better alter their actions medically. Some successful adults function well in the community (Reid et al. 2009). The Lancet Commission on Health and Climate Change emphasises that climate change initiatives must incorporate an equitable approach for all.

Furthermore, a strategy that depends on technology innovation is more comprehensive for a network of health systems. Telehealth, mHealth applications, and big data analytics enhance access and potential for disease surveillance. Solar-powered health systems for marginalised populations provide a more sustainable and resilient healthcare strategy (Kumar and Pal 2018). Moreover, in the absence of financial resources, health systems are unable to execute climate-resilient solutions. Assess the capacity to get funding from diverse sources for resilience, including global donors, public health-philanthropy collaborations, and climate-specific funds.

The Global Fund recognises the necessity of incorporating climate change considerations into health-related finance (The Global Fund 2019). Health concerns associated with climate change are addressed through interdisciplinary areas. For instance, by involving environmental considerations, agricultural practices, and zoning regulations. This intersection aids the researcher or stakeholder in comprehending the holistic nature of a preventative approach and recognising the factors contributing to vulnerability. The method to establish this link for improved health equity is derived from the Health in All Policies effort promoted by the WHO for this multisectoral requirement. Consequently, when health systems adapt to climate resilience, public health will be better protected from the impacts of climate change.

Community-Based Adaptation (CBA) Strategies

Community-based adaptation is an appropriate approach to reduce climate change impacts in vulnerable regions. CBA projects emerge from the notion that the ultimate local community is engaged and empowered and, therefore, has the resources at its disposal in other implementation forms to achieve a more suitable adaptation for their niche and socioeconomic sustainability. CBA is a bottom-up project; community involvement aids in the planning and execution and uses community resources for adaptation project sustainability. Therefore, adaptations are made that suit those who need them the most because the success and feasibility of an adaptation project depend upon culturally sensitive means of implementation (Reid et al. 2009). CBA acknowledges resources for vulnerable communities, increasing community resilience and decreasing vulnerability.

Several core components define effective CBA strategies:

1. Participatory Vulnerability Assessments (Smit and Wandel 2006).
2. Application of Traditional/Indigenous Ecological Knowledge (Berkes 2009).
3. Empowerment: This ensures that communities are educated and empowered to effectively apply and sustain adaptations (Ensor and Berger 2009).
4. Flexible and Iterative Planning (Tompkins and Adger 2004).

Examples of CBA projects include:

a. **Agriculture**. Farmers in Bangladesh have constructed floating gardens to ensure they can cultivate their crops during the rainy season, which reduces the negative impact of flooding on food supply (Huq and Reid 2007).
b. **Water**: Communities in Nepal have created rainwater harvesting systems to alleviate concern over water scarcity due to uneven rainfall amounts and patterns (Regmi and Bhandari 2013).
c. **Disaster Risk Reduction**: Communities in the Philippines have replanted mangrove forests to reduce storm surge and sedimentation, which has led to improvements in ecological and socio-cultural resiliency (Walters et al. 2008).

The benefits of CBA include:

a. **Increased Relevance and Effectiveness**: When the adaptation is for the community and by the community, it is more likely that community vulnerabilities will be addressed and more likely that adoption will be facilitated (Ayers and Forsyth 2009).
b. **Improved Social Networks**: Inclusion in the adaptation process means that people are more in tune with their bettered communities and personally empowered, thus improving climate outcomes (Dodman and Mitlin 2013).
c. **Increased Sustainability**: Community-based adaptation comes from already-known methods and community capacity, which reduces resource consumption and increases sustainable efforts (Reid et al. 2009).

National-Level Adaptation Frameworks and Planning

Although the implementation may present certain challenges, incorporating CBA as an internationally recognised standard for adaptive strategies aimed at global contributions demonstrates its efficacy in achieving effective implementation, sustainability of benefits, and community integration. Moreover, the recent advancements in institutional supervision of the CBA, the promotion of awareness and education regarding necessary resources for the CBA, and the chances for community involvement to propose potential changes are promising for a more cohesive and resilient future.

Certain nations have formulated and executed their national adaptation programs in a somewhat distinct manner:

a. **Bangladesh**: The National Adaptation Programme of Action (NAPA) for Bangladesh was instituted in 2005 in response to its status as one of the most climate-vulnerable nations globally. NAPA delineates the immediate priorities that require attention in the short term. Subsequently, the Bangladesh Climate Change Strategy and Action Plan (BCCSAP) was formulated in 2009 to focus on adaptation and mitigation efforts, as outlined by the Government of Bangladesh (2009).
b. **Germany**: The Adaptation Strategy of Germany, established in 2008, aims to mitigate susceptibility to the impacts of climate change inside the country. Adaptation will be achieved through spatial policy integration and inter-sectoral policy integration. A monitoring system is in place to assess the advancement of integrative adaptations and the overall success levels (Federal Ministry for the Environment, Nature Conservation and Nuclear Safety [BMU] 2011).
c. **South Africa**: The National Climate Change Response White Paper of South Africa, which encompasses mitigation and adaptation strategies, was published in 2011. It underscores the necessity for dynamic adaptation, driven by the socio-economic realities of South Africa (Department of Environmental Affairs 2011).

Implications

The gap signifies that climate change should be prioritized in the national development policy agenda. This necessitates policymakers formulate a holistic policy environment that fosters investment in renewable energy, enhances energy efficiency, and ensures equitable access to climate-resilient resources. Moreover, the rules must alleviate the drawbacks of private sector participation and facilitate the execution of public–private partnerships, which depend on adaptable capabilities and resources for developmental objectives that benefit all stakeholders. This research is particularly pertinent now as net zero strategies intersect with the Sustainable Development Goals (SDGs), enabling simultaneous advancements in global initiatives through targeted modifications. Consequently, the findings pertain to SDG 13: Climate Action, indicating an increased necessity for emissions reductions and the development of resilience against the impacts of climate change. The development and expansion of renewable energy and sustainable facilities will contribute to SDG 7: Affordable and Clean Energy and SDG 9: Industry, Innovation, and Infrastructure.

Ultimately, sustainable agriculture initiatives and low-carbon projects in urban environments will support this study's conclusions regarding SDG 2: Zero Hunger and SDG 11: Sustainable Cities and Communities. Future studies should assess financing methods, necessary instruments, green bonds, climate funds, and other mechanisms to address the financial deficit required for a Net Zero transition. Furthermore, assessing technical requirements by location and projected socio-economic advantages will enable effective climate change adaptation. An examination of the Net Zero efforts of other nations would enhance global awareness and keep the audience engaged in the ongoing global efforts.

Conclusion, Limitations, and Future Scope

This chapter concludes that the global initiative to attain and balance Net Zero emissions, particularly from the viewpoint of emerging nations, is a formidable challenge. This chapter concludes that renewable energy alternatives are essential, technologies must be adapted for optimal human utilisation, and regulatory frameworks should be established proactively to prevent delays in necessary climate change initiatives. Moreover, while similar tools for action originate from both developed and developing countries, the socioeconomic and infrastructural conditions in developing regions contrast with the resource availability in developed regions, necessitating a more specific approach. The knowledge acquired pertains to international collaboration for financing, technological exchange, and training prospects. The Net Zero Targets of developing countries are crucial for global climate success and provide these nations with equity and less susceptibility to climate change.

Furthermore, the focus on enhanced renewable energy, innovative sustainable initiatives, and climate-resilient infrastructure will inspire a sense of hope and progress. This chapter illustrates that worldwide cooperation and nationally led initiatives can bridge the divide between contemporary capitalist objectives and future climate-focused realities. As governments align with the broader objectives while addressing the identified requirements for international and local short-term initiatives, they will recognize that local communities are essential for the success of the Sustainable Development Goals (SDGs). This chapter illustrates how the intersections of national and local efforts might achieve emissions objectives while fostering the fairness and resilience essential for a sustainable future.

References

Apata AO, Oyenuga MO, Abeoba M, Ugom A (2024) Internet of Things (IoT) solutions for smart transportation infrastructure and fleet management. J Propuls Technol 45(4):1492–1505. https://propulsiontechjournal.com/index.php/journal/article/view/8325

Berkes F (2009) Indigenous ways of knowing and the study of environmental change. J R Soc N Z 39(4):151–156. https://doi.org/10.1080/03014220909510568

Chester MV, Allenby B (2019) Toward adaptive infrastructure: flexibility and agility in a non-stationarity age. Sustain Resil Infrastruct 4(4):173–191. https://doi.org/10.1080/23789689.2017.1416846

Department of Environmental Affairs (2011) National climate change response white paper. Government of South Africa. https://www.environment.gov.za

Fasanmi OO, Olusanya BB, Oyenuga MO (2024) Entrepreneurial training as a retirement succorance for Nigerian retirees twist 19(4):137–147. https://doi.org/10.5281/zenodo.10049652#321

Federal Ministry for the Environment, Nature Conservation and Nuclear Safety (BMU) (2011) Germany's national strategy for climate change adaptation. https://www.bmu.de

Food and Agriculture Organization of the United Nations (FAO) (2013) Climate-smart agriculture sourcebook. FAO. https://www.fao.org/climate-smart-agriculture-sourcebook/en/

Garrity DP (2004) Agroforestry and the achievement of the millennium development goals. Agrofor Syst 61(1):5–17. https://doi.org/10.1023/B:AGFO.0000028986.37502.7c

Githeko AK, Lindsay SW, Confalonieri UE, Patz JA (2000) Climate change and vector-borne diseases: a regional analysis. Bull World Health Organ 78(9):1136–1147. https://www.scielosp.org/article/bwho/2000.v78n9/1136-1147/

Gössling S, Neger C, Steiger R, Bell R (2023) Weather, climate change, and transport: a review. Nat Hazards 118:1341–1360. https://doi.org/10.1007/s11069-023-06054-2

Government of Bangladesh (2009) Bangladesh climate change strategy and action plan (BCCSAP). Ministry of Environment and Forests. https://www.climatechange.gov.bd

Hazell PB, Pomareda C, Valdés A (eds) (1986) Crop insurance for agricultural development: issues and experience. Johns Hopkins University Press. https://archive.org/details/cropinsurancefor0000unse

Hii YL, Zhu H, Ng N, Ng LC, Rocklöv J (2012) Forecast of dengue incidence using temperature and rainfall. PLoS Negl Trop Dis 6(11):e1908. https://doi.org/10.1371/journal.pntd.0001908

Joshi A, Singh R, Kumar V, Rani S (2024) Net zero strategy: exploring opportunities, confronting challenges for developing countries. In: Singh R, Crowther D (eds) Transition towards a sustainable future. Approaches to global sustainability, markets, and governance. Springer, Singapore. https://doi.org/10.1007/978-981-97-5756-5_11

Lal R (2009) Soil degradation as a reason for inadequate human nutrition. Food Secur 1(1):45–57.https://doi.org/10.1007/s12571-009-0009-z, https://link.springer.com/article/10.1007/s12571-009-0009-z

Lin BB (2011) Resilience in agriculture through crop diversification: adaptive management for environmental change. BioScience 61(3):183–193.https://doi.org/10.1525/bio.2011.61.3.4, https://academic.oup.com/bioscience/article/61/3/183/238071

Kumar K, Singh R, Sharma A, Lehri BPS (2024) Climate change adaptation strategies for achieving net-zero economy. In: Singh R, Crowther D (eds) Transition towards a sustainable future. Approaches to global sustainability, markets, and governance. Springer, Singapore. https://doi.org/10.1007/978-981-97-5756-5_8

Kumar P, Pal R (2018) India achieves WHO recommended doctor population ratio: a call for paradigm shift in public health discourse! J Fam Med Prim Care 7(5):841–844. https://doi.org/10.4103/jfmpc.jfmpc_218_18

Nwokolo SC, Singh R, Khan S, Kumar A, Luthra S (2023a) Scenarios that could give rise to an African net-zero energy transition. In: Africa's path to net-zero. CSR, sustainability, ethics & governance. Springer, Cham. https://doi.org/10.1007/978-3-031-44514-9_8

Nwokolo SC, Singh R, Khan S, Kumar A, Luthra S (2023b) Africa's awakening to climate action. In: Africa's path to net-zero. CSR, sustainability, ethics & governance. Springer, Cham. https://doi.org/10.1007/978-3-031-44514-9_9

Nwokolo SC, Singh R, Khan S, Kumar A, Luthra S (2023c) Remedies to the challenges of renewable energy deployment in Africa. In: Africa's path to net-zero. CSR, sustainability, ethics & governance. Springer, Cham. https://doi.org/10.1007/978-3-031-44514-9_3

Nwokolo SC, Singh R, Khan S, Kumar A, Luthra S (2023d) Decarbonizing hard-to-abate sectors in Africa. In: Africa's path to net-zero. CSR, sustainability, ethics & governance. Springer, Cham. https://doi.org/10.1007/978-3-031-44514-9_6

Nwokolo SC, Singh R, Khan S, Kumar A, Luthra S (2023e) Technological pathways to net-zero goals in Africa. In: Africa's path to net-zero. CSR, sustainability, ethics & governance. Springer, Cham. https://doi.org/10.1007/978-3-031-44514-9_5

Nwokolo SC, Singh R, Khan S, Kumar A, Luthra S (2023f) Global investment and development in Africa. In: Africa's path to net-zero. CSR, sustainability, ethics & governance. Springer, Cham. https://doi.org/10.1007/978-3-031-44514-9_2

Nwokolo SC, Singh R, Khan S, Kumar A, Luthra S (2023g) Influencing the scale of Africa's energy transition. In: Africa's path to net-zero. CSR, sustainability, ethics & governance. Springer, Cham. https://doi.org/10.1007/978-3-031-44514-9_4

Omale SA, Oyenuga MO, Abdullahi D, Madu I (2024) Moderating role of entrepreneurial education on innovative work behaviour and resilience of SMEs. Innovations 77(2):1166–1185 e0ce0-1166-1185.11600.pdf

Oyedele OM (2025) The cobra effect: exploring the dark side of innovation and environmental sustainability in developing nations. In: Singh R, Joshi A, Filho WL, Khan S (eds) Zero carbon industry, eco-innovation and environmental sustainability. World sustainability series. Springer, Cham. https://doi.org/10.1007/978-3-031-80220-1_15

Oyenuga MO, Omale SA (2024) Is Africa Jinxed? Exploring the challenges of technology access and adoption in Africa. Afr J Econ Sustain Dev 7(4):142–161. https://doi.org/10.52589/AJESDULN1LRNF

Regmi BR, Bhandari D (2013) Climate change governance and funding dilemmas for Nepal. South Asian Stud 31(2):11–23. https://www.southasianstudies.org

Rivera WM, Alex G (2004) Extension reform for rural development, vol 1. World Bank. https://documents1.worldbank.org/curated/en/170431468291370533/pdf/318900Extension1Reform1V1.pdf

Rockström J, Karlberg L, Wani SP, Barron J, Hatibu N, Oweis T, Bruggeman A, Farahani J, Qiang Z (2010) Managing water in rainfed agriculture—The need for a paradigm shift. Agric Water Manag 97(4)

The Global Fund (2019) Focus on climate change and health. https://www.theglobalfund.org/media/8066/publication_climatechangeandhealth_focuson_en.pdf

Tompkins EL, Adger WN (2004) Does adaptive management of natural resources enhance resilience to climate change? Ecol Soc 9(2):10. https://doi.org/10.5751/ES-00622-090210

Walters BB, Rönnbäck P, Kovacs JM, Crona B, Hussain SA, Badola R, Primavera JH, Barbier EB, Dahdouh-Guebas F (2008) Ethnobiology, socio-economics and management of mangrove forests: a review. Aquat Bot 89(2):220–236. https://doi.org/10.1016/j.aquabot.2008.02.009
Watts N, Amann M, Arnell N, Ayeb-Karlsson S, Belesova K, Berry H, Bouley T, Boykoff M, Byass P, Cai W, Campbell-Lendrum D, Chambers J, Daly M, Dasandi N, Davies M, Depoux A, Dominguez-Salas P, Drummond P, Ebi KL et al (2018) The 2018 report of the lancet countdown on health and climate change: shaping the health of nations for centuries to come. The Lancet 392(10163):2479–2514. https://doi.org/10.1016/S0140-6736(18)32594-7
World Bank Group (2023) Climate-smart agriculture. https://www.worldbank.org/en/topic/climate-smart-agriculture

Chapter 6
International Aid and Climate Finance for Developing Economies

Abstract This chapter resonates with the energy policy necessity from a developing nation's standpoint for worldwide Net Zero. Relative contributions to climate change correspond with the global adoption of renewable resources, global technology evaluations for human viability sustainability, and global legislative activities. However, emerging nations do not live on the same playing field as First World nations regarding socioeconomic position and infrastructure development. Therefore, each nation-specific lawmaker must fight for what is suitable to its requirements. However, on a worldwide compounded level, the acquisition of finance assistance supported by technical implementations and education is vital. This chapter supports the developing nation's stance for Net Zero owing to reduced susceptibility to international climate change injustices and enhanced equity.

Introduction

International aid and climate finance are essential for global development. This challenge is undoubtedly among the most significant issues confronting the globe today. Climate change poses a challenge to human and environmental systems, as well as to regional and global economy. Nonetheless, fragile developing nations contribute minimally to greenhouse gas emissions yet suffer the greatest consequences. Studies indicate that emerging island nations facing rising sea levels and agricultural districts suffering from severe droughts are more susceptible to risks, with equitable remedies presenting significant potential. Consequently, climate funding and international support are crucial interconnected domains of opportunity for equality in climate change mitigation and climate-related sustainable development.

This chapter examines international climate finance which denotes the financial resources mobilised for climate change mitigation and adaptation via public, private, and alternative channels. The principal entities overseeing climate financing are the Green Climate Fund (GCF), the Climate Investment Funds (CIF), and the Adaptation Fund. These organisations aim to allocate urgently required capital to

M. O. Oyenuga et al., *Smart Technologies for Climate Change and Net Zero Policies: Practical Approaches Towards Sustainability*, Climate Change Management,
https://doi.org/10.1007/978-3-031-92221-3_6

developing nations that frequently have significant deficiencies in human and financial resources, infrastructure, and institutional capacities necessary for attaining low-carbon and climate-resilient growth trajectories. Nonetheless, these monies are acquired through specific offers, and access necessitates a rigorous application process. In resource-constrained nations, the qualifying criteria for specific grants are daunting, compounded by the lengthy application process for prospective financing that aligns with the country's natural and developmental requirements. The GCF seeks to balance mitigation and adaptation activities, recognising the time and financial constraints faced by vulnerable nations regarding their capabilities while others offer concessional loans for large-scale renewable energy projects and adaptation initiatives requiring community-focused engagement. Nonetheless, the global disparity persists in inadequately financed initiatives, inequitable social justice, and inefficient funding and management.

Global Climate Finance Mechanisms: Overview

Climate change inflicts extensive damage on global ecosystems, economy, and society, necessitating remedial measures and adaptive strategies. Consequently, decarbonization seeks to establish a low-carbon economy that promotes diligence through minimal-impact infrastructure projects and collaborative efforts among governmental agencies. Developing nations have additional burdens in conducting business (Oyedele et al. 2023) with reduced carbon emissions and without standardized methods for achievement, each country, along with its own micro and macro cosmologies, employs distinct strategies that are most effective for its inhabitants. Numerous initiatives have been established globally that might designate leaders in worldwide efficiency and advanced low-carbon implementation (Joshi et al. 2024). For the earth to advocate for itself, humanity must promote the planet to guarantee the existence of equally healthy worlds for future generations. Consequently, Africa's shift to a net-zero energy economy is imperative, it is not a choice (Oyenuga and Omale 2024). The shift to an energy economy requires a redefined approach to energy production and use which emphasizes sustainability from both top-down and bottom-up perspectives across all societal levels. Africa, with its abundant natural resources and rich diversity, is positioning itself for economic sustainability and leadership in climate change solutions (Nwokolo et al. 2023a, b, c, d, e, f) though the infrastructural deficit necessary for generating and distributing sustainable energy resources may impede the region's socio-economic development (Kumar et al. 2024).

Climate finance denotes the financial flows originating from public, private, and alternative sources at international, national, and local levels that facilitate climate change mitigation and adaptation efforts. A global climate funding initiative has been developed to generate and distribute cash for climate change mitigation initiatives. This new finance initiative primarily pioritises poor nations that are at the highest danger from climate change. African states have benefited significantly in terms of economic advantage, enhanced infrastructure, and increased quality of life from developmental and climate funding initiatives by the global community. The financial injection has resulted in enhanced access to healthcare and education due to stabilisation, economic growth, and entry into the global marketplace, which would have otherwise not be achievable (Nwokolo et al. 2023a, b, c, d, e, f). However, Africa may be excluded from the global energy transformation without financial aid. This is because Africa has limited resources, technology, governance, and political potential, in addition to substantial capital investment requirements. Energy transition necessitates political will and teamwork to establish rules and legislation that promote it; yet, without the requisite support from all levels, stakeholders will not see the success of energy transition initiatives. Therefore, considering this, Africa may accomplish significant progress towards achieving net-zero by mid-century (Nwokolo et al. 2023a, b, c, d, e, f). Hence, international entities must provide resources through technical assistance, capacity enhancement, and climate funding to ensure that countries across Africa establish and execute sustainable energy initiatives (Nwokolo et al. 2023a, b, c, d, e, f).

The United Nation Framework Convention on Climate Change (UNFCCC) mandates that rich nations must facilitate access to financial resources for poor countries to effectively implement climate change initiatives (UNFCCC 2025). The initiatives include:

a. **Green Climate Fund (GCF)**: Founded in 2010, the GCF is a global fund that supports poor nations in executing their climate mitigation and adaptation initiatives. The objective is to facilitate the fund's transition to climate-resilient sustainable methodologies and low-emission development trajectories. The GCF obtains resources from international public and private initiatives, facilitating access to finance options through grants, concessional loans, equity, and guarantee investments for comprehensive climate intervention projects (Green Climate Fund 2020).
b. **Adaptation Fund**: The Adaptation Fund was established in 2001 under the Kyoto Protocol. According to the United Nations, the Adaptation Fund finances projects and programs to adapt to climate change in developing countries that are most vulnerable to its effects. Its funding comes from part of revenues generated through certain CDM project operations and voluntary contributions. The Fund will work to finance activities that strengthen adaptive capacity and resilience. Asian Development Bank, African Development Bank, World Bank. MDBs support climate financing on a global scale through loans, grants, and technical assistance. The ability of any department to create income via global capital markets and financial help from donor nations creates an additional useful resource pool to be used for both mitigation and adaptation efforts.

Green Climate Fund and Its Role in Developing Countries

The Green Climate Fund (GCF) is a key institution for global climate action, especially in support of developing countries. The GCF was created in accordance with the United Nations Framework Convention on Climate Change (UNFCCC) and is designated to redistribute financial resources to these nations in an effort to mitigate greenhouse gas emissions and adapt to the consequences of climate change. The GCF was formally established in 2010 during the 16th Conference of the Parties (COP 16) in Cancun, Mexico. The establishment of the Green Climate Fund was necessitated by the realization that climate change has a disproportionately adverse effect on developing countries, coupled with the need to create a special fund to assist them with finding their bearings in the global clime (UNFCCC 2015).

The GCF is one important mechanism for assistance in the climate agenda of developing countries. The GCF funds adaptation projects that help communities prepare for climate-related hazards, including floods, droughts, and rising sea levels. For example, the GCF has funded projects that have enhanced coastal defenses and early warning (Green Climate Fund 2020). The GCF also help countries to offer low-carbon economies by funding renewable energy projects, energy efficiency programs, and sustainable transportation systems. Through GCF investments, Kenya has significantly boosted geothermal energy capacity, avoiding the use of fossil fuels. Furthermore, GCF works with developing nations to strengthen the technical and institutional capacity needed to formulate, implement, and oversee robust climate policies and initiatives. This encompasses capacity building, knowledge transfer and technical assistance. Additionally, GCF by offering initial funding and risk-mitigation instruments, contributes to the crowding in of further investments from private and public sectors, scaling up climate action. This catalytic function is critical to closing the financing gap for climate projects.

Bilateral and Multilateral Aid for Climate Action

Countries offer economic assistance for climate action through bilateral and multilateral aid, two main types of mechanism that have different features, advantages and drawbacks. Knowing these differences is the key to measure how effective are they at helping us fight climate change. Bilateral foreign assistance, money from one nation to another, has been a significant source of effect finance. Bilateral aid, however, is given according to donors' strategic interests, although the global climate crisis needs to be tackled as a joint endeavour, studies show. For instance, Klöck et al. (2018) shows that, vulnerability, capacity and responsibility would be the important determinants of climate assistance allocation. They also find that countries that are well governed and that make commitments under the Paris agreement are also more likely to receive bilateral climate finance and that political and institutional contexts influence the effectiveness of aid (Chen and Han (2022)). This is in line with the findings of Michaelowa and Michaelowa (2011), which showed that bilateral aid has increasingly focused on climate-friendly interventions (particularly renewable energy or energy efficiency) since the early 1990s.

In contrast, climate finance is increasingly multilateral, characterized by the collection of diverse actors seeking to address shared challenges. Bechtel et al.

(2022) claim that multilateralism can reduce public fears about the economic costs of climate action, as it appears to distribute those costs equally across nations. This idea of fairness is crucial for wining over public support for climate measures. In addition, financing multilateral climate is the key to mobilizing resources on an adequate scale to achieve global climate goals, especially through multilateral development banks and climate funds (Aggregate Trends of Climate Finance Provided and Mobilised by Developed Countries in 2013–2020, 2022). Evaluation of multilateral aid has also been reported (Biscaye et al. 2016), showing multilateral aid is less affected by political objectives than bilateral and potentially a better indicator of development effectiveness. But both types of assistance have their challenges. Critics argue that bilateral aid can be used as a tool for the political agendas of donor countries, which may not always serve the needs of recipient countries (Wambaka 2022).

On the other hand, multilateral aid has some inherent bureaucratic inefficiencies and convoluted decision-making processes that can constrain speedy responses to climate emergencies (Rahman and Hassan 2018). While the debate between multilateralism and bilateralism continues in the academic literature, other scholars recommend moving toward minilateralism, coalitions of major emitters that are smaller and able to act more flexibly to deliver climate action (Eckersley 2012; McGee 2011).

The Role of Private Sector Finance in Climate Initiatives

There is a growing acknowledgement that private sector finance will be critical to help unlock the broad spectrum of solutions needed to combat climate change. Private sector investment plays a crucial role in the financing of mitigation efforts, such as renewable energy projects, as well as of adaptation strategies that build resilience to climate impacts. Private finance is critical for the scale-up of climate action, especially in developing countries where public funds are often scarce due to competing priorities, according to research. Although the private sector-familiarized in adapting to climate-related risks-has practical know-how in addressing climate-related vulnerabilities, awareness of political aspects of climate adaptation still has a long way to go (Pauw 2017). This fragmentation can prevent the formation of effective crews of public and private actors, which can have different effectiveness for climate action. Moreover, there is strong regional variation as to the degree to which the private sector supports climate finance. Crick et al. (2018), when meta-analyzed with studies conducted in sub-Saharan Africa, found no evidence that performance-based cash transfers improve health and education outcomes among the very poor in sub-Saharan Africa. Highlighting routes for investments to simultaneously enhance the adaptive capacity of small and medium enterprises (SMEs) while also facilitating business growth. This dual focus matters since most SMEs are confronted with multiple pressures that impede their adjustment to climate change. Actively investing in climate adaptation will not only drive investments but will also support business strategies that build climate resilience, focusing on economic gains and fostering a sustainable economy.

Also, Abdoul-Azize et al. (2021), present a case study from the Niger Republic, which shows how private sector financing can help reduce poverty and increase socio-economic resilience via targeted climate actions. Private sector finance in the context of climate is less coordinated than lending practices in emerging economies, argues Gambetta et al. (2019), and as a result, the appropriate financial instruments are not present in many emerging economies to catalyze private capital for sustainable development. This calls for legislative reform to promote synergies between financing instruments and sustainable development goals (SDGs). Furthermore, Agrawala et al. (2011) emphasises the crucial role of public sector support in equipping companies with the information they need to reliably manage climate-related risks. This entails brokering the transfer of knowledge between the scientific and business communities, which is an essential factor in facilitating evidence-based decisions in the private sphere. Despite the hurdles unraveling the roadmap for private sector finance to make a large dent in combating climate change is clear. As a market regulator national governments need to identify commercial operating models by which the private sector can invest at scale, aligning adaptation strategies with private sector objectives (Adhikari and Chalkasra 2021). That alignment is critical to generate climate-resilient results, since it incentivizes private investment in climate adaptation actions that match their private business goals. The example of the Green Climate Fund (GCF) shows how private sector engagement goes beyond co-financing and enables innovative adaptation finance (Stoll et al. 2021).

Challenges and Opportunities in Accessing Climate Finance

Accessing climate finance is pivotal for developing countries striving to mitigate and adapt to the impacts of climate change. However, numerous challenges impede the effective mobilization and utilization of these funds. Simultaneously, there are significant opportunities to enhance access and efficiency in climate finance.

Challenges in Accessing Climate Finance

The following challenges are identified as bedeviling climate financing;

1. **Complex Application Procedures**: Application processes for climate finance for developing countries are often complex and resource-intensive. To take the example of the Green Climate Fund (GCF), it follows complex application processes involving elaborate project proposals, detailed feasibility studies and rigorous fiduciary standards. The administrative capacities of many developing nations are limited and are often surpassed by these requirements resulting in serious delays or failures in accessing funding (Weikmans and Roberts 2019).

2. **Insufficient Institutional Capacity**: Developing countries also have the disadvantage of weak institutional frameworks and limited technical capacity for the design, implementation and monitoring of climate projects. The lack of project management capacity leads to poor project proposals and erodes the confidence of donors and financial institutions in the efficient implementation of funded projects (Nakhooda et al. 2013).
3. **Inadequate Access to Information**: A major barrier is a lack of access to information about available funding sources, their eligibility criteria, and how to apply for them. Several developing countries do not know about the different avenues available to climate finance or do not have access to quality guidance that allows them to navigate the maze of international climate finance (Pauw et al. 2016).
4. **Stringent Eligibility Criteria**: Often donor agencies and financial institutions impose stringent eligibility criteria such as co-financing, showing financial sustainability and alignment with certain policy priorities. Such conditions may be prohibitive for some countries with limited financial resources and policy environment (Brown et al. 2010).
5. **Political and Economic Instability**: Factors such as political instability, governance challenges and economic volatility absolutely discourage and prevent investors and other donors from committing funds in certain regions. The risks which the observed instability brings are creating hesitance with regards to the disbursement of climate finance, denying target groups of the assistance they need for climate adaptation and mitigation.
6. **Fragmentation of Funding Sources**: The multitude of sources of climate finance with different objectives, procedures and reporting requirements creates fragmentation. Such complexity can strain the administrative capabilities of developing countries, leading to inefficiencies and duplication of efforts (Bird et al. 2013).

Opportunities in Accessing Climate Finance

The climate finance landscape has become more complex than ever, with numerous new funding sources and mechanisms potentially enhancing developing countries' ability to implement climate programs. One of the clearest pathways, with potential for transformation, is through the Green Climate Fund (GCF), which has been set up to direct large sums of money into the fight against climate change in the developing world. This commitment of mobilizing US$100 billion per year for facilitating access to climate finance Samuwai and Hills (2019) hopes to make these and other vulnerable countries parties to the GCF such that they can benefit from its funding. Such funding will only be effective if allocation policies are equitable and consider the repeated needs and vulnerabilities of the countries receiving the support (Klöck et al. 2018). A second set of solutions includes innovative financing mechanisms, like the so-called green bonds recently created that direct private sector investment into climate projects. They enable investors to support environmental initiatives, and still earn a financial return. Green bonds have been shown to be effective as a tool for financing climate resilient renewable energy and infrastructure (Long 2024; Tolliver et al. 2019).

Decentralisation of the green bond market in Africa has exposed investment opportunities in Kenya, Nigeria and South Africa which have now been able to access private capital that is critical in the effort to meet climate goal targets as set out in the Paris Agreement (Ngwenya and Simatele 2020). The capacity for green bonds also signal the need for the robust development of aligned policies that facilitate cross-sectoral public–private stakeholder engagement (Ngwenya and Simatele 2020) and makes a case for the strategic leveraging of assets as it is possible to rollout climate finance into health initiatives. Recent changes in the broader international climate governance landscape illustrate interlinkages between climate change and public health and pave opportunities for health practitioners and policymakers to access climate finance, so that resilient health systems are able to be built (Borghi 2024). So, this intersection allows to reap the benefit of climate finance out there, but it is also an advocacy of those key health challenges exacerbated by climate change ensuring a tectonic shift in climate action.

Implications

The implication for policy, SDG, and future research implications suggests a more unified approach with allocated financial initiatives to enhance the efficacy of climate funding. Policy interventions should primarily concentrate on capacity building to enable poor nations to effectively analyze, implement, and sustain climate financing initiatives. When climate funds become more accessible for countries and international organizations, it entails less cumbersome application processes and transparent procedures for the acquisition and distribution of funds. Hence, regulations must be instituted for public–private partnerships to mitigate the risks associated with private sector involvement via guarantees, concessional funding, and regulatory tax incentives. Therefore, improvement in South-South cooperation is recommended to enable poor nations to collaborate on information sharing, best practices, and technical support for more effective climate initiatives.

Conclusion, Limitations, and Future Scope

International aid and climate financing tackle the many dimensions of climate change for developing countries. International initiatives, including the Green Climate Fund (GCF) and Climate Investment Funds (CIF), address financial and technical deficiencies for mitigation and adaptation projects. Additionally, international climate finance targeting the collective can be augmented through bilateral or multilateral efforts to alleviate the burden. Multilateral efforts promote more funding through aggregated resources, while diverse possibilities frequently lead to worldwide involvement. The private sector also participates via green bonds and public–private partnerships. Consequently, international climate financing generates substantial cash for renewable energy projects and critical infrastructure efforts in this domain. However, obstacles remain. Numerous climate finance options remain inaccessible. Complex application processes, institutional deficiencies, and deficiencies in poverty conditions hinder specific vulnerable communities from acquiring or accessing climate finance.

References

Abdoul-Azize H, Salé A, Soulé M (2021) Private telecommunication companies and climate finance: evidence from the Niger republic. Int J Environ Change 11:374–385. https://doi.org/10.9734/ijecc/2021/v11i1230588

Adhikari B, Chalkasra L (2021) Mobilizing private sector investment for climate action: enhancing ambition and scaling up implementation. J Sustain Financ & Invest 13(2):1110–1127. https://doi.org/10.1080/20430795.2021.1917929

Agrawala S, Carraro M, Kingsmill N, Lanzi E, Mullan M, Prudent-Richard G (2011) Private sector engagement in adaptation to climate change: approaches to managing climate risks. https://doi.org/10.1787/5kg221jkf1g7-en

Aggregate trends of climate finance provided and mobilised by developed countries in 2013–2020 (2022) Climate finance and the USD 100 billion goal. https://doi.org/10.1787/d28f963c-en

Bechtel M, Scheve K, Lieshout E (2022) Improving public support for climate action through multilateralism. Nat Commun 13(1). https://doi.org/10.1038/s41467-022-33830-8

Bird N, Brown J, Schalatek L (2013). Design challenges for the Green climate fund: a report from the ODI and Heinrich Böll Foundation. https://www.odi.org

Biscaye P, Reynolds T, Anderson C (2016) Relative effectiveness of bilateral and multilateral aid on development outcomes. Rev Dev Econ 21(4):1425–1447. https://doi.org/10.1111/rode.12303

Borghi J (2024) Climate finance opportunities for health and health systems. Bull World Health Organ 102(5):330–335. https://doi.org/10.2471/blt.23.290785

Brown J, Cantore N, Stern N (2010) Climate finance: balancing the costs of climate change for developing countries. Taylor & Francis. https://doi.org/10.4324/9780203865591

Chen S, Han X (2022) Drivers of bilateral climate finance aid: the roles of Paris agreement commitments and public governance. https://doi.org/10.21203/rs.3.rs-1809158/v1

Climate Bonds Initiative (2021) Green bonds market summary 2020. https://www.climatebonds.net

Climate Investment Funds (CIF) (2025) About the climate investment funds. https://www.climateinvestmentfunds.org/

Crick F, Gannon K, Diop M, Sow M (2018) Enabling private sector adaptation to climate change in Sub-Saharan Africa. Wiley Interdiscip Rev Clim Change 9(2). https://doi.org/10.1002/wcc.505

Eckersley R (2012) Moving forward in the climate negotiations: multilateralism or minilateralism? Glob Environ Polit 12(2):24–42. https://doi.org/10.1162/glep_a_00107

Friede G, Busch T, Bassen A (2015) ESG and financial performance: aggregated evidence from more than 2000 empirical studies. J Sustain Financ & Invest 5(4):210–233. https://doi.org/10.1080/20430795.2015.1118917

Gambetta N, Azadian P, Hourcade V, Reyes M (2019) The financing framework for sustainable development in emerging economies: the case of uruguay. Sustainability 11(4):1059. https://doi.org/10.3390/su11041059

Global Environment Facility (GEF) (2025) About the global environment facility. https://www.thegef.org/
Green Climate Fund (GCF) (2019) Rwanda Green Fund (FONERWA). https://www.greenclimate.fund
Green Climate Fund (GCF) (2020) Climate finance readiness program: Building capacity for change. https://www.greenclimate.fund
Intergovernmental Panel on Climate Change (IPCC) (2014) Climate change 2014: impacts, adaptation, and vulnerability. Cambridge University Press. https://doi.org/10.1017/CBO9781107415379
International Renewable Energy Agency (IRENA) (2019) SIDS lighthouses initiative: progress and way forward. https://www.irena.org
Joshi, A, Singh, R, Kumar, V, Rani, S (2024). Net Zero Strategy: Exploring Opportunities, Confronting Challenges for Developing Countries. In: Singh, R, Crowther, D. (eds) Transition Towards a Sustainable Future. Approaches to Global Sustainability, Markets, and Governance. Springer, Singapore. https://doi.org/10.1007/978-981-97-5756-5_11
Klöck C, Molenaers N, Weiler F (2018) Responsibility, capacity, greenness or vulnerability? What explains the levels of climate aid provided by bilateral donors? Environ Polit 27(5):892–916. https://doi.org/10.1080/09644016.2018.1480273
Kumar K, Singh R, Sharma A, Lehri BPS (2024) Climate change adaptation strategies for achieving net-zero economy. In: Singh R, Crowther D (eds) Transition towards a sustainable future. approaches to global sustainability, markets, and governance. Springer, Singapore. https://doi.org/10.1007/978-981-97-5756-5_8
Long P (2024) Financial mechanisms for energy transitions: a review article. Fulbright Rev Econ Policy 4(2):126–153. https://doi.org/10.1108/frep-07-2024-0039
McGee J (2011) Exclusive minilateralism: an emerging discourse within international climate change governance? Port J Multidiscip Int Stud 8(3). https://doi.org/10.5130/portal.v8i3.1873
Michaelowa A, Michaelowa K (2011) Old wine in new bottles? Does climate policy determine bilateral development aid for renewable energy and energy efficiency? Revue Internationale De Politique De Développement 2(2). https://doi.org/10.4000/poldev.748
Nakhooda S, Watson C, Schalatek L (2013) The 2013 climate finance update: an analysis of international support. Overseas Development Institute. https://www.odi.org
Ngwenya N, Simatele M (2020) Unbundling of the green bond market in the economic hubs of Africa: case study of Kenya, Nigeria and South Africa. Dev South Afr 37(6):888–903. https://doi.org/10.1080/0376835x.2020.1725446
Nwokolo SC, Singh R, Khan S, Kumar A, Luthra S (2023a) Technological pathways to net-zero goals in Africa. In: Africa's path to net-zero. CSR, sustainability, ethics & governance. Springer, Cham. https://doi.org/10.1007/978-3-031-44514-9_5
Nwokolo SC, Singh R, Khan S, Kumar A, Luthra S (2023b) Scenarios that could give rise to an African net-zero energy transition. In: Africa's path to net-zero. CSR, sustainability, ethics & governance. Springer, Cham. https://doi.org/10.1007/978-3-031-44514-9_8
Nwokolo SC, Singh R, Khan S, Kumar A, Luthra S (2023c) Africa's awakening to climate action. In: Africa's path to net-zero. CSR, sustainability, ethics & governance. Springer, Cham. https://doi.org/10.1007/978-3-031-44514-9_9
Nwokolo SC, Singh R, Khan S, Kumar A, Luthra S (2023d) Remedies to the challenges of renewable energy deployment in Africa. In: Africa's path to net-zero. CSR, sustainability, ethics & governance. Springer, Cham. https://doi.org/10.1007/978-3-031-44514-9_3
Nwokolo SC, Singh R, Khan S, Kumar A, Luthra S (2023e) Decarbonizing hard-to-abate sectors in Africa. In: Africa's path to net-zero. CSR, sustainability, ethics & governance. Springer, Cham. https://doi.org/10.1007/978-3-031-44514-9_6
Nwokolo SC, Singh R, Khan S, Kumar A, Luthra S (2023f) Global investment and development in Africa. In: Africa's path to net-zero. CSR, sustainability, ethics & governance. Springer, Cham. https://doi.org/10.1007/978-3-031-44514-9_2

Oyedele MO, Marcus GO, Ahungwa AI (2023) Do consumers care about green marketing practices? Insight from a developing nation. Budapest Int Res Crit Inst J (BIRCI J) 6(3):1424–1436. https://bircu-journal.com/index.php/birci/article/view/7665
Oyenuga MO, Omale SA (2024) Is Africa Jinxed? Exploring the challenges of technology access and adoption in Africa. Afr J Econ Sustain Dev 7(4):142–161. https://doi.org/10.52589/AJESDULN1LRNF
Organisation for Economic Co-operation and Development (OECD) (2015) Climate finance in 2013–14 and the USD 100 billion goal. OECD Publishing. https://doi.org/10.1787/9789264249424-en
Organisation for Economic Co-operation and Development (OECD) (2018) Blended finance for sustainable development: mobilizing private sector investment. https://www.oecd.org
Organisation for Economic Co-operation and Development (OECD) (2020) Climate finance provided and mobilised by developed countries in 2013–2018. OECD Publishing. https://doi.org/10.1787/f0773d55-en
Pauw WP, Klein RJT, Menzel C, Biermann F (2016) Public finance for climate change adaptation: the role of institutions and policymakers. Clim Change 134(1):89–103. https://doi.org/10.1007/s10584-015-1537-4
Pauw P (2017) Mobilising private adaptation finance: developed country perspectives. Int Environ Agreem Polit Law Econ 17(1):55–71. https://doi.org/10.1007/s10784-016-9342-9
Rahman N, Hassan M (2018) Climate change fiasco: what multilateral arrangements have to offer? Manag Econ Res J 4(2):223. https://doi.org/10.18639/merj.2018.04.670139
Samuwai J, Hills J (2019) Gazing over the horizon: will an equitable green climate fund allocation policy be significant for the pacific post-2020? Pac Journal Rev – Te Koakoa 25(1&2):158–172. https://doi.org/10.24135/pjr.v25i1.393
Stoll P, Pauw P, Tohme F, Grüning C (2021) Mobilising private adaptation finance: lessons learned from the green climate fund. https://doi.org/10.21203/rs.3.rs-255703/v1
Tolliver C, Keeley A, Managi S (2019) Green bonds for the paris agreement and sustainable development goals. Environ Res Lett 14(6):064009. https://doi.org/10.1088/1748-9326/ab1118
United Nations Development Programme (UNDP) (2019) Climate finance explorer: a guide for developing countries. https://www.undp.org
United Nations Environment Programme (UNEP) (2016) Adaptation finance gap report 2016. https://www.unep.org
United Nations Framework Convention on Climate Change (UNFCCC) (2015) The Paris agreement. https://unfccc.int
United Nations Framework Convention on Climate Change (UNFCCC) (2017) South-South cooperation in climate adaptation and resilience. https://unfccc.int
United Nations Framework Convention on Climate Change (UNFCCC) (2025) Climate finance. https://unfccc.int/topics/climate-finance
Wambaka K (2022) Impact of bilateral and multilateral aid on domestic savings in low and middle-income Sub Sahara African countries: mediating role of institutional quality. Int Bus Res 16(1):26. https://doi.org/10.5539/ibr.v16n1p26
Weikmans R, Roberts JT (2019) The international climate finance accounting muddle: is there hope on the horizon? Clim Dev 11(2):97–111. https://doi.org/10.1080/17565529.2017.1410087
Weiler F, Klöck C, Dornan M (2018) Vulnerability, good governance, or donor interests? The allocation of aid for climate change adaptation. World Dev 104:65–77. https://doi.org/10.1016/j.worlddev.2017.11.001
World Bank (2025) Climate change and the World Bank Group. https://www.worldbank.org

Part II
Technological Solutions for Net-Zero Emissions

Chapter 7
Technological Innovations in Renewable Energy

Abstract This chapter discusses renewable energy technologies as having the ability solve energy shortages for disadvantaged groups and places. The chapter assesses what renewable energy technologies and energy solutions are most promising based on such advances, which are known to improve rural low-resource communities' socioeconomic status. As solar energy technology advances, portable solar panels, solar micro-grids, and pay-as-you-go solar systems might power towns off the grid. Affordable and accessible electricity in rural areas increases school and hospital hours and vocational and agricultural activity. Wind energy from small and medium-sized turbines may also benefit low-income communities. Wind turbine parts must be made locally for local wind velocity, therefore skills transfer workshops must promote low-income manufacturing to ensure sustainability. Small hydropower projects are low-impact and low-cost; as well as low-head turbines are modular and portable, so they do not disrupt ecosystems or livelihoods. Communities that use solar, wind, and battery farms because enlarged central networks are too expensive, yet need dependable electricity, more strongly support off-grid hybrid solutions. Relative case studies and real successes in solar system in sub-Saharan Africa, wind in South Asia, small-scale hydropower in South America—suggest that regionally ingenious solutions to regionally complicated problems come with relative technological advances. Thus, renewable energy technology will always be the most effective tool for sustainable, fair development with solid finance, technical advances, and international aid.

Introduction

Climate change giants and net-zero policies can only ensure a sustainable future with renewable energy. International dependence on fossil fuels has always been harmful, and climatic disasters with inefficient burning and carbon emissions production no longer work, stalling socioeconomic growth. Low-carbon renewables meet disaster- and equity-based socioeconomic improvement needs. Most of this renewable energy (Oyedele 2025) small-scale hydropower, wind, and solar advances is viable and may improve energy access and resilience for marginalised, low-income, and off-grid

M. O. Oyenuga et al., *Smart Technologies for Climate Change and Net Zero Policies: Practical Approaches Towards Sustainability*, Climate Change Management,
https://doi.org/10.1007/978-3-031-92221-3_7

groups. Solar microgrids, portable, solar-driven, and pay-as-you-go systems, lower energy access obstacles. Rural schools and hospitals that need consistent energy for instruction or emergency treatment have also used solar microgrids. In developing countries with varied topography, small-scale wind turbines and hybrid solar-wind systems are common. Low-impact, cost-effective small-scale hydropower generates power in diverse locations. However, institutional barriers make long-term and widespread energy sustainability difficult. Legislation failures, hefty upfront costs, and long-term practicality up for inaccessible places prevent renewable energy from becoming mainstream.

Government choices, public awareness campaigns on renewable resources and breakthroughs, and quality of life assessments support legislative attempts to minimise such hurdles. Nonprofit and for-profit organisations working together to address community needs in energy policy and development remove financial and legal barriers. Renewable energy supports the UN Sustainable Development Goals (SDGs). Many are strengthened by renewable energy. SDG 7 (Affordable and Clean Energy) and SDG 13 (Climate Action) are met by clean, sustainable energy. SDG 2 (Zero Hunger) is supported by solar-powered agricultural irrigation, while SDG 3 (Good Health and Well-Being) is supported by reliable hospital and emergency medical power. Renewable energy is linked to more than climate change, therefore making it affordable and widespread may help many organic human development needs (Oyetunde et al. 2023). In Sub-Saharan Africa, Asia, and Latin America, renewable energies are used everywhere to solve regional problems, from micro-grids with solar panels to turbines that generate energy to small hydropower projects that use pre-existing water generation with minimal environmental impact. The pace of renewable energy technologies goes beyond fresh innovations. It implies game-changing solutions that make sustainability and equity possible. Without rapid adoption of all renewable energy, vulnerable groups will always face poverty, catastrophes, and energy inequalities.

Solar Energy Innovations for Rural Electrification

Africa is essential to achieving net-zero economies and sustainability. Nwokolo et al. (2025) delineate three levels of assistance necessary for this transformation to take place: The digitisation and connectivity of Africa drive a net-zero future, supported by three key elements: Development Support—leapfrogging, hybridisation, flexibility, decentralisation, and sustainable development; Awareness; and Global Financing. Africa may attain net-zero emissions through technical transfers, education, and funding, provided there is global awareness and equity (Nwokolo et al. 2023a, b, c, d, e; Oyenuga and Omale 2024). Nwokolo et al. (2023a, b, c, d, e) elucidate the necessity for awareness and education. Awareness and education necessary for sustainability complement current events and knowledge acquisition while an informed public understands that socio-political advancements may assist governments in securing sustainable funding. Consequently, knowledge and education on current developments are essential to facilitate sustainability. Awareness facilitates sustainability through workshops (Oyedele et al. 2024), climate change education, and community involvement, promotes intranational sustainability via research and enhanced financing for broader energy alternatives, and accelerates access to net zero in the short term for Africa (Nwokolo et al. 2024). Consequently, African governments must lead this initiative by facilitating energy awareness seminars, promoting research and development tax incentives, and formulating regulations related to renewable energy investment (Nwokolo et al. 2023a, b, c, d, e). Conglomerates may also engage in CSR efforts that mitigate greenhouse gas emissions and facilitate value-added growth. Innovations in solar technology for rural electrification deliver energy together with additional socio-economic advantages for consumers; hence, rural electrification is a global concern in developing countries. The varied technological elements and entrepreneurial advancements would improve the sustainability and efficacy of rural solar systems (IRENA 2019). Portable solar panels are economical solutions for off-grid individuals and small company proprietors. They are portable and deployable, supplying adequate power for supplementary requirements such as illumination and charging of phones/devices. Moreover, the efficacy and durability of photovoltaic (PV) technologies enhance accessibility and agricultural practices (IRENA 2019). Solar microgrids generate electricity at their installation sites, supplying power to groups of houses and businesses, making them suitable for rural households and areas where national grid infrastructure would be excessively burdensome. In Odisha, India, solar microgrids manage water irrigation pumps, grain milling equipment, and domestic enterprises such as solar farming (Bhattacharyya 2018). Consequently, advancements enhance the accessibility of solar energy. Mobile money systems enable clients to utilise pay-as-you-go (PAYG) options for their services. PAYG solar provides low-income individuals, who cannot afford a substantial upfront payment, the option to use solar technology. M-KOPA and d.light PAYG solar firms have electrified millions of off-grid rural homes in Sub-Saharan Africa, enhancing lighting, sanitation, and study duration (Climate Policy Initiative 2019). Access to solar technology enhances rural agricultural economics. Solar lights reduce reliance on kerosene, enhance irrigation capabilities, boost agricultural productivity, improve food security, and generate financial benefits. Ultimately, solar technology enhances connectedness among individuals, integrating them into broader educational and economic initiatives. In rural Kenya and Tanzania, solar-powered educational institutions and medical facilities exemplify the capacity of renewable energy to improve social initiatives (Green Climate Fund 2020). Job possibilities exist for installations, repairs, and sales. Community-trained technicians acquire specialised skills pertinent to solar technology projects, therefore reducing reliance on foreign labour and technicians. The most extensive solar initiative in Bangladesh has the Solar Home Systems (SHS) program, which has provided employment to thousands following the delivery to millions of households (IRENA 2019). However, growth initiatives were unsuccessful owing to inadequate financing from both consumer and supplier perspectives. Market expansion need time, since renewable technologies such as solar microgrids, solar panels, and PAYG solar lanterns can mitigate the rural energy deficit while simultaneously facilitating access to employment possibilities. Renewable energy sources represent the future for rising markets with sustainable prospects compared to "green" alternatives.

Kenya is a leading nation in the use of Pay-As-You-Go (PAYG) solar technology, enabling millions of off-grid households to obtain solar electricity. PAYG systems facilitate the acquisition of solar energy over time, eliminating the need for an upfront investment, hence becoming it a viable and economical energy option that promotes healthier practices via the utilisation of greener energy. This penetration has been enabled by enhanced access through mobile technology (Bhattacharyya 2018). Also, the PAYG solar home system initiative in Bangladesh revolutionised rural electrification by allowing the lowest-income people to get clean energy resources with no financial burden. PAYG solar firms, such as BRAC and Greenlight Planet, offered solar goods equipped with mobile money systems for payment integration, allowing users to acquire the items through micro-installment payments. This not only enhanced energy accessibility but also improved consumers' status in borrowing and creditworthiness for new loans and services. (Climate Policy Initiative 2019).

Bangladesh accomplished rural household electricity for millions with the Solar Home Systems (SHS) Project. The SHS initiative of Infrastructure Development Company Limited (IDCOL) emphasises concessional loans, subsidies, and training in its solar policy and execution. This initiative is among the most effective in terms of policy and global initiatives for rural electrification (IRENA 2019). Solar energy programs provide a solution for rural electrification, enhancing quality of life and fostering sustainable development in at-risk communities. Solar initiatives promote community ownership and generate local employment by circumventing conventional finance and operational challenges, using solar energy breakthroughs to align with entrepreneurial energy business models. Integrated solar startups in India, Bangladesh, and Kenya has successfully implemented solar projects through innovative solutions that align with governmental policy backing and international collaboration. Consequently, as long as sustained support and deliberate initiatives persist, solar energy projects will address the disparity in energy access while promoting a more sustainable and equitable world for all inhabitants.

Wind Energy Projects in Low-Income Countries

Wind energy is an essential renewable resource for mitigating electricity shortages in low-income nations, where millions lack dependable access to power. Technological innovations, regulatory structures, and global cooperation have rendered wind energy progressively feasible for sustainable electricity. Its scalability, cost-efficiency, and ecological advantages provide it a crucial alternative for alleviating energy poverty and fostering economic development (Bhattacharyya 2018).

Wind energy provides a clean, low-emission substitute for fossil fuels, while tackling economic and environmental issues. In low-income nations, using wind power may diminish dependence on imported fuels, improve industrial output, and promote rural electrification. Research indicates that several low-income countries, especially those with coastline and elevated terrains, have considerable unexploited wind energy potential (IRENA 2019). Small-scale wind initiatives, especially decentralised mini-grids, offer localised energy solutions, obviating the necessity for expensive grid extensions while guaranteeing inexpensive and sustainable electricity availability.

Recent advancements have enhanced the viability and efficiency of wind energy in economically disadvantaged areas:

1. **Small-Scale Wind Turbines**: Engineered for rural areas, these economical turbines supply electricity to residences, educational institutions, and small enterprises (IRENA 2019).
2. **Hybrid Wind Systems**: The integration of wind energy with solar or hydro solutions guarantees a reliable power supply, particularly in regions characterised by variable wind patterns (Climate Policy Initiative 2019).
3. **Smart Turbine Technology**: The incorporation of AI and IoT for real-time performance monitoring and predictive maintenance has diminished operating expenses and enhanced turbine efficiency (Green Climate Fund 2020).
4. **Local Manufacturing**: Cultivating wind technology domestically diminishes expenses and creates jobs, promoting economic sustainability (Bhattacharyya 2018).

Besides, wind energy initiatives generate job opportunities and promote economic growth. The Lake Turkana Wind Project in Kenya generated thousands of employment opportunities and bolstered local enterprises (IRENA 2019). Mini-grids in Ethiopia have provided electricity to off-grid populations, enhancing healthcare and education while decreasing dependence on paraffin (Climate Policy Initiative 2019). Wind energy enhances sustainability by reducing carbon emissions and water consumption, rendering it suitable for dry areas (Green Climate Fund 2020). By surmounting financial, technological, and governmental obstacles, wind energy may significantly contribute to energy security and economic resilience in low-income countries.

Notwithstanding its benefits, many challenges impede the proliferation of wind energy in low-income nations:

a. **Significant Initial Expenditures**: The capital-intensive characteristics of wind infrastructure render finance a substantial obstacle (OECD 2020).
b. **Policy Barriers**: Discrepant laws and insufficient incentives hinder investment in wind projects (UNDP 2011).
c. **Technical Capacity**: The deficiency of qualified workers for installation and maintenance impacts long-term sustainability (IRENA 2019).
d. **Infrastructure Limitations**: Inadequate grid systems and insufficient storage options result in intermittency challenges (Green Climate Fund 2020).
e. **Community Acceptance**: Social opposition stemming from land use issues and aesthetic considerations necessitates inclusive decision-making (Bhattacharyya 2018).

Hydropower: Small-Scale Solutions for Local Communities

In recent years, small-scale hydropower has emerged as a viable alternative for rural electrification and community sustainable development, as it is a successful renewable energy source with zero negative emissions. Furthermore, small-scale systems micro and mini-hydropower projects are economically viable, straightforward to install, and expandable alternatives that are also environmentally non-intrusive. This article examines the social, economic, and environmental advantages of small-scale hydropower systems as a viable choice for enhancing electricity access in developing countries and fostering community development when resources are available for such enhancements.

Small-scale hydropower produces power outputs between few kilowatts and a few megawatts, rendering it appropriate for modest generation requirements. It utilises the kinetic energy of flowing water from smaller sources, like rivers, streams, and canals. It necessitates reduced building and infrastructure compared to traditional hydropower, possesses a diminished environmental impact, and may be located at or near the consumption site to minimise transmission losses (IRENA 2019). Numerous off-grid regions exist where grid expansion initiatives are constrained or costly. Small-scale hydropower produces electricity without constraints, harnessing the naturally flowing water sources available to rural communities. Moreover, small-scale hydropower is adaptable and can be implemented in virtually any geographic area to meet diverse community energy requirements, ranging from residential and educational use to microenterprise initiatives, as demonstrated in the article "Technological Pathways to Net-Zero Goals in Africa" (Nwokolo et al. 2023a, b, c, d, e). Recent breakthroughs in small-scale hydropower encompass improvements in efficiency and viability. Recent developments in turbine technology, namely low-head turbines, augment energy output in low-flow conditions or low-head water bodies, hence expanding prospective siting locations in places previously deemed unsuitable for hydropower.

Furthermore, recent innovations in modular system designs and water flow regulation enhance viability (Green Climate Fund 2020). Moreover, the use of prefabricated systems facilitates the expedited and cost-effective deployment of energy solutions, which is crucial for developing nations with limited financial resources for such initiatives (Nwokolo et al. 2023a, b, c, d, e). Intelligent hydropower systems also exist there. Digitised projects, along with IoT-based monitoring and AI-driven fault detection for maintenance, ensure these projects function at optimal capacity with little downtime, hence reducing LCOE (IRENA 2019).

Small-scale hydropower projects offer a reliable energy source for residential, commercial, and institutional applications, hence promoting socio-economic development in developing and rural regions compared to other forms of alternative energy. Small-scale hydropower initiatives closely correspond with SDG 1: No Poverty and SDG 2: Zero Hunger, since rural electricity enables farmers to energise their irrigation systems, operate grain mills, and preserve perishable goods through freezing or refrigeration (Nwokolo et al. 2023a, b, c, d, e). Moreover, SDG 4: Quality Education and SDG 3: Good Health and Wellbeing depend on enhanced access to sustainable energy, as such initiatives facilitate increased resources; educational institutions addressing SDG 4 could utilise surplus energy to illuminate classrooms for prolonged instructional hours and to operate computers for technological education; healthcare facilities could employ surplus energy to power refrigeration systems for vaccines and x-ray apparatus (IRENA 2019). Moreover, small-scale hydropower initiatives financed by donors generate employment opportunities for construction, operation, and maintenance. Skills training equips community members with the necessary technical expertise, fostering a sense of ownership in operation and maintenance, thereby mitigating project failure due to donor withdrawal if the technical knowledge is acquired but not imparted to the residents (Nwokolo et al. 2023a, b, c, d, e). Small-scale hydropower also possesses beneficial environmental potentials. For example, small-scale initiatives do not disrupt aquatic ecosystems as significantly as large-scale projects. Numerous projects are run-of-the-river, indicating that water is not required after electricity generation (Nwokolo et al. 2023a, b, c, d, e). Secondly, it endorses international community objectives for climate impact by generating clean energy independent of fossil fuels and diminishing carbon emissions (Green Climate Fund 2020). Nevertheless, despite an abundance of advantageous resources, it remains a seldom-utilized technique. Small-scale hydropower incurs several expenses. Initially, the building prerequisites for optimal installation necessitate certain auxiliary components, resulting in elevated one-time installation expenses (Nwokolo et al. 2023a, b, c, d, e). Operational costs are minimal if energy production does not anticipate supplementary expenses; yet, the initial expenditure serves as a disincentive for many. Moreover, adequate installation necessitates proficient labour, and local training is essential for installation, both of which demand community capacity-building initiatives (IRENA 2019). Moreover, if water levels vary seasonally, energy output diminishes, necessitating supplementary energy production solutions or reserves to fulfil requirements (Green Climate Fund 2020). Consequently, although small-scale hydropower may be advocated by local regulatory bodies and supported by favourable policies where deemed beneficial, overcoming financial cost barriers requires a sunk cost, and training is essential to guarantee that all low-income groups reap the benefits.

Off-Grid Renewable Energy Technologies: A Path to Energy Access

Many low-income rural people in Africa and Asia lack electricity. Off-grid renewable energy solutions are conceivable since extending the conventional electricity infrastructure to these communities is complicated, expensive, inefficient, and time-consuming. Off-grid renewable energy solutions produce clean, reliable electricity without large-scale aggregation and contribute to socioeconomic growth and global social and climatic commitments. Off-grid renewable energy solutions include solar home systems, solar mini-grids, grid-tied and hybrid systems, and biomass energy help fight energy poverty and achieve the UN Sustainable Development Goals (Nwokolo et al. 2023a, b, c, d, e). Thus, these methods enable local energy independence, helping low-income rural populations survive power system breakdowns (Nwokolo et al. 2023a, b, c, d, e). Innovative technologies can solve off-grid renewable energy access difficulties. The main off-grid technology is solar home systems (SHS), which use photovoltaic panels, batteries, and LED illumination. Recently developed energy-efficient solar cells and lithium-ion batteries make solar home systems (SHS) dependable and cost-effective. Pay-as-you-go (PAYG) systems are distributed to low-income families in Sub-Saharan Africa, allowing mobile money payments over time (IRENA 2019).

Regional networks of solar, wind, and hydroelectric min-grids with energy storage and smart-grid technology provide reliable generation and delivery to whole communities. Min-grids are a sustainable alternative to solar home systems and can encourage small enterprise and community electrification (Nwokolo et al. 2023a, b, c, d, e). Durable hybrid off-grid systems use renewable resources and energy storage. Regions with only one renewable resource may benefit from integrating solar and wind energy to avoid redundancy. India and Bangladesh have hybrid solar-wind systems (Nwokolo et al. 2023a, b, c, d, e). Biomass and biogas convert organic waste into electricity. Off-grid settlements have small biomass plants that generate electricity from waste. Small biomass plants offer reliable energy in agricultural Asia, whereas biogas plants reduce rural firewood dependency and indoor air pollution from wood burning (Green Climate Fund 2020).

Off-grid technologies boost the economy and provide access to education, healthcare, and jobs raises family earnings. Off-grid solar home systems (SHS) and mini-grids power household appliances without kerosene or gasoline. Power allows businesses to operate longer and more efficiently. East African solar mini-grids boost small-scale agriculture and trade (Nwokolo et al. 2023a, b, c, d, e). Educational institutions can operate longer with illumination and computers. Solar energy can prolong clinic hours, improving medical diagnostics, immunisation preservation, and vital testing. Diagnostic and therapeutic solar energy reduces maternal, foetal, and newborn mortality in Rwanda (IRENA 2019). However, off-grid solutions need substantial development and sustained operations since each item must be installed and maintained. Solar technicians are trained in formal and informal settings, creating a trustworthy local network (Nwokolo et al. 2023a, b, c, d, e).

However, off-grid energy projects have challenges. Despite PAYG, solar panels, batteries, and inverters require a large initial investment. Micro-financing and grants would make them more feasible (Nwokolo et al. 2023a, b, c, d, e). Second, changing laws, bureaucratic hurdles, and insufficient incentives for off-grid development complicate matters. Governments must establish renewable energy frameworks with subsidies and tax incentives to encourage private sector participation (Green Climate Fund 2020). Third, rural areas may lack the human resources for off-grid projects, thus training and supply chain improvements are needed (Nwokolo et al. 2023a, b, c, d, e). Fourth, weather conditions affect solar and wind energy availability, requiring storage and hybrid systems for consistent energy access (IRENA 2019). Successful case studies support off-grid options. East Africa's M-KOPA and d.light pay-as-you-go solar solutions provide rural electrification using mobile money payback systems that allow households to buy solar home systems. Bangladesh has solar mini-grids for off-grid enterprises and communities, and foreign financing shows the promise of large-scale decentralised renewable energy projects. Rural India promotes hybrid solar-wind systems to reduce energy intermittency, electrify and educate communities (IRENA 2019). Thus, the government, private investors, and international organisations may increase financial feasibility for off-grid energy systems through subsidised initiatives, concessional loans, and grants to reduce end-user costs. To ensure sustainable off-grid system installation, use, and maintenance, local professionals must attend capacity-building seminars and training. A government-mandated legal and regulatory framework can reduce bureaucratic barriers to private sector renewable project participation. To improve off-grid system efficiency and reliability, the government may fund hybrid, energy storage, and smart-grid research. Off-grid renewable energy technologies can improve electricity access in disadvantaged nations. Solar home systems (SHS), mini-grids, hybrid energy systems, and biomass technologies turn rural problems into sustainable energy sources with socioeconomic advantages and climate change mitigation potential. Politicians and the global community can solve development's costs, technological limits, and governance issues. This comparative house case study of Africa and Asia shows that off-grid renewable energy is crucial for fair and sustainable development everywhere.

Case Studies of Successful Renewable Energy Projects: Transforming Energy Landscapes Globally

Renewable energy initiatives combat climate change globally, diminish dependence on fossil fuels, and enhance sustainable development. International project case studies offer socio-economic and technological empirical insights for the implementation of renewable energy objectives in other projects. The Solar Home Systems (SHS) project in Bangladesh is one of the most successful renewable energy initiatives in South Asia, enhancing electricity access for several off-grid people. The Infrastructure Development Company Limited (IDCOL) oversees the project, which has delivered over four million solar home systems (SHS) nationwide, benefiting about 20 million individuals. Microfinance and a subsidised pay-as-you-go scheme enable even the lowest-income households to get solar energy at little expense. The SHS initiative has enhanced livelihood prospects for participants by providing access to lights and phone charging, reducing reliance on paraffin and freeing up time formerly spent on energy generation for income-generating activities.

Furthermore, the initiative generated employment opportunities in the production, installation, and maintenance of SHS (IRENA 2019). The Lake Turkana Wind generating Project (LTWP) in Kenya is the biggest wind generating facility in Africa, supplying up to 15% of the country's electricity needs. This project was a public–private partnership aimed at assessing and utilising sustained wind in the region to provide a stable energy supply for a large-scale wind farm, while also promoting awareness of the necessity for future local infrastructure development (Nwokolo et al. 2023a, b, c, d, e). In instances when interconnected grids are impractical or prohibitively costly, initiatives have facilitated electrification nonetheless. In Nepal, micro-hydro plants have enabled distant mountainous communities to get electrification. Nepal possesses more than 3,000 micro-hydro projects, predominantly successful due to community ownership, multifunctional uses (household/agricultural), and negligible environmental consequences achieved by run-of-the-river technology (IRENA 2019).

In India, hybrid wind-solar mini-grids integrate solar panels, wind turbines, and battery storage to establish a reliable electricity supply for off-grid rural applications. Mini-grids provide reliable energy to residences, educational institutions, and enterprises, hence diminishing dependence on costly diesel generators used for rural electrification (Nwokolo et al. 2023a, b, c, d, e). In Cambodia, rice husk power plants implement a novel waste-to-energy system that utilises agricultural waste to generate electricity, benefiting farmers and reducing dependence on fossil fuels (Africa's Awakening to Climate Action, Nwokolo et al. 2023a, b, c, d, e). Offshore projects have commenced the implementation of wind energy, exemplified by the Hornsea Wind Farm in the United Kingdom, which employs some of the most advanced wind turbines currently available, in conjunction with high-voltage undersea cables to convey renewable energy to the national grid (IRENA 2019).

Geothermal advancements offer energy stability. The Olkaria geothermal facility in Kenya supplies about 50% of the nation's energy requirements and reduces reliance on foreign energy sources (Nwokolo et al. 2023a, b, c, d, e).

Notwithstanding these achievements, obstacles to the use of renewable energy persist. Firstly, there is a sporadic initial expense; secondly, there are policy and regulatory challenges; thirdly, there is a scarcity of well skilled personnel. The necessity for capital investment during pre-construction and installation is a challenge in low-resource regions; however, initial cost obstacles can be mitigated through blended finance and public–private partnership initiatives (Green Climate Fund 2020). Likewise, in instances where supportive policies are in place, bureaucratic delays hinder project execution, highlighting the necessity for regulatory environments that enable investors to promptly and efficiently initiate projects (Nwokolo et al. 2023a, b, c, d, e). A limited number of qualified specialists available to operate renewable systems hinders sustainable usage; hence, sustainability necessitates the implementation of training initiatives to ensure ongoing operational capabilities (IRENA 2019).

Moreover, the challenge of intermittency with renewable sources solar and wind necessitates the implementation of hybrid systems or sophisticated storage solutions to ensure reliable energy availability. These stories demonstrate the imperative for community engagement, financial resources, technical progress, and global attempts to enhance renewable energy programs. Community involvement in building and implementation fosters a sense of ownership among individuals, hence reducing the likelihood of neglecting maintenance and sustainability initiatives. Moreover, finance mechanisms like pay-as-you-go in Bangladesh and concessional solar loans for low-income people in India facilitate the feasibility of such initiatives. Moreover, technical innovations like as India's wind-hybrid systems, battery storage solutions, and solar smart-grid technologies enhance the efficacy of these initiatives. Finally, international initiatives from governmental and business sectors, together with foreign aid from developed nations through the UN, allocate cash for extensive projects and collaborative collaborations. Policymakers must ensure the feasibility of energy technology by suitable subsidies, tax incentives, or effective regulatory frameworks. The inclusion of renewable energy projects in national or regional development plans is essential for national rural electrification initiatives and comprehensive economic growth.

Implications

Numerous Sustainable Development Goals, in addition to SDG 7 and SDG 13, are relevant, including SDG 8 (Decent Work and Economic Growth), as initiatives generate employment opportunities. As energy storage technology progresses, next study can evaluate ideal hybrid systems by evaluating alternate energy combinations and energy requirements. Furthermore, evaluating variability can enhance the dependability and predictability of renewable energy sources. An evaluated situational evaluation of SDG indicators throughout time would facilitate the incorporation of renewable energies in select projects, enabling multidisciplinary specialists to devise improved solutions for under-resourced areas in the future. Ultimately, evaluating many funding mechanisms throughout time—pay-as-you-go, blended finance, public–private partnerships—will result in enhanced chances for impact in the future. By accurately evaluating the obstacles and implementing the essential insights, renewable energy may be expedited for a worldwide shift to net zero in a timely manner.

Conclusion, Limitations and Future Scope

The chapter on technological innovations in renewable energy underscores the transformative potential of these advancements in addressing global energy challenges, particularly in underserved regions. By leveraging innovations in solar, wind, and hydropower technologies, the chapter demonstrates how renewable energy solutions can promote socio-economic development, foster sustainability, and mitigate climate change. Key innovations such as portable solar panels, solar micro-grids, hybrid systems, and small-scale hydropower projects have proven effective in providing electricity to rural and low-income areas, enhancing education, healthcare, and local economic activities. These technologies not only address energy poverty but also contribute to global sustainability goals by reducing greenhouse gas emissions and reliance on fossil fuels.

Moreover, real-world case studies from regions such as Sub-Saharan Africa, South Asia, and Latin America illustrate the feasibility and impact of renewable energy initiatives, highlighting the importance of localized and community-driven approaches. However, the chapter also identifies critical barriers to scaling these technologies, including high upfront costs, technical and policy challenges, and limited infrastructure. Addressing these obstacles requires targeted interventions such as capacity-building programs, supportive policies, and innovative financing mechanisms.

Therefore, the chapter emphasizes the pivotal role of technological innovation and international collaboration in advancing renewable energy adoption. By fostering an integrated approach that combines innovation, policy support, and strategic investment, renewable energy technologies can bridge the energy access gap and contribute to a more sustainable and equitable future.

References

Bhattacharyya SC (2018) Energy economics: concepts, issues, markets, and governance. Springer. https://doi.org/10.1007/978-3-319-69658-5

Climate Policy Initiative (2019) Global landscape of climate finance 2019. https://www.climatepolicyinitiative.org

Green Climate Fund (GCF) (2020) Climate finance readiness program: Building capacity for change. https://www.greenclimate.fund

International Renewable Energy Agency (IRENA) (2019) SIDS lighthouses initiative: progress and way forward. https://www.irena.org

Nwokolo SC, Singh R, Khan S, Kumar A (2024) Technological pathways for Africa's net-zero economy: technology solutions to unlock Africa's sustainable future. https://www.sciencedirect.com/book/9780443314865/technological-pathways-for-africa-s-net-zero-economy

Nwokolo SC, Singh R, Khan S, Kumar A (2025) Role of digitalization and connectivity for achieving a net zero economy in Africa. In: Technological pathways for Africa's net-zero economy, pp 175–192. https://doi.org/10.1016/B978-0-443-31486-5.00006-9

Nwokolo SC, Singh R, Khan S, Kumar A, Luthra S (2023a) Impacts of climate change in Africa. In: Africa's path to net-zero. CSR, sustainability, ethics & governance. Springer, Cham. https://doi.org/10.1007/978-3-031-44514-9_7

Nwokolo SC, Singh R, Khan S, Kumar A, Luthra S (2023b) Influencing the scale of Africa's energy transition. In: Africa's path to net-zero. CSR, sustainability, ethics & governance. Springer, Cham. https://doi.org/10.1007/978-3-031-44514-9_4

Nwokolo SC, Singh R, Khan S, Kumar A, Luthra S (2023c) Scenarios that could give rise to an African net-zero energy transition. In: Africa's path to net-zero. CSR, sustainability, ethics & governance. Springer, Cham. https://doi.org/10.1007/978-3-031-44514-9_8

Nwokolo SC, Singh R, Khan S, Kumar A, Luthra S (2023d) Africa's awakening to climate action. In: Africa's path to net-zero. CSR, sustainability, ethics & governance. Springer, Cham. https://doi.org/10.1007/978-3-031-44514-9_9

Nwokolo SC, Singh R, Khan S, Kumar A, Luthra S (2023e) Remedies to the challenges of renewable energy deployment in Africa. In: Africa's path to net-zero. CSR, sustainability, ethics & governance. Springer, Cham. https://doi.org/10.1007/978-3-031-44514-9_3

Organisation for Economic Co-operation and Development (OECD) (2020) Climate finance provided and mobilised by developed countries in 2013–2018. OECD Publishing. https://doi.org/10.1787/5f8326b5-en

Oyedele OM (2025). The cobra effect: exploring the dark side of innovation and environmental sustainability in developing nations. In: Singh R, Joshi A, Filho WL, Khan S (eds) Zero carbon industry, eco-innovation and environmental sustainability. World sustainability series. Springer, Cham. https://doi.org/10.1007/978-3-031-80220-1_15

Oyedele MO, Sunday AO, Abuh AI (2024) Fostering technological-enhanced training and development for business survival and performance in the new normal. J Propuls Technolo 45(3):1858–1869. https://propulsiontechjournal.com/index.php/journal/article/view/7479

Oyenuga MO, Omale SA (2024) Is Africa Jinxed? Exploring the challenges of technology access and adoption in Africa. Afr J Econ Sustain Dev 7(4):142–161. https://doi.org/10.52589/AJESDULN1LRNF

Oyetunde TO, Oyenuga MO, Adoga GJ (2023) Green management practices and organisational efficiency in higher institution. Budapest Int Res Crit Linguist Educ (BirLE) J 6(3):276–290. https://bircu-journal.com/index.php/birle/article/view/7716

United Nations Development Programme (UNDP) (2011) Blending climate finance through national climate funds. https://www.undp.org

Chapter 8
Digital Technologies and Climate Action

Abstract The evolution of digital technologies to enable greenhouse gas emissions to become more sustainable across sectors is becoming an increasingly important complement to climate action. This chapter explores the use of digital technologies (i.e. blockchain; artificial intelligence and big data analytics) in aiding climate actions. And, yes, there are case studies and real-world examples to help facilitate discussion of how these technologies might take emissions tracking and management up a notch in terms of transparency, traceability and efficacy. It also addresses challenges and opportunities tied to each of these green digitalization pathways in the context of climate action, seeking to provide a holistic perspective of how digital technologies can fundamentally change the way we approach climate change and our engagement in sustainable development.

Introduction

With the world facing the daunting challenges posed by climate change, there is an urgent imperative. Even more so, digital technologies today are a potentially powerful means to combat climate change (Nwokolo et al. 2024a, c), by promoting opportunities for sustainability, increasing efficient use of natural resources, and decreasing greenhouse gases. The implementation of digital solutions in climate action is changing how industries operate, how governments frame policies, and how communities connect with sustainability activities. Digital technologies range from simple tools and systems such as blockchain, artificial intelligence, the Internet of Things, and big data analytics. Real-time monitoring and data collection become possible, helping to make emission reporting more transparent. For instance, this technology is taking the limelight for traceability in supply chain management where supply chain stakeholders have the opportunity to verify their carbon footprint and as a result ensure alignment with sustainability principles (Guo 2023). Here, blockchain serves to establish decentralised and immutable ledgers as a way to build trust, hence motivating partnerships in reducing emissions.

M. 0. Oyenuga et al., *Smart Technologies for Climate Change and Net Zero Policies: Practical Approaches Towards Sustainability*, Climate Change Management,
https://doi.org/10.1007/978-3-031-92221-3_8

Artificial intelligence further plays a significant role in the action of climate through the ability to provide predictive analytics and optimise available resources. Algorithms for AI can analyze vast of data that may help spot patterns and trends, thus aiding businesses and governments in making sound decisions on which strategies will be effective in reducing emissions. For example, artificial intelligence can optimize industrial processes, such as energy use. This optimizes operations and then reduces carbon emissions dramatically (Engvall and Flak 2022). Tap into Data-Driven Insights The development of efficient climate policies and strategies hinges largely on data-driven insights. Exploiting big data analytics also allows for a holistic assessment of climate impacts and vulnerabilities. Data integration from various sources, such as satellite imagery, weather patterns, and socio-economic indicators, provides stakeholders with the ability to develop targeted adaptation strategies based on a better comprehension of the risks. The Digital Earth platform, for example, offers essential data on climate variables and serves as a base for Earth system models in the prediction of climate; this would assist researchers and policymakers in informed decisions (Jia et al. 2019). While these digital technologies show much promise for climate action, a few challenges remain. Data privacy, cybersecurity, and the digital divide are a few of these that would hinder the effective use of these technologies. Only with a strong partnership between governments, businesses, and communities to build toward an environment that encourages innovation and sustainability can pandemic breaches be effective through digital solutions (Guo 2023). We also need to consider how digital technologies impact the environment, with energy use from data centres and blockchain infrastructures that contribute to emissions if not properly managed.

The Role of AI in Climate Prediction and Adaptation

The advancements in this field will create powerful weapons against climate change, and provide best use scenario solutions for climate prediction and adaptation. It can be the greatest division of data that can help anything work out the total dynamic situation help better forecast, and enable more proactive decisions about climate impact issues by creating CSS using large data and well-structured algorithms. One of the biggest areas of application for AI for the climate has been to identify the complex, interleaved data produced by climate models, satellite images and past weather observations. Machine learning algorithms are capable of detecting correlations or patterns that are not visible using some other analytical techniques. For example, machine learning techniques have been widely used to mirror the efficiency that emerges from the analysis of big datasets at certain spatial and temporal scales for weather prediction and climate analyses (Bochenek and Ustrnul 2022). This ability helps researchers and policymakers develop stronger strategies for responding to climate transformations.

AIs also make climate models much more accurate. Reichstein et al., were optimistic about utilizing deep learning methods to better represent these complex climate processes, such as storm processes, precipitation, and overall temperature variability. They would then introduce A.I. into these climate models to make more dynamic predictions of climate scenarios years in the future, giving that information so regions at risk can plan and redistribute resources. Focusing on AI's potential role in climate adaptation, the technology can be used to address the potential vulnerabilities the stakeholders in ecosystems and communities face, such vulnerabilities can envelop the entire life cycle of adaptation (Cheong et al. 2022). AI can use data relating to climate effects, like vegetation change, water resources and extreme weather events.

AI can also give communities valuable information that can help them consider resilience against climate change conditions in the development of targeted interventions. On the other hand, the use of AI technologies could provide for real-time climate-related risk monitoring and management. Employing IoT's need for AI–integration allows better data continuity for constant polling of environmental settings and responses to threat signatures when they become visible. In agriculture, for example, Artificial intelligence power tools that determine how to manage irrigation and crops based on changing weather have guaranteed real-time responses to farmers to climatic changes with between the least input of fertilizers, pesticides and water (Nakalembe and Kerner 2023) In this manner, food security can be addressed proactively without harming the ecosystem. This promise is exciting for instance for AI for climate prediction and adaptation, so there are many challenges to crowd in. Data quality and availability are arguably one of the most significant factors that could determine how successful AI applications will be. AI-driven solutions in agriculture greatly rely on access to qualitatively good data and the capability to analyse it proficiently (Callaghan et al. 2022). Integrating AI into the existing climate landscape comes with its challenges, but that's one of the areas it can be strongest, working with a wide range of stakeholders, from governments and researchers to industry leaders. As stated by Manish et al. (2023), dissemination of knowledge and forming networks is a pivotal task that countries need to address to fully harness the potential of AI for climate change. Different players can and should work together to create interconnected approaches to harnessing the power of AI technologies to support climate resilience and sustainability.

The Role of the Internet of Things (IoT) in Real-Time Environmental Monitoring

Internet of Things (IoT) has been identified as a state-of-the-art technology with a tendency to transform the field of environmental monitoring (Apata et al. 2024) whereby real-time data can be acquired by networking multiple sensors, instruments and systems in a vast network. It can provide knowledge about the environment and thus can be used for more satisfactory decision processes. Environmental monitoring is one of the major IoT applications that are used mainly for real-time environmental conditions. For example, Udrea et al. (2020) hint that an IoT solution that captures indoor climate data in open space offices shows the possibility of generating healthier indoor conditions through the usage of IoT. These systems facilitate the acquisition of real-time data, allowing stakeholders to act promptly in response to environmental alterations and to mitigate any potential threats. IoT technologies are used extensively in agriculture for precision farming, pest control etc.

According to Gao et al. (2020) argue that the fusion of IoT with UAV allows integrated monitoring of crop health and pest occurrence. So farmers utilize real-time data about present weather and soil moisture to make better decisions on irrigation, fertilization, and pest control, which ultimately improves productivity and reduces impact on the environment. This has increased resource efficiency and hence agriculture has become sustainable. Further, the Internet of Things is significantly contributing to sustainable water resource management. The smart water management system allows for efficient water resource management by using IoT sensors to monitor water quality and usage in real time. For example, a Smart Water Monitoring System using IoT has been established in the cities, which helps alert the authorities about water supply network leaks and prevent the loss of water and misuse of water. These practices are crucial for environmental sustainability (Oyedele et al. 2025), to counter the concern of water scarcity in regions. IoT enablement with renewable energy systems presents immense prospects for environment monitoring.

Through the use of IoT devices monitoring energy consumption and production, all stakeholders can benefit by optimizing the use of renewable energies like solar and wind. Optimizing the exploitation of renewable resources would reduce carbon emissions and thus diminish reliance on fossil fuels. Further, with the help of IoT, energy demand response mechanisms will be brought about, empowering consumers to tailor their energy consumption according to real-time scenarios. While there are too many benefits that IoT provides for environmental monitoring, several challenges need to be overcome. Some of the concerns against data privacy and security must be addressed in the collection and transmission of sensitive data on the environment. According to Jebur (2022), sustainable IoT solutions which have minimal effects on the environment but protect the data are urgently required. The digital divide may also mean that IoT technology is not widely available in specific regions, mainly in developing countries where infrastructure and resources are often scarce (Oyenuga and Omale 2024; Joshi et al. 2024). Collaborative efforts of stakeholders are a must to take full advantage of IoT in monitoring the environment. Governments, companies, and research institutions must all work together for the development of standards, policies, and frameworks that support IoT technology deployment. In addition, public awareness and education will contribute to greater acceptance and understanding of IoT applications in environmental monitoring.

Data Analytics for Optimizing Energy Use

The integration of data analytics into energy management systems is revolutionizing how industries and consumers optimize energy use. With advanced analytical techniques, organizations can gain insight into patterns of energy consumption, pinpoint inefficiency hotspots, and develop strategies that lead to significant reductions in energy waste and greenhouse gas emissions. One large application to utilize optimization with data analytics is the generation of machine learning algorithms to estimate the possible energy intake. Home energy management systems that apply machine learning and IoT techniques to monitor energy consumption for smart homes are further discussed here (Sudha 2024). Such systems recommend to the user a specific action to save energy based on an analysis of historical data and usage patterns at this specific time, which could involve a modification in the heating and cooling settings or better optimization of user usage of certain types of appliances in the house. Not only does this technology provide an advancement in energy utilization efficiency but empowers the end-user (Oyedele et al. 2023) with decision autonomy over consumption (Marcus et al. 2020).

In terms of electric vehicles, data analytics will play a vital role in the optimization of energy management.

This study by Padmavathy et al. (2023) represents the likelihood that intelligent machine learning algorithms can analyze driving patterns, traffic conditions, and weather to predict energy consumption and optimize schedules for charging. These systems can mitigate energy waste from energy management solutions while providing personalized solutions for efficient ranges of driving for electric vehicles. This optimization is crucial because the adoption of electric vehicles is increasing, and efficient energy use is required to support the transition to sustainable transportation. Additionally, the use of data analytics in smart grids improves the efficiency of energy distribution and consumption. As Mocanu et al. (2019) have pointed out, smart grids make use of big data analytics; monitoring energy utilization in real time helps to improve the demand side of management and load forecasting. By optimising the distribution process for utilities, peak demand could be curtailed. Overall grid reliability also improves.

This is also important in helping harness renewable power supplies into the grid while providing stable access to energy supply. Data analytics is also applicable in industrial processes, as can be seen in this aspect. As Nugrahanto (2024) notes, predicting and optimising CNC machining processes regarding energy consumption had to depend on machine learning. Manufacturers (Oyenuga et al. 2023) can identify patterns by analyzing operational data and optimising parameters to enhance energy efficiency and thus lessen associated operational costs and emissions. This application of data analytics is a way to enhance productivity and, indirectly, adds to the sustainability of manufacturing operations despite the many benefits of data analytics in optimizing energy use. Several challenges need to be overcome. Issues such as data privacy, cybersecurity, and the digital divide can influence the effectiveness of these technologies. Challenges in the management and integration of data will limit the application of data analytics in smart grids (Bhattarai et al. 2019). Moreover, the requirement for skilled manpower (Singh et al. 2024) in terms of data analysts or interpreters is very vital in unlocking the potential of data-driven energy optimization methods.

Blockchain Technologies for Tracking Emissions

Blockchain technologies turned out to be a game-changer because blockchain enabled visibility, traceability, and accountability far greater than anything that was experienced until 23 years ago, at least in the look of many industries from supply chain management to emission tracking. A decentralized and time-stamped ledger allows emissions data to be recorded in real-time use cases which results in more cooperation and trust among the actors involved.

One of the significant benefits of utilizing blockchain technology for tracking emissions is the potential for enhanced transparency in the supply chain. Blockchain technology enables stakeholders in the green supply chain to cooperate and build trust among each other through the provision of an immutable record of emissions data (Guo2023). These contributions minimize information asymmetry and enable stakeholders to collectively work toward meeting sustainability objectives. For example, it could allow companies to offer emissions data to regulators, consumers and other stakeholders, so long as claims of sustainability are verifiable and credible. Carbon trading systems could also be applied through blockchain technology, to have a strong and orderly means of tracking carbon credits and emissions allowances. Rogerson and Parry (2020) state that the transparency offered by the blockchain will reduce supply chain friction and improve the trust between participants thus leading to more efficient carbon markets. It can also help simplify trading and make it less vulnerable to fraud and manipulation in carbon markets by providing real-time tracking of emissions and carbon credits.

Blockchain has its role as traceability of emissions that occur in product manufacturing, transportation, and distribution (Zhang 2024) within low-carbon supply chains. Such an integration between the IoT and blockchain technology would assist companies in knowing their emissions through different stages in the supply chain, thereby achieving more accurate reporting and accountability, especially in complicated supply chains involving industries where it is challenging to track and prove emissions. Such companies can engage consumers with systems based on blockchain technology to determine the carbon footprints of the products they have purchased for further sustainable consumption habits (Wang et al. 2020). Such an approach not only empowers the consumer to make informed choices but also encourages the companies to change their ways to adopt more sustainable practices and improve environmental performance. Blockchain technology has the potential for tracking emissions, but various challenges are also associated with this technology. Technology is still very early, and issues such as scalability, interoperability, and regulatory uncertainty would be major drawbacks to its broader adoption (Nwokolo et al. 2024b, d). According to Helo and Hao, blockchain integration requires planning as well as joint efforts by stakeholders to achieve harmonization with current systems and processes (Helo and Hao 2019). Moreover, the energy consumption of blockchain networks, especially those relying on proof-of-work consensus mechanisms, poses environmental concerns about this emerging technology.

Case Studies of Digital Technology in Climate Initiatives

Digital technologies are increasingly playing a central role in climate initiatives, including innovative means of monitoring and managing climate impacts (Nwokolo et al. 2023). A classic example is blockchain technology in carbon credit trading. Some of the international eco-digital projects discussed are where blockchain is applied to give more transparency in carbon markets, and Maksymova et al. (2023) demonstrate how decentralized ledgers that cannot be tampered with allow stakeholders to make real-time tracks of carbon credits and emissions in real-time. This builds trust and cooperation between participants. This innovation could help make the carbon trading processes more efficient, thus motivating higher involvement in activities that help to reduce emissions for climate neutrality (Kumar et al. 2024). Souza (2023) further points out that digital technologies can boost energy efficiency and climate protection through the COVID-19 pandemic (Omale et al. 2021). In education and public awareness, educational institutions applied online platforms in the spread of information and resources (Oyedele and Iember 2021), involving students and communities in climate action. The successful implementation of digital tools for climate education proves the potential of technology in raising awareness and empowering people to practice sustainability.

Another important case study is the use of artificial intelligence (AI) in climate change adaptation strategies. Lee and Mwebaza (2021) emphasize the role of AI in enhancing climate technology transfer from developed to developing countries. AI can find a pattern or trends in the masses of climate data, allowing policymakers and businessmen to make decision-making based on the information acquired. This way, it improves the effectiveness and resource allocation through climate initiatives. It will be a change for vulnerable areas.

An instance worthy of mention is also the integration of IoT technologies for real-time monitoring of environmental conditions. Narayana mentions the creation of smart monitoring systems that are capable of employing IoT sensors to monitor environmental parameters, including air quality and temperature, in real-time (Narayana 2024). These kinds of systems yield critical information for informing climate action decisions and providing information for informing policy actions that make stakeholders respond to environmental change and mitigate the potential risks associated with it. This includes the concept of "smart forests" that Gabrys developed (Gabrys 2020). Using digital networks (Nwokolo et al. 2024b, d) to analyze and manage environmental processes is one way through which cities can increase their adaptive capacity to climate change. The use of such digital technologies for climate resilience is improving urban resilience and enhancing community engagement and stewardship of natural resources.

Implications

The integration of digital technologies into climate action initiatives holds enormous implications for enhancing sustainability, improving efficiency, and promoting innovation in every sector. First, digital technologies enable the collection, analysis, and reporting of data in a more accurate and timely manner. It serves as a pillar for good monitoring and decision-making on climate change issues. For example, IoT devices can be used to monitor environmental parameters in real-time, which helps stakeholders understand climate adaptation strategies. This capability increases the ability of governments, businesses, and communities to respond proactively to climate-related challenges. Digital technologies also make climate action efforts more transparent and accountable. For instance, blockchain technology is likely to strengthen traceability within carbon markets: carbon credits can be tracked better, and therefore emissions can also be tracked much more efficiently.

It encourages stakeholders to commit and take part in climate neutrality and compliance with regulatory frameworks and international agreements as the data can be validated and verified using these digital platforms. The evolution of digital touches every part of everyday life, from working from home to business transformation. In a low-carbon economy, digital innovation can open up new business channels and strengthen competitiveness in the green sectors of tomorrow. Every incremental adoption of one set of digital health technologies can lead to new value, resource allocation, and efficiency benefits being adopted in a wide range of sectors, particularly energy and transportation.

Such a transition is also paramount to climate action and increasing economic resilience and sustainability. However, leveraging digital technologies effectively for climate initiatives is no small task. The efficiency of their successful use can be prevented, due to factors like data privacy and cyber security, etc. Environmental costs are also a must, they indicate emissions unless managed otherwise. Proper use can come only by integrating technological innovation within social and economic considerations.

Conclusion, Limitations, and Future Scope

Examining the role of digital technologies in climate action will involve improved collection and transmission of data through tools such as the Internet of Things (IoT), blockchain, and artificial intelligence that enhance transparency and efficiency in supporting solutions to climate challenges. The effectiveness of these technologies in enhancing resilience in climate actions and sustainability efforts is illustrated through the case studies in the chapter. There can, however, be challenges related to the adoption of digital technology; it will be critical to overcome these challenges to realize the full potential of technology in climate initiatives. Such joint action involves coordinated efforts by governments, businesses and communities to develop enabling policies, investment in research and development and sectoral coordination. Therefore, as countries determine how to rebuild following COVID-19, they must prioritize how digital technologies are integrated into climate action initiatives so that the world moves forward in tackling climate change with a more sustainable, resilient future. By nurturing ideas that are capable of solving the pressing challenge of climate change, we can make the world a more sustainable planet.

References

Apata AO, Oyenuga MO, Abeoba M, Ugom A (2024) Internet of Things (IoT) solutions for smart transportation infrastructure and fleet management. J Propuls Technol 45(4):1492–1505. https://propulsiontechjournal.com/index.php/journal/article/view/8325

Bhattarai BP, Paudyal S, Luo Y, Mohanpurkar M, Cheung K, Tonkoski R, Hovsapian R, Myers KS, Zhang R, Zhao P, Manic M, Ahang S, Zhang X (2019) Big data analytics in smart grids: state-of-the-art, challenges, opportunities, and future directions. IET Smart Grid 2(2):141–154. https://doi.org/10.1049/iet-stg.2018.0261

Bochenek B, Ustrnul Z (2022) Machine learning in weather prediction and climate analyses—applications and perspectives. Atmosphere 13(2):180. https://doi.org/10.3390/atmos13020180

Callaghan M, Schleussner CF, Nath S, Lejeune Q, Knutson TR, Reichstein M, Hansen G, Theokritoff E, Andrijevic M, Brecha RJ, Hegarty M, Jones C, Lee K, Lucas A, van Maanen N, Menke I, Pfleiderer P, Yesil B, Minx J (2022) Machine learning-based evidence and attribution mapping of 100,000 climate impact studies. https://doi.org/10.21203/rs.3.rs-783398/v2

Cheong S, Sankaran K, Bastani H (2022) Artificial intelligence for climate change adaptation. Wiley Interdiscip Rev Data Min Knowl Discov 12(5). https://doi.org/10.1002/widm.1459

Engvall T, Flak L (2022) The state of information infrastructure for global climate governance. Transf Gov People Process Policy 16(4):436–448. https://doi.org/10.1108/tg-05-2022-0064

Gabrys J (2020) Smart forests and data practices: from the internet of trees to planetary governance. Big Data Soc 7(1):205395172090487. https://doi.org/10.1177/2053951720904871

Gao D, Sun Q, Hu B, Zhang S (2020) A framework for agricultural pest and disease monitoring based on internet-of-things and unmanned aerial vehicles. Sensors 20(5):1487. https://doi.org/10.3390/s20051487

Guo W (2023) The impact of blockchain technology on integrated green supply chain management in china: a conceptual study. J Digit Real & Mast (Dream) 2(02):58–65. https://doi.org/10.56982/dream.v2i02.112

Helo P, Hao Y (2019) Blockchains in operations and supply chains: a model and reference implementation. Comput Ind Eng 136:242–251. https://doi.org/10.1016/j.cie.2019.07.023

Jebur T (2022). Greening the internet of things: a comprehensive review of sustainable iot solutions from an educational perspective. Indones J Educ Res Technol 3(3):247–256. https://doi.org/10.17509/ijert.v3i3.57428

Jia G, Zhang L, Zhu L, Xu R, Liang D, Xu X, Bao T (2019) Digital earth for climate change research 473–494. https://doi.org/10.1007/978-981-32-9915-3_14

Joshi A, Singh R, Kumar V, Rani S (2024) Net zero strategy: exploring opportunities, confronting challenges for developing countries. In: Singh R, Crowther D (eds) Transition towards a sustainable future. Approaches to global sustainability, markets, and governance. Springer, Singapore. https://doi.org/10.1007/978-981-97-5756-5_11

Kumar K, Singh R, Sharma A, Lehri BPS (2024) Climate change adaptation strategies for achieving net-zero economy. In: Approaches to global sustainability, markets, and governance, pp 163–183. https://doi.org/10.1007/978-981-97-5756-5_8

Lee W, Mwebaza R (2021) Digitalization to achieve technology innovation in climate technology transfer. Sustainability 14(1):63. https://doi.org/10.3390/su14010063

Maksymova I, Mietule I, Kulishov V (2023) Digital solutions for a climate neutral economy: international framework of eco-digital projects. In: Environment technology resources proceedings of the international scientific and practical conference, vol 1, pp 123–127. https://doi.org/10.17770/etr2023vol1.7291

Manish K, Sree G, Karthikeya G, Kedara G, Reddy C, Packiaraj H (2023) Climate change analysis using machine learning. Int Res J Mod Eng Technol Sci. https://doi.org/10.56726/irjmets47177

Marcus GO, Oyenuga MO, Ahungwa AI (2020) Effects of sales promotion on consumer buying behaviour of food seasoning among Nigerian households: a case study of nestle Maggi Naijapot. Budapest Int Res Crit Inst J (BIRCI J) 3(1):134–139. https://doi.org/10.33258/birci.v3i1.724

Mocanu E, Mocanu D, Nguyen P, Liotta A, Webber M, Gibescu M, Slootweg J (2019) Online building energy optimization using deep reinforcement learning. IEEE Trans Smart Grid 10(4):3698–3708. https://doi.org/10.1109/tsg.2018.2834219

Nakalembe C, Kerner H (2023) Considerations for ai-eo for agriculture in sub-saharan Africa. Environ Res Lett 18(4):041002. https://doi.org/10.1088/1748-9326/acc476

Narayana T (2024) Advances in real time smart monitoring of environmental parameters using iot and sensors. Heliyon 10(7):e28195. https://doi.org/10.1016/j.heliyon.2024.e28195

Nugrahanto I (2024) Innovative approaches to sustainable computer numeric control machining: a machine learning perspective on energy efficiency. Sustainability 16(9):3569. https://doi.org/10.3390/su16093569

Nwokolo SC, Singh R, Khan S, Kumar A (2024a) Role of digitalization and connectivity for achieving a net zero economy in Africa. Elsevier eBooks, pp 175–192. https://doi.org/10.1016/b978-0-443-31486-5.00006-9

Nwokolo SC, Singh R, Khan S, Kumar A (2024b) Key players in net zero technology. Elsevier eBooks, pp 193–215. https://doi.org/10.1016/b978-0-443-31486-5.00007-0

Nwokolo SC, Singh R, Khan S, Kumar A (2024c) Technological pathways for Africa' s net-zero economy. Elsevier eBooks. https://doi.org/10.1016/c2023-0-52499-1

Nwokolo SC, Singh R, Khan S, Kumar A (2024d) Net zero technology and circular economy. Elsevier eBooks, pp 237–247. https://doi.org/10.1016/b978-0-443-31486-5.00009-4

Nwokolo SC, Singh R, Khan S, Kumar A, Luthra S (2023) Technological pathways to net-zero goals in Africa. In: CSR, sustainability, ethics & governance, pp 93–210. https://doi.org/10.1007/978-3-031-44514-9_5

Omale SA, Oyenuga MO, Oriaku C (2021) Effect of organizational culture on the performance of hospitality industry in covid-19 era. Abhigyan Manag J 39(3):24–32. https://doi.org/10.56401/abhigyan/39.3.2021.24-32

Oyedele OM (2025). The cobra effect: exploring the dark side of innovation and environmental sustainability in developing nations. In: Singh R, Joshi A, Filho WL, Khan S (eds) Zero carbon industry, eco-innovation and environmental sustainability. World sustainability series. Springer, Cham. https://doi.org/10.1007/978-3-031-80220-1_15

Oyedele OM, Iember AA (2021) Covid-19 and the future of higher education. IEEE Technol Policy Ethics 6(4):1–3. https://doi.org/10.1109/ntpe.2021.9778140

Oyedele MO, Marcus GO, Ahungwa AI (2023) Do consumers care about green marketing practices? Insight from a developing nation. Budapest Int Res Crit Inst J (BIRCI J) 6(3):1424–1436. https://bircu-journal.com/index.php/birci/article/view/7665

Oyenuga MO, Omale SA (2024) Is Africa Jinxed? Exploring the challenges of technology access and adoption in Africa. Afr J Econ Sustain Dev 7(4):142–161. https://doi.org/10.52589/AJESDULN1LRNF

Oyenuga MO, Andah RA, Nduji R (2023) Employee participation and organisational performance in an emerging economy. Konfrontasi j: Cult Econ Soc Changes 10(2):91–105. https://doi.org/10.33258/konfrontasi2.v10i3.281

Padmavathy R, Jeya Prakash K, Greeta T, Divya K (2023) A machine learning-based energy optimization system for electric vehicles. In: E3s web of conferences, vol 387, p 04008. https://doi.org/10.1051/e3sconf/202338704008

Rogerson M, Parry G (2020) Blockchain: case studies in food supply chain visibility. Supply Chain Manag Int J 25(5):601–614. https://doi.org/10.1108/scm-08-2019-0300

Singh R, Crowther D, Ravesangar K (2024) Harnessing human capital for a sustainable future: HR's pivotal role toward net zero policies. In: Approaches to global sustainability, markets, and governance, pp 27–47. https://doi.org/10.1007/978-981-97-5756-5_2

Souza R (2023) University outreach during the pandemic: digital technologies to teach energy efficiency and climate protection. Educ Pesqui 49:e264343. https://doi.org/10.1590/s1678-4634202349264343en

Sudha N (2024) Hems using machine learning and IoT in energy management. Int Res J Adv Eng Hub 2(06):1652–1658. https://doi.org/10.47392/10.47392/irjaeh.2024.0227

Udrea I, Gheorghe V, Cartal L, Duminică D, Petrache S, Apostolescu T, Kraus V (2020) IoT solution for monitoring indoor climate parameters in open space offices. In: E3s web of conferences, vol 180, p 02012. https://doi.org/10.1051/e3sconf/202018002012

Wang M, Wang B, Abareshi A (2020) Blockchain technology and its role in enhancing supply chain integration capability and reducing carbon emission: a conceptual framework. Sustainability 12(24):10550. https://doi.org/10.3390/su122410550

Zhang C (2024) Blockchain traceability adoption in low-carbon supply chains: an evolutionary game analysis. Sustainability 16(5):1817. https://doi.org/10.3390/su16051817

Chapter 9
Green Infrastructure and Smart Urbanization

Abstract This chapter links sustainable development and green infrastructure with intelligent urban development. This shows the necessity of avoiding development mistakes since other more vulnerable places may suffer the environmental, social, and economic consequences. The notions of green infrastructure argue that natural systems should survive city building. For instance, urban resilience projects that reduce climate change include animals, vegetation, ecosystems, sustainable runoff management, and urban tree canopy correction. Thus, planning should accommodate these when and when needed. Smart cities increase urban living technology and sustainability. IoT, big data, and AI are used to build new cities from the bottom up to optimize resource usage, energy use, and public service accessibility. This chapter examines the creation of an accessible low-carbon transportation network that includes electric automobiles, alternative public transit, and nonmotorized activities like walking and biking. African and Asian smart cities use renewables, garbage management, and skyline digitalization. The same goes for underdeveloped nations. However, socioeconomic difficulties, insufficient policy advice, and a lack of funding still hinder sustainable urbanization.

Introduction

Urbanisation constitutes a global crisis as it requires fundamental design and application viability for sustainable development in a globalised context. As population and industrial centres grow, the proliferation of urban areas signifies that mankind is currently experiencing heightened energy consumption and escalating levels of greenhouse gases and other externalities that adversely affect ecosystems. This aspect complicates reality, both in the current situation and as much as necessary for the future. Simultaneously, reality progresses at a consistent rate. Consequently, significant potential exists as green construction and smart urbanisation offer innovative solutions to counteract the negative consequences of previous urbanisation, while enhancing social efficacy and economic viability. Grey infrastructure refers to

M. O. Oyenuga et al., *Smart Technologies for Climate Change and Net Zero Policies: Practical Approaches Towards Sustainability*, Climate Change Management,
https://doi.org/10.1007/978-3-031-92221-3_9

constructions that are developed without regard for the natural environment, necessitating subsequent retrofitting of sustainable or health-oriented measures that should have been incorporated initially. This constitutes grey development—grey urban rivers, roofs, and construction projects are executed without regard for water. Consequently, WSUD represents an ambiguous domain for Water-Sensitive Urban Design, urban forestry, green roofs, and development initiatives where watermarking, storm resilience, and air quality are prioritised from the outset. WSUD is a planning and design technique that incorporates the urban water cycle into development.

In contrast to a conventional development strategy that incorporates grey infrastructure, Water Sensitive Urban Design (WSUD) offers a development approach that augments or complements natural flow and interactions. This encompasses enhanced rainfall collection, natural aquifer recharge, and superior water quality. Consequently, WSUD refers to the execution and development of urban infrastructures—the edifices and waterways—with water-related issues at a micro level. WSUD is committed to enhancing micro-level environmental aesthetics while simultaneously addressing micro-level floods and stormwater challenges, as well as macro-level air quality concerns. Consequently, the anticipation from a system that accommodates such an offset depends on resilience improvements, as well as enhanced carbon sequestration, augmented human health, and expanded biodiversity. Green urbanisation utilises sustainable practice from olden days to address challenges of the subsequent era (Oyetunde et al. 2023). While traditional urbanisation depletes natural resources, green urbanisation advocates for their preservation for future generations via environmental stewardship and sustainable development. This relates to the concrete multidisciplinary aspects of urban forestry, biophilic design, and urban carbon offsets. These cities contribute significantly to the planet's naturalisation beyond merely promoting a healthy living within nature.

Moreover, there can be interplay between architects and urban planners at a larger level to evaluate design from a generalised, rather than an industry-specific, viewpoint. This chapter positions green infrastructure and smart development within the sustainability initiatives promoted by the UN on a global scale. Consequently, it aligns with SDG 11 (Sustainable Cities and Communities) to ensure that urban projects are affordable, equitable, and secure, with SDG 7 (Affordable and Clean Energy) for the implementation of clean, renewable energy systems, SDG 6 (Clean Water and Sanitation) for water management infrastructures, and SDG 13 (Climate Action) due to its intrinsic connection to climate change adaptation initiatives. Consequently, the many crossings indicate that in the absence of this convergence of attainable achievements at this stage of execution and success, the global intersection of prospects arising from more prevalent initiatives to foster growth would not materialise.

Green Infrastructure and Urban Planning: Principles and Applications

The convergence of urbanised regions and reality fosters a synergy between green infrastructure and urban design, promoting an efficient, practical, and productive coexistence with nature for transformed, sustainable, resilient, and accessible integrated communities. Particularly as global urbanisation intensifies—with an increasing number of emerging nations opting for urbanization—the ultimate practical applications advocate for the principles of urban planning and green infrastructure. The theoretical foundations of green infrastructure and urban planning derive from nature-based solutions, multidisciplinary benefits for anthropomorphised environments, and sustainable research for future generations. Consequently, solutions for heat islands, heightened floods, and diminished air quality necessitate the implementation of green infrastructure, which should be expanded beyond its current scope to promote resilient communities that support human growth while acknowledging the realities of coexistence with the natural environment. Central to green infrastructure is the preservation and repair of natural systems. When urban settlements develop in regions already occupied by natural landscapes, biodiversity is absent, and ecosystem services are lacking. Consequently, the necessity for green belts, parks, and urban forests arises, as these woods function as carbon sinks that absorb surplus atmospheric carbon and lower ambient temperatures in metropolitan areas. Enhanced air quality is a secondary benefit, since pollution is mitigated by trees and forests, leading to improved atmospheric conditions (Nwokolo et al. 2023a, b, c, d, e, f, g, h).

Africa is at the forefront of urban forestry, demonstrating the extensive impact of ecological restoration on development. For instance, with facilities designed for burgeoning urban centers—which, in the context of the Sahel, entails tree planting and projected water conservation to mitigate desertification trends—there is also a responsible reduction of water usage to enhance climate resilience and quality-of-life advantages. Moreover, Water Sensitive Urban Design (WSUD) is congruent with green infrastructure, particularly in susceptible, flood-prone urban areas. The predominant focus of green infrastructure is water management, particularly through Water Sensitive Urban Design (WSUD) elements like permeable pavements, vegetated roofs, and rain gardens, which mitigate surface runoff and promote groundwater percolation, hence addressing flood issues in highly urbanised areas (IRENA 2019). Southeast Asia hosts Jakarta, Indonesia, which has successfully used WSUD principles to mitigate its persistent flooding issues. Implementing green or vegetated roofs on business and residential structures facilitates the collection and diversion of rainfall, while permeable pavements mitigate surface runoff.

Moreover, rain gardens and bioswales contribute to the mitigation of stormwater and flooding problems that deteriorate homes and urban projects over time. This implementation fosters sustainable infrastructure for both current design and future habitation. Urban agriculture represents a burgeoning concept of green infrastructure, which involves modern agriculture on vacant plots and roofs within the city to cultivate food locally, reducing the necessity for transportation across the country or globe. Agriculture serves as a sustainable mechanism for climate resilience, and food security relies on the concept of urban agriculture, particularly in densely populated regions where access to fresh produce is often limited. Urban agriculture is present in African cities like Nairobi and Lagos, where urbanisation intersects with agricultural, promoting green spaces and economic growth. Consequently, urban agriculture can enhance biodiversity in densely urbanised regions and reduce concrete heat absorption when exposed to sunshine.

As urban areas advance, circular economy programs are increasingly integrated to mitigate trash. A circular economy encompasses reclamation initiatives, energy efficiency, and renewable energy strategies in metropolitan areas. Consequently, they correspond with the sustainability aspect of ESG activities, promoting efficient and environmentally friendly growth (Green Climate Fund 2020). Reclaimed recyclable garbage serves as resources for urban construction. In India, destroyed structures were repurposed to supply recycled components for new construction projects. This initiative has reduced carbon emissions at building sites by 50%. Furthermore, advancements in renewable energy technologies, such as solar panels and wind turbines, are implemented in urban environments to accommodate building and development requirements while reducing dependence on non-renewable resources. However, barriers to implementation persist. For example, previous policy deficiencies and their exclusion from earlier evaluations impede the effective implementation of incremental Green Infrastructure in urban settings in subsequent years. Moreover, there is insufficient finance to support this new initiative, particularly in marginalised communities. Ali et al. (2024) assert that incremental adjustments are inadequate to effect meaningful impact. There are also technological impediments.

Moreover, the social dimension is lacking—the extent to which politicians and the public exhibit obstinacy, ignorance on the need for change, and uncertainty about the relevance of sustainability is substantial. The gradual solution arises from a synergy of government, business, and independent activities. Public–Private Partnerships (PPP) provide the necessary finance and expertise in green technologies for extensive green infrastructure initiatives. For example, in the United States and Kenya, tree-planting initiatives may be implemented in metropolitan areas, while solar water diversion devices might be installed on private properties. The focus is on completion. Moreover, cross-collaboration is essential. The Green Climate Fund and the United Nations Development Programme in Kenya offer financial and technical support to facilitate urban sustainability; these collaborations address financing deficiencies and guarantee appropriate measures are implemented in foreign nations.

Smart Cities: Integrating Sustainability into Urbanization

The concept of a smart city, as articulated by Nwokolo et al. (2023a, b, c, d, e, f, g, h), is merely an upgraded version of human life through urbanization and contemporary technological requirements while reducing negative environmental effects.

The smart city employs IoT (Internet of Things), big data, and AI (Artificial Intelligence) to establish an increasingly precise relationship between advanced technology and sustainable benefits, resulting in a situation of energy-efficient resource distribution, integrated public services, and minimized carbon emissions all under one populated roof. Renewable energy systems that decrease dependence on fossil fuels sound like only one aspect of the urban ecology concerning smart cities. A smart city's urban energy grid is primarily dependent on renewable energy through solar and wind. The smart grid penetrates devices and applications that assess real-time usage and apply AI-driven algorithms to predict and facilitate energy transfer better, helping with stabilization and proper supply and demand. Thus, for example, in Copenhagen, Denmark, energy smart grid systems are constructed to develop energy storage devices related to temperature control to maintain an average renewable energy supply when necessary; this average use is simpler to accomplish due to total carbon remission facilitation. (Byrne et al. 2024). Yet transportation is among the largest contributors to greenhouse gas emissions, so the answer is to establish a low-carbon system to promote better sustainability,which developing countries have different needs. For instance, Oyenuga and Omale (2024) faults underdeveloped road systems, while relying on fossil fuels at a high rate. Yet some low-carbon transportation systems are already in operation.

E-mobility is becoming more common across Africa and parts of Asia. For instance, solar-powered charging stations and low-cost electric two-wheelers have revolutionized transportation from rural to urban environments (Nidhi et al. 2024). The government of Rwanda has instituted e-mobility initiatives to remove petrol-driven vehicles and substitute them with electric alternatives to reduce emissions while simultaneously creating green job opportunities. Second, proper public transportation systems exist globally, whether bus rapid transit (BRT) or per-railed systems. These systems prevent the need for personal vehicular traffic; furthermore, artificial intelligence decreases heat production by suggesting the best routes, preventing overcrowding and excessive fuel consumption, which promotes increased efficacy (Krishnan et al. 2024). Third, considerations for non-motorized transportation reduce the need for passthroughs where adequate developments for biking and walking exist. New technology is required for low-carbon transportation. IoT sensors placed within transportation infrastructures can optimize congestion and therefore carbon output through digital traffic lights (Byrne et al. 2024; Apata et al. 2024). Yet even with options presented, many challenges abound. From expensive capital investments to regulatory policies that fail to support low-carbon initiatives, government intervention and transparency with the general public are required to implement and subsidize low-carbon solutions. Therefore, when citizens are unaware or resistant to change, socio-cultural issues emerge. A necessity for community stakeholders and education to alleviate such issues should exists, calling for partnership and collaboration to be established. (Ali et al. 2024; Omale et al. 2024).

Additionally, poor compliance with policies and regulations also posit challenge. Many developing nations suffer from subpar institutions and governance. Urbanization suffers from corruption, bureaucracy, and incompetence (Byrne et al. 2024).

Therefore, such problems should be addressed by establishing better institutions and providing more accountability.

E-mobility is an emerging route that developing countries are following as it supplants fossil-fuel generating transportation. E-mobility encompasses any type of electric vehicle (EV)—e-buses, two-wheelers, and e-passenger vehicles—and is made possible by the potential for renewable energy. For instance, two-wheelers are inexpensive and multi-functional, making them a widely adopted means of transportation in India and Kenya. Such a means of transportation in the developing world allows developing countries to get around with this inexpensive, sustainable means of transportation for its endeavors (Nwokolo et al. 2023a, b, c, d, e, f, g, h). E-mobility projects such as electric motorcycle taxi services in Rwanda have reduced emissions and provided actual jobs within the green economy. They spin out of solar charging stations which guarantee that an electric diversion does not wander off away from renewable energy efforts (Ali et al. 2024). In addition, government projects in these areas like Sub-Saharan Africa have included tax credits and rebates for those transitioning to EVs, a commendable effort of decarbonizing the transportation sector.

Within the developing world, low carbon transport occurs with public transportation as it generates less reliance on personal vehicles, and therefore, personal vehicle emissions. One such component is the Bus Rapid Transit (BRT) system that spans from Colombia to Nigeria. BRTs have improved efficiency for high capacity transit, which decreases emissions and promotes more densely populated, less congested urban centers (Byrne et al. 2024). Digital transportation is just as relevant in Cape Town, in South Africa too. The MyCiTi BRT operation uses digital functionalities for better routing and tracking of passengers. For example, data analytics via AI can reduce average wait time and fuel consumption, which means a more efficient transit operation and happier travelers (Liinamaa et al. 2024). In addition, to reduce the public transit carbon footprint even more, the options for the BRT fleet of vehicles now include electric buses. NMT, like walking and biking, are some of the most sustainable, low-carbon transportation solutions there are. Developing nations are forming NMT-related infrastructure with bike lanes and extended areas for pedestrian sidewalks to encourage active transportation without relying on fossil-fuel-dependent transportation. For example, Kigali, Rwanda has a municipality biking initiative where citizens are encouraged to bike for short-distance travel; the municipality subsidizes bicycles for residents, holding civic engagement meetings to discuss the health benefits and sustainable nature of this transportation option (IRENA 2019). Likewise, the Masala Aloo initiative in India possesses walkable developments for pedestrians under the Smart Cities Mission, which champions greenery and green corridors.

IoT and AI are transforming transportation systems to be more efficient and sustainable. IoT sensors in traffic systems, for instance, may detect jams; the information gained can lead to real-time adjustments based on traffic patterns. Fewer delays at stoplights or stop signs mean less idling and increased progress through an intersection; this is critical in city environments with many stop signs/red lights, thereby decreasing unnecessary fuel consumption and reducing emissions in the atmosphere

(Krishnan et al. 2024; Nwokolo et al. 2023a, b, c, d, e, f, g, h). Nairobi, Kenya. It even has something called Bolt and Uber Green, carpooling services where the cars are hybrid or electric. These socially and environmentally aware services are possible due to big data creating carpooling paths that use less fuel and are cheaper for patrons (Green Climate Fund 2020).

Many developing countries are building solar charging stations and solar, low-energy consuming transportation hubs to aid the cause. (Nwokolo et al. 2023a, b, c, d, e, f, g, h) This is true in India, where solar panels atop bus shelters generate power for BRT systems to operate independently of grid power. This is also true in Senegal, where electric buses operate via renewable energy for a pilot project on replicable decarbonization of public transportation efforts (Sevindik 2024). These efforts show the ability to collaborate between renewable energy and clean transportation to allow cities to fit into larger efforts for climate change mitigation.

However, despite the potential for low-carbon transportation, challenges still exist in these emerging markets. For example, electric vehicles and renewable energy infrastructure are capital intensive. Many emerging economies do not have the applicable GDPs to have extensive electric vehicle fleets or extensive solar/wind farms, although international aid and public–private partnerships provide necessary offsets (Ali et al. 2024). In addition, policies and governance are precarious. Either too much is gained through noncompliance with emission standards, or there is too little legislative support for alternative, greener transportation solutions. Finally, socio-cultural factors are at play. There are many who are satisfied in their ways and do not want to change their behavior, or there are those who are misinformed about what low-carbon transportation options could provide (Nwokolo et al. 2023a, b, c, d, e, f, g, h).

Case Studies of Smart City Initiatives in Developing Countries

Innovative citywide initiatives from developing nations demonstrate how sustainable development and new technology may change urban life. From boom cities to minimal resources to substantial positive climate change advancements, these citywide projects improve citizens' everyday lives through digital emphasis, green resources, and sustainable development through realistic urban design. F This technology is essential in an area with an annual water problem (IRENA 2019). Smart city developments in the region help the Smart Bus Rapid Transit System (BRTS) decide the optimal routes and how to serve commuters. Thus, the BRTS employs AI-informed algorithms to cut fuel usage and maintain its timetable for public transit users. Digital changes for merchant public transit access improve efficiency and inclusivity (Nwokolo et al. 2023a, b, c, d, e, f, g, h).

Digital traffic management uses artificial intelligence to adjust stoplights to reduce idling and reroute traffic (Ali et al. 2024). Electric vehicle charging stations are solar-powered, and ridesharing apps offer green transportation. The TransMilenio Bus Rapid Transit (BRT) system in Bogotá solves numerous transportation issues and is another popular smart city project. System management for dedicated BRT lanes and on-board riders get real-time updates and routing via digital interventions. Finally, it lowers traffic and supports public transit. Additionally, hybrid and electric BRT buses reduce greenhouse gas emissions (Byrne et al. 2024). The mayor and his staff offer bike-shares and miles of paved bike lanes to promote non-motorized transportation as part of Bogotá's sustainability goals for accessibility for all and to reduce human impact on the environment.

Cape Town uses solar energy from municipal rooftop solar panels and community solar projects to reduce coal-generated electricity and achieve carbon neutral (Liinamaa et al. 2024). Electric public transit is also available in this city. Flooding is common in Jakarta. Knowing the experience, it utilised smart technology to improve catastrophe resiliency. The Flood Early Warning System in Jakarta uses IoT sensors and AI predictive analytics to monitor water levels in real time. Permeable pavement and green roofs in new builds absorb water and reduce run-off, and digital mapping for flooded at-risk regions shows how smart city technologies reduce natural disaster vulnerability (Green Climate Fund 2020).

Nairobis are also using technology to enhance service delivery and public services. Example: Huduma Kenya is a digital centre for government services including licensing, taxes, and utility bills. Programmes like these decrease bureaucratic redundancy and improve service (Sevindik 2024). Nairobi is also introducing electric buses and motorbikes through public–private partnerships. AI-based ridesharing apps also reduce fuel usage and wasteful driving by providing better instructions, supporting Nairobi's low-emission urban transportation plans (Krishnan et al. 2024). For public safety and energy efficiency, IoT solar-powered streetlights are being built around the city. Dakar's smart city energy initiatives use renewable and sustainable energy. Solar street lighting in the city cut energy costs and pollution. The Dakar waste-to-energy complex converts organic waste into biogas, which satisfies energy needs and solves trash disposal problems.

Thus, these developed-world case studies demonstrate technological and urban planning potential. Durability and built-in sustainability can benefit Cape Town's renewable efforts and Toronto's Edge waterfront development, which uses sustainable materials that could be threatened by shoreline erosion. Technology-related citizen safety is also a major problem. Thus, developed global case studies create the groundwork for safe technology urban development.

Barriers to Sustainable Urbanization in Developing Economies

Urbanisation may boost the economy and development, but its flaws make it unsustainable. Thus, poor infrastructure, funding, policy, and disparities hinder urbanization's sustainability. Many sustainable urbanisation initiatives are viable but waiting for implementation, and others may be in developing nations where infrastructural modifications are needed (Joshi et al. 2024). Waiting until public transit, water, and sewage facilities are developed makes sense. In certain places of South Asia and

Africa, where public transit is usually reliable, one must either take their own car or find informal transport, which most people do not want to do. More cars on the road increase congestion and GHG emissions (Nwokolo et al. 2023a, b, c, d, e, f, g, h). Sustainable urbanisation is hindered by resource shortages. Green infrastructure, renewable energy, and climate-resilient housing are too expensive for many communities. The necessity and prioritised prioritisation of health, education, and poverty reduction strain public funds (IRENA 2019). Additionally, many emerging economies have significant governance gaps (Nwokolo et al. 2023a, b, c, d, e, f, g, h).

Sustainable urbanisation requires consistent urbanisation activities. Many rising nations have severe circumstances and ignore high-efficiency resource cycles owing to weak governance. In Abidjan, Côte d'Ivoire, the informal waste management system uses too much plastics, creating an intolerable nexus that might be prevented with better land use rules. Government fails to prioritise land use due to inefficient land usage.

Social inequality complicates sustainable urbanisation. Among the most obvious is informal settlements. Millions of people live in informal settlements across developing Asia, Africa, and South America, whether they name them slums, favelas, or shantytowns. Without government or NGO action in housing restrictions, these areas would grow badly, filthily, and without adequate water and power. They generate environmental and public health concerns (Nwokolo et al. 2023a, b, c, d, e, f, g, h). One of the largest slums is Kibera in Nairobi, Kenya. Geographic isolation from wealthier, higher socioeconomic places encourages carelessness and indifference, degrading environmental, health, and social conditions. Development does not happen since residents do not own the land and displacement/development expenses are high.

Needing technical competence also hinders sustainable urbanisation. To succeed, development areas need specialisation and government or private sector bureaucracy. Without project management skills and a detailed knowledge, sustainability measures would fail (Ali et al. 2024). Thus, globalisation must address gaps and increase capacity for such scenarios (INRENA 2019; Nwokolo et al. 2023a, b, c, d, e, f, g, h). Other issues include global warming, deforestation, and air pollution (Nwokolo et al. 2023a, b, c, d, e, f, g, h); socio-cultural and psychological issues (Byrne et al. 2024; Nwokolo et al. 2023a, b, c, d, e, f, g, h); and urbanisation.

Implications

This policy change is nationwide required. National policy reforms are needed to build green infrastructure and smart cities. Federal green infrastructure policy should encourage ecological urban landscapes and nature-based solutions including urban forestry, water-sensitive urban design, and energy projects. Public–private partnerships (PPPs) must also be included in legislative changes to solve finance gaps and private equity contributions. To improve urban sustainability, transportation, housing, and energy must improve. Poor nations need money and international climate change funds to build green infrastructure. The program aligns with various Sustainable Development Goals, creating a development framework. SDG 11 (Sustainable Cities and Communities): Green infrastructure, flexible designs, and technological integration promote resilience. Access to services and resources improves quality of life over scattered developments. SDG 7: Affordable and Clean Energy: Decades of urban development anticipated access and cohesion, necessitating energy system integration. SDG 6: Adaptive urban design optimises water use and access to potable water, mitigates overdevelopment-induced floods, and increases clean water availability. SDG 13 (Climate Action): Climate-oriented urbanisation aids natural adaptation and mitigation, emphasising climate change requires global action.

Conclusion

Sustainable urbanisation in emerging nations may increase population, protect the environment, promote economic justice, and promote social equity. As urbanisation rises, developing nations face limited resources, infrastructural pressure, greenhouse gas emissions, and informal housing building. Sustainable housing developments are in areas prone to transportation collisions (if they comply with regulations), lacking waste management or collection routes, and with limited water access; therefore, population growth cannot accommodate more unsustainable developments. However, impoverished nations lack the means to use green technology like renewable energy to build sustainable infrastructure and promote sustainable development. Unfortunately, antiquated norms, weak regulatory frameworks, and socio-cultural impediments hamper sustainable technology resources and advancement. All issues may be resolved. Governments, private investors, and international groups must collaborate. This chapter examines international resource opportunities, particularly the Green Climate Fund, financial options for urbanisation, and public–private partnerships (PPPs) for essential resources and guidance, such as HICN and World Bank Purchasing Power Parities, Sustainable Urban Development in Nairobi, Kigali, and Jakarta. Fair policies must prioritise vulnerable individuals and satisfy all demands, including the underserved. The Global South can lead the next generation of urbanism with resilient, egalitarian, and sustainable solutions with inventiveness, communal duty, and the ability to duplicate successful efforts from other places.

References

Ali S, Gondro C, Yan Q, Banzhaf W (2024) End-to-end decentralized tracking of carbon footprint using IoT and distributed databases. In: Proceedings of the 7th conference on cloud and internet of things (CIoT 2024), pp 1–6. IEEE. https://ieeexplore.ieee.org/document/10757144

Apata SB, Oyenuga MO, Abeoba M, Ugom A (2024) Internet of Things (IoT) solutions for smart transportation infrastructure and fleet management J Propuls Technol 45(4):1492–1505. https://propulsiontechjournal.com/index.php/journal/article/view/8325

Byrne NM, O'Mahony C, Hayes M (2024) Validating decarbonisation strategies of climate action plans via digital twins: a limerick case study. Front Sustain Cities. https://www.frontiersin.org

Green Climate Fund (GCF) (2020) Climate finance readiness program: building capacity for change. https://www.greenclimate.fund

International Renewable Energy Agency (IRENA) (2019) SIDS lighthouses initiative: progress and way forward. https://www.irena.org

Joshi A, Singh R, Kumar V, Rani S (2024) Net zero strategy: exploring opportunities, confronting challenges for developing countries. In: Singh R, Crowther D (eds) Transition towards a sustainable future. Approaches to global sustainability, markets, and governance. Springer, Singapore. https://doi.org/10.1007/978-981-97-5756-5_11

Krishnan C, Sahdev SL, Paul GW (2024) Blockchain-based traceability solutions for environmental sustainability. In: Puri N, Singh H, Mishra N (eds) Sustainability reporting and blockchain technology, pp 291–310. Taylor & Francis. https://doi.org/10.4324/9781003378341-18

Nidhi P, Ahmed T, Singh R (2024) Cloud IoT in healthcare transforming patient monitoring and data analytics. Int J Innov Sci Res Technol. https://ijisrt.com

Nwokolo SC, Singh R, Khan S, Kumar A, Luthra S (2023a) Technological pathways to net-zero goals in Africa. In: Africa's path to net-zero. CSR, sustainability, ethics & governance. Springer, Cham. https://doi.org/10.1007/978-3-031-44514-9_5

Nwokolo SC, Singh R, Khan S, Kumar A, Luthra S (2023b) Influencing the scale of Africa's energy transition. In: Africa's path to net-zero. CSR, sustainability, ethics & governance. Springer, Cham. https://doi.org/10.1007/978-3-031-44514-9_4

Nwokolo SC, Singh R, Khan S, Kumar A, Luthra S (2023c) Impacts of climate change in Africa. In: Africa's path to net-zero. CSR, sustainability, ethics & governance. Springer, Cham. https://doi.org/10.1007/978-3-031-44514-9_7

Nwokolo SC, Singh R, Khan S, Kumar A, Luthra S (2023d). Scenarios that Could Give Rise to an African Net-Zero Energy Transition. In: Africa's Path to Net-Zero. CSR, Sustainability, Ethics & Governance. Springer, Cham. https://doi.org/10.1007/978-3-031-44514-9_8

Nwokolo SC, Singh R, Khan S, Kumar A, Luthra S (2023e) Africa's awakening to climate action. In: Africa's path to net-zero. CSR, sustainability, ethics & governance. Springer, Cham. https://doi.org/10.1007/978-3-031-44514-9_9

Nwokolo SC, Singh R, Khan S, Kumar A, Luthra S (2023f) Remedies to the challenges of renewable energy deployment in Africa. In: Africa's path to net-zero. CSR, sustainability, ethics & governance. Springer, Cham. https://doi.org/10.1007/978-3-031-44514-9_3

Nwokolo SC, Singh R, Khan S, Kumar A, Luthra S (2023g) Decarbonizing hard-to-abate sectors in Africa. In: Africa's path to net-zero. CSR, sustainability, ethics & governance. Springer, Cham. https://doi.org/10.1007/978-3-031-44514-9_6

Nwokolo SC, Singh R, Khan S, Kumar A, Luthra S (2023h) Global investment and development in Africa. In: Africa's path to net-zero. CSR, sustainability, ethics & governance. Springer, Cham. https://doi.org/10.1007/978-3-031-44514-9_2

Omale SA, Yusuf SO, Oyenuga MO, Ikemefuna M, Ojo SS, Momodu ID (2024) Organizational learning in the post-covid 19 era: a prerequisite for stakeholder satisfactions. Twist 19(3):265–272 https://twistjournal.net/twist/article/view/339

Oyenuga MO, Omale SA (2024) Is Africa Jinxed? Exploring the challenges of technology access and adoption in Africa. Afr J Econ Sustain Dev 7(4):142–161. https://doi.org/10.52589/AJESDULN1LRNF

Oyetunde TO, Oyenuga MO, Adoga GJ (2023) Green management practices and organisational efficiency in higher institution. Budapest Int Res Crit Linguist Educ (BirLE) J 6(3):276–290. https://bircu-journal.com/index.php/birle/article/view/7716

Sevindik S (2024). The role of big data and digitalization in energy transition and business sustainability. In: Spataru C, Lv X, Carvalho P, Nowbuth MD, Ameli N (eds) The Elgar companion to energy and sustainability, pp 90–107. Edward Elgar Publishing. https://doi.org/10.4337/9781035307494.00012

Chapter 10
Circular Economy and Waste Reduction Technologies

Abstract The circular economy focuses on reducing waste, improving resource efficiency, and designing systems that restore natural ecosystems. The circular economy replaces the traditional linear take-make-dispose model with one that emphasizes the need to recycle, reuse, and remanufacture to support sustainable growth. At the core of this framework is innovations in waste management. However, advanced recycling technologies and upcycling systems allow the same material to be effectively turned back into available resources, thus reducing landscape waste and reliance on precious virgin raw materials. Improvements in the efficiency of recycling processes, for instance, stemming from AI-base sorting technologies and bio-based treatments, will allow greater flexibility in recycling systems that can respond to different waste streams. Cleaner production is one of the most important methods that can be implemented to minimize environmental effects. Other concepts, such as closed-loop systems and industrial symbiosis, allow manufacturers to recycle by-products, reducing waste and greenhouse gas emissions. This chapter references circular economy initiatives via the manufacturing sector and the potential this has for resilient supply chains.

Introduction

Following growing global attention to environmental issues and climate change, sustainability and the circular economy have become urgent topics on the international political agenda and they have become increasingly important to economic growth in many booming economies. For millennia, the economy has been linear: people identify natural resources, extract them, produce and consume goods and services, create waste and eventually repeat the process. The take-make-dispose paradigm of that has complicated things and further strained nature's limited resources while liberally spewing excess carbon into the atmosphere. As the entire economy depends on the resources, waste, and carbon output need to be minimised or eliminated, the implementation of a circular economy is the most viable solution for a safe economy and a safe environment.

M. 0. Oyenuga et al., *Smart Technologies for Climate Change and Net Zero Policies: Practical Approaches Towards Sustainability*, Climate Change Management,
https://doi.org/10.1007/978-3-031-92221-3_10

The circular economy contrasts with the linear model, emphasising resource efficiency, waste reduction, and natural systems renewal. It encompasses reduction, reuse, recycling, remanufacturing, and systems reliant on goods and materials that are never newly created, rather than always utilising virgin resources. The technology explosion significantly enhances the feasibility of a circular economy. AI-driven sorting systems, biomimicry recycling systems, and industrial symbioses enable companies to leverage technology for enhanced efficiency, waste reduction, and the attainment of sustainability goals. Industries are evolving, particularly those with the greatest material throughput: automobile, electronics, and construction. The circular potential in these industries includes remanufacturing and closed-loop production. Altering the duration of a product's lifecycle not only diminishes the sector's carbon footprint but also enhances economic efficiency derived from increased throughput and improved durability. Communities are significant since these developing regions require trash reduction to promote initiatives like community recycling and intrastate sustainability programs.

This research is justified by global climate change phenomena, exemplified by the Paris Agreement's objective to mitigate climate change, and the United Nations' Sustainable Development Goals, which recognise these aims while allowing for enhancement. For instance, SDG 12 represents Responsible Consumption and Production, while SDG 13 pertains to Climate Action; the former is closely linked to insights from the circular economy, while the latter emphasises the necessity of resource utilisation and consumption to foster a more sustainable world. The objective of this study endeavour is to advocate for this initiative. This is a critical period for the globe, necessitating new and scalable strategies to promote economic growth and environmental sustainability.

The Principles of a Circular Economy

The circular economy is an innovative paradigm for societal resource utilization and consumption before disposal. The systems-thinking paradigm of the circular economy serves as an alternative to the linear economy's "take, make, discard" model. The circular economy seeks to maximize resource utilization indefinitely inside a closed loop, minimize superfluous waste and pollution, and endeavour to renew and repair natural ecosystems. The circular economy principles correspond with the world's visions toward sustainable development as well as the international actions undertaken to tackle shortages of resources, environmental disasters, and socioeconomic imbalances (Nwokolo et al. 2023a, b, c, d, e, f, g). The conceptual principles of a circular economy enhance logistics for attaining a balance between the requirements of the environment and economic growth. Circular economy concepts rely on product uniqueness, which is about durability, reparation, and recyclability. Products are designed with longevity, ease of maintenance, and, when rendered unusable, straightforward recyclability to recover valuable resources. Many household appliance and electronics sectors rely on modular design to ensure durability, facilitate repair, and enhance recyclability. The circular economy seeks to move dependence from non-renewable resources to renewable resources. One of the primary inputs needed is renewable energy resources. Circular production activities will rely on solar and wind power, reducing dependence on fossil fuels and reducing carbon footprint (Nwokolo et al. 2023a, b, c, d, e, f, g). In addition, biobased products formed from renewable resources are used as raw materials for producing them. Agricultural waste is preferred as the origin of bioplastics in packages over conventional plastics because it affords an opportunity to extract raw materials that are renewable. Circular systems apply to agriculture also, where crop rotation and organic composting enhance the quality of the soil and reduce dependence on artificial fertilizers. These initiatives inherently align with the circular economy, improving the resilience against climate change in the farm (Sevindik 2024; Nwokolo et al. 2023a, b, c, d, e, f, g).

A circular economy is only possible if there is digital technology. The Internet of Things, blockchain technology, and artificial intelligence all make supply chains more transparent and efficient. Blockchain technology makes it possible to track the lifecycle of a product fully, so it is recycled rather than dumped during downturns. This responsibility compels the manufacturer to be responsible both to the consumer and to himself (Oyedele et al. 2023) which drives the manufacturer towards sustainability (Ali et al. 2024). Besides, IoT may monitor in real-time, actual usage of resources such as energy usage or industrial production, which may enable detection of anomalies against the predicted efficiency (Apata et al. 2024). In the construction industry, IoT can indicate the on-site presence of excess materials or variance in the energy output that could indicate a future problem (Kumar et al. 2024). In manufacturing, IoT can be used in closed spaces to detect excessive energy that can be rectified today for a better tomorrow. Artificial intelligence can predict the utilization of spare resources and can detect the potential problems associated with back-ordered commodities in a supply chain (Green Climate Fund 2020).

There are economic and environmental advantages to these circular economy initiatives. For instance, while materials remain in motion, organizations may reduce their reliance on production and virgin resources, which can enhance profit margins and improve competitive positioning in the marketplace. The electronics industry conserves billions annually through take-back and repair initiatives for refurbished items (IRENA 2019). The circular economy conserves non-renewable resources, diminishes carbon emissions, mitigates pollution, and aids in achieving sustainable objectives established by global climate change initiatives (Kumar et al. 2024). The need for recycling, remanufacturing, and repair services generates new employment opportunities. This demonstrates how contemporary international initiatives may also assist individuals on a socioeconomic level (Byrne et al. 2024). Despite the viability of a circular economy with accessibility, several practical challenges exist. The deployment incurs high costs, with initial expenses discouraging several organizations from proceeding. Many small and medium-sized firms (SMEs) (Nwokolo et al. 2023a, b, c, d, e, f, g; Joshi et al. 2024) cannot facilitate the substantial reworks in supply chain transformation required for modular manufacturing and renewable inputs. Furthermore, implementation would be limited without stakeholders being informed and educated on circular economy strategies. While individuals adhere to a linear economy, consuming products and producing resources and waste, circular initiatives may seem overly radical or superfluous; thus, educational institutions and enterprises must promote this transition to foster a shift from a resource-intensive and wasteful mentality (Oyetunde et al. 2023).

Innovations in Waste Management: Recycling and Reuse

As the circular economy evolves, waste management trends emerge that facilitate recycling and reuse initiatives to convert trash into a resource. These activities alleviate landfill concerns, promote responsible natural resource utilization, reduce greenhouse gas emissions, enhance sustainability for both government and commercial sectors, and generate innovation. Emerging recycling opportunities will progressively complicate initiatives as traditional waste management systems struggle to adapt. The incorporation of new machinery into current sorting facilities, with machine learning, represents a significant advancement. Robots employed in sorting facilities are outfitted with sensors enabled by machine learning, which enhance their visual and tactile capabilities beyond those of humans for assessing quality and type (Krishnan et al. 2024). Near-infrared sensors provide improved differentiation of plastic polymers. Chemical recycling is a novel development in the realm of polymers. Mechanical recycling, often including the shredding of polymer materials, results from prolonged use of these materials, whereas chemical recycling decomposes some polymers back to their original monomers. Consequently, chemical recycling achieves the quality of virgin plastics. Consequently, this approach mitigates plastic waste and enhances recycling opportunities for composite materials that were previously non-recyclable. For instance, Japan and EU facilities employ chemical recycling to produce new feedstock from mixed discarded plastics or plastics deemed unrecyclable (IRENA 2019). Moreover, reuse systems refurbishment and remanufacturing are essential to the circular economy. These strategies seek to prolong a product's lifespan, reducing the necessity for virgin input resources and preventing unnecessary exploitation. Manufacturers are establishing take-back schemes in the consumer electronics sector that enable makers to reclaim outdated devices for remanufacturing or refurbishment, facilitating lower-cost resale. This promotes e-waste reduction while also addressing the demand for more affordable technology (Nwokolo et al. 2023a, b, c, d, e, f, g). The lithium-ion battery is another extremely recyclable component. These are present in automobiles and consumer goods. Closed-loop recycling solutions for these batteries are developed, mainly in the EU. Advanced technology can recover lithium, cobalt, and nickel, ensuring these metals remain inside the economic cycle. This reduces virgin mining activities and decreases the carbon emissions associated with battery manufacture (Green Climate Fund 2020). Additionally, the building provides recycling opportunities through modular design. Structures utilizing modular and prefabricated components may be readily disassembled and repurposed. Components can be altered to construct essential structures without surplus waste. This is beneficial in areas with high population density where sustainable construction solutions are desired (IRENA 2019). Decentralized waste management systems enable communities to oversee their recycling and reuse initiatives. Moreover, municipalities are implementing IoT-enabled waste management systems in which garbage receptacles are equipped with sensors that relay their fullness, thereby optimizing collection routes to conserve fuel and labour hours. Moreover, AI systems in smart cities monitor regional garbage production to inform regulators about required diversion regulations and initiatives (Ali et al. 2024). Consequently, this circular economy necessitates innovation, community engagement, and policy implementation. The government must subsidize the proposed development of more recycling centres and implement laws that promote a circular economy.

Cleaner Production: Advancing Sustainable Industrial Practices

Cleaner production is a proactive strategy for reducing environmental impact through better resource use, reduced waste generation, and newer sustainable technologies. Cleaner production plays a critical role in global concerns about industrial effluents amounting to trillions of GHG emissions. The industrial components of cleaner production: energy reductions, processing efficiencies, material substitutions, and digitalization are how the industrial world can achieve sustainability goals. Instead of managing waste and emissions after production, cleaner production emphasizes the avoidance of waste and emissions during production. For example, industrial sites can choose lower emission practices instead of those based on energy, which are more prone to create emissions. For instance, if heating during production is energy-based, shifting to a more low-emission heating option reduces reliance upon specific energies (Sevindik 2024). However, another cleaner production method comes from closed-loop systems, which take industrial byproducts and reinsert them into production. The steel industry would be an excellent collaborator. Steel is created from a byproduct, slag, used to create cement. Using excess slag not only reduces industrial waste but also reduces the carbon footprint needed to create new cement (Krishnan et al. 2024). Similarly, fabric manufacturers use waterless dyeing to create fabrics instead of requiring a dyeing process that uses gallons of water and subsequent energy efforts (Byrne et al. 2024). In addition, industrial symbiosis is another new concept associated with cleaner production. It concerns the exchange of wastes or by-products between industries such that wastes from one are materials for another. Thus, this supports more productive efficiencies while minimizing efforts that run counter to environmental reduction. Cleaner production has become more feasible than ever with digital advancements such as artificial intelligence, the Internet of Things, and big data. For instance, with IoT sensors, industries can gauge, on a worldwide scale, how much they use in terms of resources, how much they emit, and how much waste they create. Thus, these IoT sensors can identify inefficiencies or losses and immediately suggest changing operations in real-time. For instance, IoT metering on energy consumption can record how much energy is used across assembly lines and enable IoT feedback to machine operators to suggest re-calibration decisions to combat energy waste (Ali et al. 2024). Second, artificial intelligence can support cleaner production through predictive analytics regarding machine maintenance and determining appropriate activity levels. For example, suppose AI recognizes that a mixing machine in a pharmaceutical company was used on specific chemicals under specific conditions. In that case, it can predict how hot and extended the following reaction should be, saving energy and avoiding drastic by-product development (Krishnan et al. 2024). Finally, using blockchain technology adds transparency to supply chains and peer-to-peer accountability. Thus, organizations can determine if their third-party vendors are reputable and sustainable and trace their products and parts from generation to assembly to waste.

Manufacturing as a Catalyst of the Circular Economy

The industrial sector is a fundamental player in the global marketplace. Its input toward a circular economy and sustainable, transformative efforts development is great. Innovations in processes that generate efficiencies, reductions in detrimental impacts, and incremental levels of improvement allow the industrial sector to lower its carbon footprint to benefit industrial growth effectively. Circular economy efforts in manufacturing concern material use, longevity increase, energy use, and efficient and effective endeavours with overlapping efforts like exhaust sharing. A component of circularity in production is to create with future recoverability and durability in mind. This implies that when a product is at the end of its lifecycle, parts can be reclaimed, reused, or recycled so that subsequent creation does not necessitate virgin materials.

One great avenue ahead is with modular products. Modular products are easier to take apart and easier to reassemble, and they have interchangeable parts. Brands (Oyenuga et al. 2021) focus on disassembling and reassembling their pieces, which makes for easier repair. The consumer electronics and automotive manufacturing industries are actively applying this design element for greater product sustainability due to increased durability (Nwokolo et al. 2023a, b, c, d, e, f, g). For example, within the consumer electronics industry, with the highest production level of cellular devices, certain models are produced with greater repair feasibility to avoid e-waste and empower consumers to replace non-functioning parts and aged components instead of relying on the manufacturer. Additionally, within the automotive industry, a popular approach is remanufacturing vehicle parts to original standards as they age. This provides decreased demand for materials, decreased energy consumption, and efforts to manufacture; therefore, remanufacturing is typically a cost-saving endeavour as opposed to starting anew. Due to the circular economy package, the

EU has long kept track of transformative statistical analysis relative to this approach in the last few years (Krishnan et al. 2024).

Manufacturing with renewable energy is a natural shift to circularity as it balances the manufacturing need with a sustainability effort. Thus, manufacturing with increased renewable energy occurs more often because plants use solar, wind, and other sourced efforts to manufacture the energy needed for their operations without using fossil fuels. For example, many manufacturing facilities in India have rooftop solar (IRENA 2019) to create this energy in a quasi-cheap format to satisfy on-site energy needs. This reduces energy costs and provides energy independence and reliability for citizens of regions with compromised power access (IRENA 2019). Ultimately, even importing renewable energy benefits the manufacturers by providing extra green branding options as consumers require more sustainable solutions to their inquiries. Therefore, even multinational enterprises (MNEs) have renewable energy aspirations, many of which are planned to go 100% by 2030. These countries exert micro-level directed initiatives to comply (Byrne et al. 2024). Thus, it is creating investment in macro-level renewable projects that replenish the sustainability of similar efforts. Industrial symbiosis and the eco-industrial park are operational strategies that make circular manufacturing possible. When industries depend on each other for waste streams, energy by-products, and outflows, they leverage resources even better. For example, a manufacturing company with a surplus of waste heat output may supply that surplus as an energy input to the company next door, creating an intentionally designed closed-loop system (Krishnan et al. 2024).

These closed-loop systems exist in China. China's eco-industrial parks take advantage of the in-between enterprises. They facilitate the exchange of waste products and surplus waste streams, waste energy, waste thermal energy, and surplus resources; what may be chemical by-products for one single enterprise may be the raw materials/ energy inputs for the enterprises next door. A fertilizer enterprise can use the chemical company's waste output as chemical by-products; this simultaneously reduces transportation and emissions. Ultimately, such enterprises reduce costs for these companies and encourage better ecological stewardship, which means China's industrial symbiosis efforts can be exported elsewhere (Green Climate Fund 2020). Another important development in manufacturing for the circular economy is a supply chain that transforms itself yet operates circularly. For instance, circular supply chains advocate remanufacturing, placing products back into the supply flow, and reducing waste creation during the manufacturing phase.

Therefore, new technologies, such as blockchain systems and IoT, have been developed to create better traceability and transparency regarding keeping materials in supply chains for potential reuse (Ali et al. 2024). For example, through blockchain, manufacturers can follow the proper steps to alert once a material has been reclaimed or remanufactured so it does not unknowingly go down the waste stream. This also creates avenues for sustainability efforts. In textiles, for instance, clothing brands have blockchains that show the textiles used are remanufactured scraps and not virgin materials. In terms of electronics, there are IoT chips that tell the initial manufacturer when a piece of technology is being used or if it no longer works (Nwokolo et al. 2023a, b, c, d, e, f, g).

Finally, if consumers refuse to purchase upcycled or refurbished goods, there is no market access and demand for circular solutions. The answer lies within a collaboration of policymakers, the energy industry, and international institutions. For example, policymakers can promote circular creation via tax credits, loans, or subsidies for renewable technology. Extended producer responsibilities—legally requiring that producers must take accountability for the effects of their product during its life and after its life—work effectively to promote and incentivize reuse and recycling (IRENA 2019). A second approach to achieving circularity in manufacturing on a much larger scale involves public–private partnerships. Anything from eco-industrial parks to beyond can be created when resources and expertise combine through a PPP. For instance, clean energy companies partnering with manufacturing companies have similar clean energy processes established within the multicultural manufacturing atmospheres for the greater good of circularity efforts expedited in the forefront (Krishnan et al. 2024).

Case Study Projects on Circular Economy and Waste Reduction

The following are case studies where the circular economy has taken place.

Singapore Zero Waste Masterplan: A Circular Integrated Masterplan

Perhaps the most cohesive effort of a circular economy at one of the furthest levels comes from Singapore's Zero Waste Masterplan, which extends to efforts like a nationalized waste management system. For the circular economy to sustain Singapore in the future, it aims to reduce waste in landfills by 30% by 2030. This is done through systematic sorting of waste by technological and human assets, community awareness efforts, and high-level recycling facilities. For example, food waste is recycled through anaerobic digestion, which creates biogas and fertilizer. The former reduces the chances of methane production from decaying organic matter, and the latter supplies nutrients for rapid development efforts in sustainable agriculture (Green Climate Fund 2020). Another component of Singapore's circular economy involves reclaimed e-waste. Through e-waste collection drives and manufacturing-relevant material recovery, e-waste can reclaim precious gold, silver, and copper and reenter the supply chain. Therefore, international standards for reducing carbon output levels related to excess metal production are supported. Ultimately, this effort is deemed successful based on extensive governmental efforts, community buy-in, and reliance on high-level technologies to promote a closed-loop system.

Rwanda's Decentralized Waste Recycling Initiatives

In Rwanda, community-led recycling initiatives of waste exemplify how local solutions work towards the advancement of the circular economy. The decentralized composting plants transform the organic waste into fertilizers; hence, better soil health while minimizing synthetic application. This therefore helps solve the two major challenges: low productivity in agriculture and landfill methane emission. The availability of cheap sustainable fertilizers supports the farmers. On the other hand, the accumulation of waste is reduced in urban places (Ali et al. 2024). The third aspect is the recycling of non-organic wastes like plastics and metals, which the Rwandans set out to reclaim. Through community cooperatives, these materials are collected, and processed, and, therefore, sustain employment opportunities while encouraging entrepreneurship. This shows that circular economy principles can be tailored to attain specific needs at the regional level while delivering social, economic, and environmental benefits.

South Africa's Waste-To-Energy Program: Closing Loops in Urban Waste Management

South Africa's Waste-to-Energy, which intends to feed some of this urban waste directly to incinerators, or modern incinerators that help make electricity by using waste mixed as feed, in the interest of solving this municipal waste menace at hand will directly impact South Africa to move towards her renewable energy ambition (IRENA 2019). Beyond environmental benefits, the WTE program creates job opportunities in the collection of wastes, sorting, and plant operation. However, the project encounters challenges such as high initial capital costs and awareness campaigns to the public to implement segregation at source. South Africa's experience brings out the point that funding support and public education are crucial when scaling up the waste-to-energy solution.

European Union's Circular Plastics Alliance: Cross-Sectoral Collaboration

The Circular Plastics Alliance (CPA) is within the European Union. This alliance is one example of cross-sectoral collaboration toward circular economy objectives. This initiative, launched to address plastic waste, brings together manufacturers, recyclers, and policymakers to increase the intake of recycled plastics in various industries. The aim here is to introduce 10 million tons of recycled plastics into newly produced products by the year 2025 (Nwokolo et al. 2023a, b, c, d, e, f, g). The CPA aims to enhance recycling infrastructure, develop material recovery technologies, and standardize recycled plastic quality. In this regard, the alliance helps in ensuring a stable supply chain, thereby providing a solid market for circular plastics. Some of the success stories of the CPA include the adoption of recycled polymers by the automotive industry for vehicle interiors and the use of recycled plastics in insulation materials by the construction sector.

India's Urban Waste Management Projects

India has taken up several projects of urban waste management, in tandem with the principles of a circular economy. The Swachh Bharat Abhiyan (Clean India Mission) is dedicated to segregation at the source and recycling and composting. Indore and Pune have become role models as decentralized waste-processing units convert organic waste into biogas and compost. These projects have both eliminated reliance on landfills and leveraged renewable electricity and fertilizers (Ali et al. 2024). Joint ventures have also spurred creative solutions, such as waste-to-energy plants and plastic-to-fuel conversion technologies. In a recycling initiative that converts unserviceable waste into new natural resources industrial oil from plastics is extracted by processing low-grade plastics in the city. In this banner, I have placed some examples of waste management that can be made short-term and long-term, when you join government initiatives with private-sector innovation.

China's Eco-Industrial Parks: Synergizing Industrial Circularity

China boasts the world's most concerted industrial circular economy approach through its Eco-Industrial Parks (EIPs). The factories within them collaborate to use best all consumption and residuals what is waste for one firm is energy or input for another. The excess energy produced by a power plant can be used for heating purposes in neighbouring firms, and the excess chemicals produced by one form of production can be used as inputs for fertilizers (Krishnan et al. 2024). Suzhou and Tianjin each have EIPs with exceptional resource inputs and very few waste outputs. These parks are the perfect theoretical underpinning for the circularity of the replicated industrial version. It indicates what is good for all in a particular geographical space where industries commit is good for like-resourced industries that could work together to reduce the environmental impact and at the same time increase the savings on the costs of operations. Denmark's Kalundborg Symbiosis has been the best example of industrial symbiosis, as well as the circular economy, worldwide. For example, an entire industrial network operates in which firms have a surplus that others can exploit such that a closed-loop and value from waste is created. The surplus heat from the energy plant is utilized in heating public housing; the gypsum from the flue gas desulfurization plant goes to the cement plant (Green Climate Fund 2020).

Implications

This project connects to several UN Sustainable Development Goals. For instance, implementing circular economy efforts champions SDG 12 (Responsible Consumption and Production) as it creates a more palpable level of consumption and reduces waste. Moreover, the ensuing creations and innovations from better recycling and reusing efforts champion SDG 9 (Industry, Innovation, and Infrastructure), while renewable energies in a circular fashion champion SDG 7 (Affordable and Clean Energy).

In addition, the idea of a circular agenda has pieces of sustainable agriculture efforts and community-based reduction of waste efforts, which champion SDG 13 (Climate Action) and SDG 11 (Sustainable Cities and Communities). Using such efforts levels the playing field among resources and prevents more vulnerable regions from having better access to a failing environment, thus championing SDG 10 (Reduced Inequalities). Future research opportunities will include whether implementing a circular economy occurs on a different scale regionally developing countries may not have the resources or infrastructure to accommodate such large-scale efforts and whether certain technological innovations are better. For example, blockchain technological innovation for supply chain transparency or AI for sorting may facilitate a more streamlined process. In these cases, the data should show how these technologies improve efficiency and similar findings with similar efforts. There are plenty of comparative case studies: China's eco-industrial parks, eco-parks in Indonesia, and Kalundborg, Denmark's industrial symbiosis come to mind. A study of the financial implications behind circular economy efforts—from green bonds to carbon credits may provide a financial hurdle easily surmountable in one region and not another. Finally, studying customer satisfaction (Oyenuga and Fredrick 2023) and marketing may reveal public relations efforts championing the circular economy.

Conclusion

This chapter illustrates how international sustainability problems may be solved through circular economic approaches and waste minimization technologies and where policy, industry, and research should go in the future. As far as policy is concerned, the findings of this article reveal that a circular economic implementation will need to be done at the national and local levels. First, policy must encourage waste minimization and resource use efficiency alongside sustainable development of new, conservative, and efficient recycling technologies. In addition, tax credits, grants, and subsidies should come from the most efficient regulatory bodies to promote a manufactured mindset of circular components from day one modularity, feasibility of remanufacturing, and industrial symbiosis so that compliance down the road is easy; compliance should also be encouraged. Furthermore, standards should be mandated that coincide with reusability and recyclability, the efficiency of recycling, and decontamination and discharge requirements for transparent environmental integrity to encourage intra-industry awareness; if one industry's waste stream is another's feedstock, that should be known. Ultimately, policy needs to connect both public and private enterprises with substantial public resources and private financing will help disproportionately large waste minimization efforts and pay for researched feasibility into large-scale application of suggested circular economic technologies.

References

Ali S, Gondro C, Yan Q, Banzhaf W (2024) End-to-end decentralized tracking of carbon footprint using IoT and distributed databases. In: IEEE international conference on internet of things. https://www.ieee.org

Apata AO, Oyenuga MO, Abeoba M, Ugom A (2024) Internet of Things (IoT) solutions for smart transportation infrastructure and fleet management. J Propuls Technol 45(4):1492–1505. https://propulsiontechjournal.com/index.php/journal/article/view/8325

Byrne NM, Khoudja M, Desmond R (2024) Validating decarbonization strategies of climate action plans via digital twins: a Limerick case study. Front Sustain Cities. https://doi.org/10.3389/fsc.2024.00012

Green Climate Fund (GCF) (2020) Climate finance readiness program: building capacity for change. https://www.greenclimate.fund

International Renewable Energy Agency (IRENA) (2019) SIDS lighthouses initiative: progress and way forward. https://www.irena.org

Joshi A, Singh R, Kumar V, Rani S (2024) Net zero strategy: exploring opportunities, confronting challenges for developing countries. In: Singh R, Crowther D (eds) Transition towards a sustainable future. Approaches to global sustainability, markets, and governance. Springer, Singapore. https://doi.org/10.1007/978-981-97-5756-5_11

Krishnan C, Sahdev SL, Paul GW (2024) Blockchain-based traceability solutions for environmental sustainability. Energies. https://doi.org/10.3390/en15040839

Kumar K, Singh R, Sharma A, Lehri BPS (2024) Climate change adaptation strategies for achieving net-zero economy. In: Singh R, Crowther D (eds) Transition towards a sustainable future. Approaches to global sustainability, markets, and governance. Springer, Singapore. https://doi.org/10.1007/978-981-97-5756-5_8

Nwokolo SC, Singh R, Khan S, Kumar A, Luthra S (2023a) Impacts of climate change in Africa. In: Africa's path to net-zero. CSR, sustainability, ethics & governance. Springer, Cham. https://doi.org/10.1007/978-3-031-44514-9_7

Nwokolo SC, Singh R, Khan S, Kumar A, Luthra S (2023b) Technological pathways to net-zero goals in Africa. In: Africa's path to net-zero. CSR, sustainability, ethics & governance. Springer, Cham. https://doi.org/10.1007/978-3-031-44514-9_5

Nwokolo SC, Singh R, Khan S, Kumar A, Luthra S (2023c) Scenarios that could give rise to an African net-zero energy transition. In: Africa's path to net-zero. CSR, sustainability, ethics & governance. Springer, Cham. https://doi.org/10.1007/978-3-031-44514-9_8

Nwokolo SC, Singh R, Khan S, Kumar A, Luthra S (2023d) Africa's awakening to climate action. In: Africa's path to net-zero. CSR, sustainability, ethics & governance. Springer, Cham. https://doi.org/10.1007/978-3-031-44514-9_9

Nwokolo SC, Singh R, Khan S, Kumar A, Luthra S (2023e) Remedies to the challenges of renewable energy deployment in Africa. In: Africa's path to net-zero. CSR, sustainability, ethics & governance. Springer, Cham. https://doi.org/10.1007/978-3-031-44514-9_3

Nwokolo SC, Singh R, Khan S, Kumar A, Luthra S (2023f) Decarbonizing hard-to-abate sectors in Africa. In: Africa's path to net-zero. CSR, sustainability, ethics & governance. Springer, Cham. https://doi.org/10.1007/978-3-031-44514-9_6

Nwokolo SC, Singh R, Khan S, Kumar A, Luthra S (2023g) Global investment and development in Africa. In: Africa's path to net-zero. CSR, sustainability, ethics & governance. Springer, Cham. https://doi.org/10.1007/978-3-031-44514-9_2

Oyedele MO, Marcus GO, Ahungwa AI (2023) Do consumers care about green marketing practices? Insight from a developing nation. Budapest Int Res Crit Inst J (BIRCIJournal) 6(3):1424–1436. https://bircu-journal.com/index.php/birci/article/view/7665

Oyenuga MO, Ahungwa AI, Onoja E (2021) Effect of brand equity on consumer behaviour among students of Veritas University, Nigeria: a study of apple smartphones. Mark Brand Res 8:48–64. https://doi.org/10.33844/mbr.2021.60329

Oyenuga MO, Fredrick E (2023) Managing product features and customer satisfaction via product development, product branding, and product packaging: evidence from CWAY table water. Mark Brand Res 10(1):28–44 https://doi.org/10.32038/mbr.2023.10.01.03

Oyetunde TO, Oyenuga MO, Adoga GJ (2023) Green management practices and organisational efficiency in higher institution. Budapest Int Res Crit Linguist Educ (BirLE) J 6(3):276–290. https://bircu-journal.com/index.php/birle/article/view/7716

Sevindik S (2024) The role of big data and digitalization in energy transition and business sustainability. Sustain Bus Rev. https://doi.org/10.1007/sustainable2024

Chapter 11
Energy Efficiency Technologies for Low-Income Countries

Abstract Technology is crucial for enhancing access to cheap, dependable, sustainable energy, climate regulation, and economic stabilization. The chapter addresses energy efficiency technologies for home applications, including energy-efficient cookstoves, solar lamps, and appliances that consume minimal or no energy. These extend beyond a mere aspiration for those seeking to minimize energy expenses or alleviate chronic ailments associated with overexertion; they may be seamlessly integrated into daily routines, resulting in reduced energy consumption and enhanced quality of life. For commercial and industrial purposes, energy-efficient appliances and waste-to-energy technologies yield energy savings from both a cost reduction and pollution mitigation standpoint, enabling enterprises and manufacturers in financially constrained situations to establish their energy efficiency requirements. Furthermore, microgrids essentially advanced smart grid technology—are examined as innovative approaches to energy transmission. Increased focus on distributed energy resources (DER) and energy storage technologies addresses several infrastructural deficiencies related to nanotechnology. Ultimately, these nanotechnological solutions are substantiated by case studies in suboptimal income contexts demonstrating effectiveness through energy efficiency applications, yielding both possible solutions and tangible benefits. Ultimately, expected governmental engagement—such as credits and deductions—reinforces the viability of these nanotechnological developments through public and private collaborations. Low-income nations may achieve an energy-efficient future alone through a collective effort involving informed technology developments, legislative direction, and private community participation.

Introduction

Energy efficiency technologies matter relative to climate change stabilization efforts and the immediate energy needs of low-income countries. As the world searches for affordable, reliable, and sustainable energy, considerations become energy efficient, economically feasible, socially equitable, and environmentally effective because energy access and energy use for development in resource-poor nations come with

M. O. Oyenuga et al., *Smart Technologies for Climate Change and Net Zero Policies: Practical Approaches Towards Sustainability*, Climate Change Management,
https://doi.org/10.1007/978-3-031-92221-3_11

a learned cost for sustainable development (Oyenuga and Omale 2024). When the world acknowledges energy efficiency, it shows that it is necessary to satisfy energy supply and demand with the least costly environmental concern. For example, energy efficiency upgrades will comprise nearly 40% of the reduction efforts required to achieve net zero by 2050 (IRENA 2019). This is true for resource-poor nations that are energy-poor but can be energy-rich. The access to electric and off-grid technologies redirects energy access away from zealous, expensive, and polluting resources and creates the opportunity for more sustainable resources. For example, in millions of homes worldwide, energy-dependent household technology is required concerning minimum-viable technology.

From a dependency perspective, households that rely on biomass fuels are connected to deforestation and its resulting negative indoor air quality prospects. Energy-efficient improved cookstoves reduce fuel consumption by up to 60%, which means less threat to indoor air quality and more significant economic and health advantages through community applications (Joshi et al. 2024). In addition, solar lighting systems replace the need for kerosene lamps across many rural areas; the quality of education, safety, and economic growth potential have improved in off-grid regions due to the availability of this technology (Nwokolo et al. 2023a, b, c, d, e, f).

Furthermore, energy-efficient technologies would also benefit low-income countries' industrial sectors. Energy-saving developments like waste heat recovery and mechanisms that work below optimal capacity reduce energy intensity within various sectors from textiles to cement (Krishnan et al. 2024; Ali et al. 2024). While innovative solutions and technological advancements appear to bridge the gaps for energy-saving energies to be effectively implemented in low-income nations, barriers exist (Oyedele 2025). First, the cost to implement is generally high on top of the necessary know-how and effective supply chains required for on-site implementation. As such, public–private partnerships have sought to pave the way to eliminate barriers to entry as well as government incentivization mechanisms that decrease financial barriers; PAYG solar financing systems and some subsidies dramatically reduce installation costs for greater influx of implementation (Green Climate Fund 2020). Lastly, government incentivization is critical for energy efficiency planning endorsements and engagement. Taxes, subsidization, and regulatory actions go a long way in providing the necessary landscape for widespread opportunities. For example, in India, the PAT scheme promotes tradable energy savings certificates to sectors attempting to integrate energy efficiency; massive emissions reductions were assessed in heavily-inducing sectors; similarly, in Germany, those retrofitting older housing schemes with apartments that received energy-efficient upgrades as their efforts toward retrofitting were appreciated by regulatory action (IRENA 2019). While it shows that this chapter focuses on an assessment of energy efficiency technologies based on low earnings/low-income demand—household dependence, electro-gridding, and governmental concern, it also discovers this theme of significant concern that energy efficiency could be the portal through which climate actions become sustainable stabilization efforts. Low-income developing nations could emulate this effort of integrating energy efficiency technologies as their minor climate change corrective action toward stabilization of growth for energy in poverty reduction while also satisfying international expectations.

Energy-Saving Technologies for Households

Home energy efficient technologies are required to increase quality of life, reduce energy expenditures, and promote sustainability growth in developing economies (Nwokolo et al. 2023a, b, c, d, e, f). Such state-of-the-art, transformative technologies are created and adopted to address issues like low accessibility to energy, low resource availability for energy and fuel expenditures, and ecological pressures to create a better world.

For homes relying on traditional biomass fuels, energy-saving cookstoves are one of the most successful interventions in the world. Typical cooking situations arise around a fire or basic range, signifying more fuel burned in the form of firewood or charcoal, which are linked to deforestation. In addition, burning wood and its natural materials or charcoal contribute to reduced air quality within a home (Oyetunde et al. 2023), impacting respiratory and cardiovascular functioning for anyone in the vicinity typically women and children, as they spend the most time in the cooking environment. Therefore, improved cookstoves reduce such occurrences via reduced fuel usage (up to 60% less) and reduced toxic emissions (Joshi et al. 2024; IRENA 2019). In Kenya and India, governmental subsidized units or NGO efforts have successfully spread such stoves, where acceptance of use is high, and respiratory outcomes decades later demonstrate extreme improvements, such as reductions in deforestation. Solar lighting systems can transform the lives of these off-grid families likely living in rural settings where additional electricity is too expensive or unavailable. Solar lighting systems include solar lanterns, in-home solar light kits, and miniature solar grids. With solar lighting systems, these families will not need to depend on kerosene lamps or candles; solar lighting systems have no variable costs like fossil fuels and their health implications inhaling chemicals while one sleeps has many problems. As climate disasters escalate alongside the increasing darkness of Sub-Saharan Africa, solar lighting provides a two-fold solution. Increased hours of study for children after school and increased hours for entrepreneurs led to boosted productivity and safety for those working at night (Nwokolo et al. 2023a, b, c, d, e, f).

Solar Sister and Lighting Africa have provided affordable solar solutions to millions of households through creative pricing and outreach in Uganda. In addition, low-power appliances are an effective energy-saving option domestically. Energy-saving refrigerators, fans, televisions, and others use less wattage, rendering them appropriate for low-income families and areas with limited grid access. For example, a fully loaded energy-efficient refrigerator utilizes over 50% fewer gallons of electricity than a regular refrigerator due to increased insulation and energy-saving compressors (Green Climate Fund 2020; Nwokolo et al. 2023a, b, c, d, e, f). In addition, LED light bulbs replace incandescent bulbs as LEDs need less wattage while providing more lumens. For example, appliance initiatives in Bangladesh and Nepal indicate sizable decreases in household electricity consumption and associated emissions over the years. However, community-based programs are critical to bringing energy-efficient technology directly to low-income families. For example, many of these programs run micro-funding solutions to motivate family conversion to technology ownership, from solar kits to energy-efficient devices to better cookstoves. PAYG or Pay-As-You-Go solar solutions enable families to buy in with small incremental payments over time. Such a device makes affordability far more manageable. In Tanzania, PAYG has enabled thousands of families to obtain solar lighting with a sustainable payback system for providers and microfunding solutions for customers (Ali et al. 2024). Therefore, in addition to the funding, an element of the expected education and outreach would be required to ensure the efficacy of these energy-saving devices (Nwokolo et al. 2023a, b, c, d, e, f). For instance, educating families on the care and use of solar panels and cookstoves/appliances would ensure proper use and durability. In addition, women are the primary end-users of these devices and are active in community outreach and training sessions, facilitating their adoption, acquisition, and proper sustainable use.

Digital technology is used, too, to bring efficiency to energy use at home. For instance, IoT devices allow for household monitoring and better regulation of energy use to prevent inefficient habits that waste energy (Nwokolo et al. 2023a, b, c, d, e, f). In addition, smart thermostats and smart meters allow families to adjust their heating, cooling, and energy-generating systems with real-time responses to family occupancy, reducing such expenses from energy bills (Krishnan et al. 2024). In addition, low-income areas of cities find more and more prepaid electric meters so families can access energy without worrying about excess debt payment problems. However, obstacles to accessibility for low-income families remain. For instance, the initial investment still poses a huge problem. Families lack the liquid cash for the down payment necessary for solar lighting or energy-efficient systems. Although microfinancing and subsidization options reduce the cash needed for an impactful initial investment, these options come about rarely and too late. Another issue for energy-efficient offerings is supply and distribution chains that are not always dependable. For example, they will never get them if there is no connected way to supply such items to these underserved regions in rural, off-the-grid areas. In addition, with no chance for post-repair and questioning, people are less inclined to invest in something they might have to ask about or seek assistance with down the line. However, there is an adoption restriction due to cultural and social considerations. Some clans do not want to upset the natural balance or go against what has been done for them and known for generations before so this technology could be an unwelcome obstacle. Therefore, it will take persuasive town hall meetings and seminars to educate, shift perspectives, and demonstrate how such energy reductions would be beneficial down the line. **A panacea** to such problems will emerge from collaboration between tiers of government, the private sector, and international **endeavours**. For instance, many energy-saving technological advancements will be created and deployed through public–private partnerships (PPPs) as the various stakeholders will all possess various resources and expertise to enhance the projects. In addition, the government will be involved with anticipated subsidies or tax credits and the projects that create reasonable legislation for improved creation and consumption of energy-saving products (Nwokolo et al. 2023a, b, c, d, e, f; Green Climate Fund 2020).

Industrial Energy Efficiency Solutions

Industrial energy efficiency solutions are pivotal in fostering sustainable development by reducing energy consumption and greenhouse gas (GHG) emissions while enhancing operational productivity (Nwokolo et al. 2023a, b, c, d, e, f). In low-income countries, where energy costs can be prohibitive and access to modern infrastructure is limited, these solutions offer a pathway to sustainable industrial growth. Implementing energy efficiency measures not only reduces environmental impacts but also lowers production costs, making industries more competitive in global markets.

Process optimization is a foundational strategy for industrial energy efficiency, focusing on refining operations to minimize energy wastage without compromising productivity. In energy-intensive industries like cement manufacture, new technologies of preheating kilns by exhaust heat from others can reduce energy demand. These systems conserve production quality while diminishing energy requirements as they capture and redirect heat (Byrne et al. 2024). In the food processing industry also, heat exchangers recover energy in cooling and heating operations that result in reduced overall energy use. Such small, significant modifications as increasing the insulation on pipelines or even just scheduling machinery runtimes in low-income countries can result in some massive energy savings. These measures are particularly effective in manufacturing sectors like textiles, where energy-intensive processes like dyeing and finishing can be optimized through better scheduling and resource management (IRENA 2019).

Investing in energy-efficient machinery is a cornerstone of industrial energy efficiency. Modern equipment, such as high-efficiency motors and variable-speed drives, can reduce energy usage by 20–30% compared to traditional machinery (Nwokolo et al. 2023a, b, c, d, e, f). In textile industries, advanced looms and dyeing machines not only consume less power but also enhance production output, supporting sustainability and profitability (IRENA 2019). For example, in Sub-Saharan Africa, several garment factories have adopted automated, energy-efficient looms that have halved their electricity costs while increasing output. Similarly, in the agricultural processing sector, energy-efficient milling machines and cold storage units have gained popularity. These technologies reduce post-harvest losses while consuming significantly less energy, enabling farmers and cooperatives to maximize profits. The initial cost of upgrading machinery remains a challenge, but long-term savings and environmental benefits make these investments highly desirable (Alam et al. 2024).

Waste heat recovery is transformative in terms of industrial efficiency. Waste heat recovery captures, and reuses industrial process-generated waste heat. A steel or glass manufacturing plant now uses waste heat to produce electric power. Because of this electricity generation, waste heat recovery also reduces energy dependence on outside means. For example, steel plants in India and China have installed waste heat recovery boilers that convert excess heat into power, saving millions of dollars annually and reducing carbon emissions (Krishnan et al. 2024). The integration of digital technologies, including the Internet of Things (IoT) and artificial intelligence (AI), is transforming industrial energy efficiency. Sensors, enabled by IoT, track energy consumption across facilities in real time, revealing inefficiencies and opportunities for improvement. These sensors provide actionable data that allow industries to optimize energy consumption dynamically (Ali et al. 2024; Nwokolo et al. 2023a, b, c, d, e, f).

AI-driven algorithms predict equipment maintenance needs to prevent unnecessary downtime and associated energy losses. Companies, for example, should have predictive maintenance systems in automotive and electronics manufacturing where engine data is analyzed for potential faults before they occur and this ensures production is not slowed up. This saves energy, but more significantly extends the life of equipment, thereby reducing replacement costs.

Digital twins would thus be cutting-edge solutions offering a virtual model of physical systems within the industry. Industries use these digital platforms to test efficient strategies without their practical processes; therefore, optimizing the energy consumption in extractions and refined products in the oil and gas sector has minimized the cost expenditure while reducing pollutants emissions (Sevindik 2024).

Despite the well-identified benefits in industrial energy efficiency solutions, it has been widely identified that such solutions face the barriers of entry in low-income countries. Probably, the two most significant discouraging factors would be the considerable capital costs entailed in installing modern systems, and the massive investments required for machinery upgrades. The backbone of most economies includes SMEs that do not afford such investments-Oyedele et al. 2024. Another critical challenge is the lack of technical expertise required to implement and maintain these solutions. To overcome these barriers, financial and policy interventions are crucial (Nwokolo et al. 2023a, b, c, d, e, f). Governments can incentivize energy efficiency through tax breaks, subsidies, and low-interest loans for purchasing advanced machinery. For example, several countries have established energy audit programs under which they give incentives to the industries in identifying inefficiencies and taking corrective action (Green Climate Fund 2020). PPPs may also have an important role in scaling industrial energy efficiency solutions. With the utilization of private sector expertise and public sector funding, PPPs may help to roll out leading-edge technologies in resource-constrained environments.

Smart Grids and Energy Distribution Efficiency

Smart grids are a revolutionary approach in energy distribution and have the potential to increase efficiency, reliability, and integration of renewable energy. For low-income countries, they present a good opportunity to address energy access challenges and optimize resource utilization.

The intelligent grid makes use of AMI and real-time data analytics in monitoring and controlling energy flows. AMI has the potential to track electricity consumption, identify outage points, and minimize energy theft, a very common phenomenon in low-income communities (Byrne et al. 2024). The same system can allow for demand response programs wherein the consumers control energy usage at certain peak hours by balancing supply with demand. The smart grids have integration of renewable energy. Energy sources like solar and wind power are not continuous and thus require sophisticated management to stabilize them. Smart grids have used AI-based forecasting and energy storage systems that can help the grid integrate as much renewable energy as possible to minimize dependence on fossil fuels. Microgrids powered by smart technology offer decentralized energy solutions for rural areas (IRENA 2019; Nwokolo et al. 2023a, b, c, d, e, f). The grids connect communities to renewable sources of energy; hence, people are able to have power in off-grid regions. For example, in Kenya, microgrids that include solar panels and battery storage have improved access to energy while lowering costs (Nwokolo et al. 2023a, b, c, d, e, f).

Case Studies on Implementing Energy Efficiency in Low-Income Settings

Rwanda: Agro-Processing Industries

Agro-processing industries in Rwanda have successfully embraced energy-efficient technologies, such as modern drying and milling equipment, reducing energy costs by up to 40% while improving the quality of their products. This has been facilitated by microfinancing schemes, enabling small-scale processors to invest in these technologies. The cost savings are further supplemented by improvements in productivity and reductions in environmental impacts, thereby demonstrating how financial support mechanisms can spur industrial efficiency in resource-constrained settings (Nwokolo et al. 2023a, b, c, d, e, f).

India: Perform, Achieve, and Trade (PAT) Scheme

India's PAT scheme is one of the flagship programs in market-based energy efficiency. The scheme encourages industries such as cement, iron, and steel to use more energy-efficient means by giving them tradable energy-saving certificates, which are tradable for exceeding their target on energy saving. Industries can sell these certificates to others, which creates a competitive market for gains in efficiency. This approach has led to massive energy savings and innovation in industrial processes (IRENA 2019).

China: Eco-Industrial Parks

China's eco-industrial parks are the best examples of industrial symbiosis where waste heat, materials, and by-products of one facility are reused by the neighboring industries. Such parks as in Tianjin and Suzhou have been designed with advanced waste heat recovery systems and shared utility infrastructure that help achieve considerable energy savings and emission reductions. The parks thus present an economic and environmental potential for collaborative energy efficiency models (Krishnan et al. 2024).

Vietnam: Process Optimization in Manufacturing

In energy efficiency programs focusing on process improvements and waste heat recovery, Vietnam has taken successful measures. Implementation of these procedures has led the factories to lower energy costs up to 15–20% by increasing the competitiveness and providing national environmental outcomes. Some critical steps include developing high-efficiency boilers, effective insulation, and enhanced manufacturing machineries. Such results suggest that targeted programs from the government can play an essential role in enhancing industrial sustainability (Byrne et al. 2024).

Kenya: Pay-As-You-Go Solar Solutions

M-KOPA, a company in Kenya, has dramatically improved access to energy through pay-as-you-go solar-based solutions. This opens up kerosene to villagers and small industries for clean energy alternatives. Solar kits provide light, radios, and other small appliances that would otherwise be available only with expensive kerosene options. The model also supports micro-enterprises by providing reliable and affordable power for their work (IRENA 2019).

South Africa: Industrial Waste-To-Energy Initiatives

Industrial WTE projects have featured innovative ways in which waste can power South Africa. These include examples such as the conversion of agricultural residues and municipal waste to biogas and electricity. These helps reduce dependence on landfills, minimize energy generation costs, and achieve set renewable energy targets. It also illustrates the role public–private partnerships play in sustainable industrialization (Green Climate Fund 2020).

Bangladesh: Textile Sector Efficiency

Bangladeshi textile manufacturing has found energy-efficient technology to reduce the cost of its products and environmental issues in the entire system. This sector also began to use advanced dyeing machines and also started using waterless dyeing technology, which reduces its water and energy consumption. Heat recovery systems are also implemented in this sector; it captures the waste heat from the production process (Sevindik 2024).

Brazil: Sugarcane Ethanol Plants

Brazil's sugarcane ethanol plants are a shining example of how bioenergy is integrated into industrial processes worldwide. The plants consume bagasse as a fuel for producing steam and electricity, and the energy is self-sufficient with minimal external dependence. The system also helps the circular economy. The extra electricity produced in these plants is usually sold to the national grid, adding to Brazil's renewable energy mix (IRENA 2019).

Ethiopia: Cement Industry Optimization

The cement industry in Ethiopia has embraced process optimization strategies, such as preheating kilns and using alternative fuels like biomass. These measures have reduced energy consumption and operational costs to meet global sustainability standards. International organizations have supported capacity-building programs that have driven these improvements (Nwokolo et al. 2023a, b, c, d, e, f).

Philippines: Energy Efficiency in SMEs

Small and medium enterprises (SMEs) in the Philippines have installed energy-efficient lighting and cooling systems that reduce their costs. The government's energy audit program helps assess areas for improvement and sends proposals for low-cost adjustments, which proves it possible to execute interventions effectively in resource-constrained settings. Energy savings from the businesses that participated in these programs reached as high as 30% (Green Climate Fund 2020).

Germany: Advanced Heat Recovery in Steel Plants

Advanced heat recovery systems are seen in German steel plants that allow these to capture the excess of the production processes to produce electricity and therefore dramatically lower the costs and emissions generated from energy sources. Integration with digital monitoring has improved efficiency due to the operations that have the potential to maximize them in real-time conditions as reported by Krishnan et al. 2024.

Indonesia: Palm Oil Industry Efficiency

The palm oil companies in Indonesia promote cleaner production technology, such as methane capture for waste ponds, and use generated methane to fire power plants thereby reducing greenhouse gas emissions by using renewable energy sources. Therefore, meeting the climate ambition of Indonesia also showcases efficiency in energy-intensive agriculture-based enterprises (Byrne et al. 2024).

Turkey: Glass Manufacturing Energy Efficiency

All these plants in Turkey used energy-efficient furnaces as well as installed automatic systems reducing their energy input by 25%. Waste heat recovery systems helped save even further excess heat energy in preheating raw material feed. Improvements in environmental legislation have given firms a possibility of competitiveness both locally and across borders (Sevindik 2024).

The Role of Government Incentives in Promoting Energy Efficiency

Key to pushing the adoption of energy efficiency technologies across the globe is government incentives. These tools that address financial and technical barriers assist countries in developing and developed statuses, encourage private sector participation, and ensure that national energy goals align with global sustainability goals. The use of subsidies, tax incentives, regulatory mandates, capacity building, and international collaboration have all facilitated important advances in energy efficiency across the globe.

Subsidies and grants are great means of energy efficiency due to the financial barriers erected towards adoption. In Bangladesh, there was a collaboration between the government and the World Bank to subsidize solar home systems that enabled more than four million rural households to turn to clean energy. The system not only resulted in improved access to energy but also stimulated the growth of local solar businesses and brought about economic development. (Nwokolo et al. 2023a, b, c, d, e, f). For example, in Kenya, the provision of subsidies on fuel-efficient cookstoves has resulted in decreased consumption of household energy, reduced deforestation, and improved indoor air quality with more benefits to women and children (Krishnan et al. 2024).

The Federal Ministry for Economic Affairs and Energy of Germany provides grants for homeowners and businesses as it is one of the leading countries that are efficient in energy—retrofitting, enhancing better insulation and sophisticated heating systems. These programs are a part of the overall policy strategy to reach carbon neutrality by 2045. One of the examples is the Energy Efficiency Program of Brazil, which provides subsidies for energy-saving technologies in low-income communities to stabilize the grid and reduce energy poverty (IRENA 2019). Tax incentives also spur demand in the market for energy-efficient technologies. For example, the Section 12L Tax Incentive in South Africa rewards businesses with rebates for achieving measurable energy savings through efficiency projects. This has led industries to invest in a high-efficiency motor and more optimized production systems, which reduce the energy cost and the emission of greenhouse gases (Green Climate Fund 2020). The United States provides tax credits to both homeowners and businesses for upgrades that are made on systems thought of as energy efficient, such as erecting solar panels, installing a heating system, and energy-efficient windows. Incentives for clean technologies have increased their adoption significantly, cutting energy bills and greenhouse gas emissions across the nation. In India, tax exemptions for energy-efficient equipment under the Perform, Achieve, and Trade (PAT) scheme have allowed industries like steel, aluminium, and cement to move to optimal positions in terms of energy consumption while generating tradable energy-saving certificates (IRENA 2019). In the PAT scheme in India, specific energy efficiency targets are set for energy-intensive industries. Facilities which exceed these goals are rewarded by tradable energy-saving certificates. Those that do not comply will face penalties. This market-based mechanism has produced significant energy savings in sectors like cement, power, and textiles (IRENA 2019). In the European Union, member states are to ensure that, by 2030, the energy efficiency of the Union shall be improved by at least 32.5%. The requirements include mandatory energy audit for large enterprises and retrofitting measures for public buildings.

Capacity building is the need of the hour for local technicians to ensure long-term sustainability of energy efficiency projects. In Ethiopia, government-led training programs have improved technical skills in installing and maintaining energy-efficient technologies, such as solar water heaters and advanced irrigation systems

(Krishnan et al. 2024). In Vietnam, industrial energy-efficiency workshops trained factory managers how to implement both waste heat recovery systems and optimum process techniques while cutting energy by up to 20% on their costs: Byrne et al. 2024.Public awareness programs will reinforce these by creating a culture of energy savings. For example, the "Switch to Save" program in Australia is educating households on the advantages of LED lights and energy-efficient appliances and encouraging adoption with direct information and monetary motivation. For instance, in Uganda, "Energy Efficiency Week" emphasizes how common solutions to consume less energy in household settings can be promoted through community events and live demonstrations (Nwokolo et al. 2023a, b, c, d, e, f).

Energy efficiency incentives have multiplied their effects in collaboration with governments, international organizations, and the private sector. For instance, Rwanda has had partnerships with UNDP that offer financing and technical support to programs for energy-efficient cookstoves, which in turn reduce fuel consumption and enhance public health outcomes (Krishnan et al. 2024).

Implications

This study provides the policy consequence for the circular economy-related initiatives that endorse the use of initiatives for a circular economy, which should be reinforced by national and regional policies; however, waste reduction technology to all universality is ancillary. Tax credits, grants, and further subsidies should be established in order to invite business involvement in circular initiatives like remanufacturing and modularity, as well as industrial symbiosis. At the same time, regulatory practices need to be followed in order to harmonize the quality recycling, reuse, and environmental disclosure obligations of the businesses. National endeavors need to be spearheaded by a multidisciplinary consensus on integrated waste management, recycling, and technological infrastructure. Universal waste reduction technologies should have solid support for improvement and solid lobbying for the scaling efforts. This endeavour promotes the achievement of many UN SDGs. The activities have a cyclical structure, and this aids in achieving SDG 12 through responsible consumption and production with increased resource efficiency and waste reduction. The incremental growth of the recycling and reusing sectors will comply with the criteria of SDG 9 regarding industry, innovation, and infrastructure. Use of renewable energy in circular economies is in line with SDG 7 concerning affordable and clean energy. Further, the successful execution of such circularity via the agricultural projects and community-based waste reduction businesses fosters both SDG 13, Climate Action, and SDG 11, Sustainable Cities and Communities.

Conclusion

The role of energy efficiency technologies and strategies in addressing both the challenges of energy access and sustainability is underscored in this chapter. It is through technologies such as household energy-saving products, including solar-powered lighting systems and efficient cookstoves, that quality of life is enhanced while energy cost and environmental degradation are reduced. These innovations are critical in low-income settings, where access to reliable energy is a challenge that remains unbroken. Industrial energy efficiency solutions are huge opportunities for increasing productivity and reducing greenhouse gas emissions. Such techniques as process optimization, waste heat recovery, and use of energy-efficient equipment not only facilitate sustainable industrial development but also save companies on operation costs. Digital technologies such as IoT and AI enhance the same by real-time monitoring of energy consumption and improvements in the process.

References

Alam MN, Turi JA, Bhuiyan AB, Kharusi SA, Oyenuga M, Zulkifli N, Iqbal J (2024) Factors influencing intention for reusing virtual reality (VR) at theme parks: the mediating role of visitors satisfaction. Cogent Soc Sci 10(1). https://doi.org/10.1080/23311886.2023.2298898

Ali S, Gondro C, Yan Q, Banzhaf W (2024) End-to-end decentralized tracking of carbon footprint using IoT and distributed databases. In: IEEE international conference on internet of things. https://www.ieee.org

Byrne NM, Khoudja M, Desmond R (2024) Validating decarbonization strategies of climate action plans via digital twins: a limerick case study. Front Sustain Cities. https://doi.org/10.3389/fsc.2024.00012

Green Climate Fund (GCF) (2020) Climate finance readiness program: building capacity for change. https://www.greenclimate.fund

International Renewable Energy Agency (IRENA) (2019) SIDS lighthouses initiative: progress and way forward. https://www.irena.org

Joshi A, Singh R, Kumar V, Rani S (2024) Net zero strategy: exploring opportunities, confronting challenges for developing countries. In: Singh R, Crowther D (eds) Transition towards a sustainable future. Approaches to global sustainability, markets, and governance. Springer, Singapore. https://doi.org/10.1007/978-981-97-5756-5_11

Krishnan C, Sahdev SL, Paul GW (2024) Blockchain-based traceability solutions for environmental sustainability. Energies. https://doi.org/10.3390/en15040839

Nwokolo SC, Singh R, Khan S, Kumar A, Luthra S (2023a) Impacts of climate change in Africa. In Africa's path to net-zero: exploring scenarios for a sustainable energy transition, pp 237–262. Springer Nature Switzerland, Cham. https://doi.org/10.1007/978-3-031-44514-9_8

Nwokolo SC, Singh R, Khan S, Kumar A, Luthra S (2023b). Scenarios that could give rise to an African net-zero energy transition. In: Africa's path to net-zero. CSR, sustainability, ethics & governance. Springer, Cham. https://doi.org/10.1007/978-3-031-44514-9_8

Nwokolo SC, Singh R, Khan S, Kumar A, Luthra S (2023c) Africa's awakening to climate action. In: Africa's path to net-zero. CSR, sustainability, ethics & governance. Springer, Cham. https://doi.org/10.1007/978-3-031-44514-9_9

Nwokolo SC, Singh R, Khan S, Kumar A, Luthra S (2023d) Remedies to the challenges of renewable energy deployment in Africa. In: Africa's path to net-zero. CSR, sustainability, ethics & governance. Springer, Cham. https://doi.org/10.1007/978-3-031-44514-9_3
Nwokolo SC, Singh R, Khan S, Kumar A, Luthra S (2023e) Decarbonizing hard-to-abate sectors in Africa. In: Africa's path to net-zero. CSR, sustainability, ethics & governance. Springer, Cham. https://doi.org/10.1007/978-3-031-44514-9_6
Nwokolo SC, Singh R, Khan S, Kumar A, Luthra S (2023f) Global investment and development in Africa. In: Africa's path to net-zero. CSR, sustainability, ethics & governance. Springer, Cham. https://doi.org/10.1007/978-3-031-44514-9_2
Oyedele OM (2025) The cobra effect: exploring the dark side of innovation and environmental sustainability in developing nations. In: Singh R, Joshi A, Filho WL, Khan S (eds) Zero carbon industry, eco-innovation and environmental sustainability. World sustainability series. Springer, Cham. https://doi.org/10.1007/978-3-031-80220-1_15
Oyedele MO, Sunday AO, Abuh AI (2024) Fostering technological-enhanced training and development for business survival and performance in the new normal. J Propuls Technol 45(3):1858–1869. https://propulsiontechjournal.com/index.php/journal/article/view/7479
Oyenuga MO, Omale SA (2024) Is Africa Jinxed? Exploring the challenges of technology access and adoption in Africa. Afr J Econ Sustain Dev 7(4):142–161. https://doi.org/10.52589/AJESDULN1LRNF
Oyetunde TO, Oyenuga MO, Adoga GJ (2023) Green management practices and organisational efficiency in higher institution. Budapest Int Res Crit Linguist Educ (BirLE) J 6(3):276–290. https://bircu-journal.com/index.php/birle/article/view/7716
Sevindik S (2024) The role of big data and digitalization in energy transition and business sustainability. Sustain Bus Rev. https://doi.org/10.1007/sustainable2024

Part III
Policy and Governance for Net-Zero Transitions

Chapter 12
Global Net-Zero Commitments: International Agreements

Abstract The Paris Agreement, which came into force in 2015 under the UNFCCC is founded on different commitments and provides a framework that countries will work toward to keep global warming well below 2 °C. This chapter postulates what net-zero targets represent at different national commitment stages with corporate aspirations. It also looks into the necessary interventions in all industries around technology and sustainable practices to make a net zero emissions world. This study illuminates the key elements of the Paris Agreement as well as the strategies for effective implementation.

Introduction

The Paris Agreement: An Overview

On 12th December 2015, all the countries took it upon themselves to deal with the atmosphere through the first-ever climate pact, otherwise known as the Paris Agreement The Agreement established a common commitment among states to slow down the pace of global warming as follows, well below 2 °C, and endeavour to limit an increase above pre-industrial levels (Maslin et al. 2023). This requires the all but eradication of net greenhouse gas (GHG) emissions, leading to radical changes needed throughout global climate mitigation policy and practice. It is a landmark pact to ensure that global temperatures do not rise by more than 2C above pre-industrial levels and make efforts to limit it while aiming for no less than "1.5 C". Yet to achieve such visions will require a mix of political resolve, improved transparency and greater ambition in every nation. Given the climate emergency and unparalleled world crisis, it is a given that any future mitigation action will have to draw inspiration from the Paris Agreement if sustainable development is at all possible.

M. 0. Oyenuga et al., *Smart Technologies for Climate Change and Net Zero Policies: Practical Approaches Towards Sustainability*, Climate Change Management,
https://doi.org/10.1007/978-3-031-92221-3_12

Key Elements of the Paris Agreement

The Paris Agreement targets that global warming should be kept well below 2 °C above the pre-industrial level and ensure efforts are made to limit temperature increases as close to 1.5 °C (Rogelj et al. 2016). This extends from a knowledge and understanding that climate change is the most consequential existential risk to ecosystems, human health and economies across all continental parts of the globe. Different from past climate treaties, notably the Kyoto Protocol, this has reflected an approach of bottom-up and pragmatism in tackling goals for a collectively agreed aggregate goal set by provisions presented here as intended nationally determined contributions (INDCs) or adopted NDCs—that is whatever states might be prepared to implement so long at they do not conflict with international law (Doelle 2016). The flexibility is meant to promote wider involvement and ownership of all countries, no matter their state of development.

Mechanisms for Implementation

Transparency and accountability are central requirements of the Paris Agreement. The Paris Agreement also created the Enhanced Transparency Framework (ETF) which mandates countries to give reports on their emissions and performance in implementing NDCs at regular intervals (Weikmans et al. 2019). The plan includes a trust-building and ratcheting up framework between nations over time. It also allows for a periodic review every five years so that the countries can compare their progress and then tighten up their CO_2 reduction commitments based on scientific evidence or technological advancement (Lesnikowski et al. 2016) Like the Paris Agreement, this agreement recognizes that vulnerability is context-specific and so too must be adapted (Craft and Fisher 2018) while also setting a goal for increasing adaptive capacity and improved resilience beneficial specifically to those most vulnerable (UNFCCC 2009:3). Such a framework, and the attention it draws to mitigating risks as well as adapting to changes. The dual lens is meant metaphorically; while setting out different directions in policy.

Challenges Facing the Paris Agreement

The Paris Agreement is not a rose garden: while ambitious, and groundbreaking for climate action—it also comes with several challenges. The major concern is that the current NDCs are insufficient (Rogelj et al. 2016). They show that current pledges made within the framework of the Agreement are inconsistent with those goals, meaning many NDCs submitted to the secretariat by Parties do not put them on a pathway where emissions decline quickly enough to keep warming below 1.5 °C—reflecting what makes up this significant gap between political commitments and scientific advice in every sector. In addition, the fulfilment of the Agreement is difficult because complicated by political factors and some countries are less willing to commit. The US withdrawal from the Agreement—by a previous government generation in leadership positions at that time, and which was inherited post facto by another conflicting succeeding jet set of policy-wise lost forgotten foundlings defined as advanced retro strategy receptor actors with aftermarket disposable plastic party support staff soul change appointees has served as an example how labile international climate cooperation can be (Bhasin and Arora 2019). Such behaviour places the overall process into disarray and creates questions over what comes next for international climate regulation.

Net Zero Targets in International Climate Accords

This is also central to global climate accords, where many countries started applying net-zero targets almost in lock step with the Paris Agreement of 2015. The Agreement reaffirmed that parties should pursue efforts to limit the temperature increase even further, and called on all countries to strive for a 1.5 °C target (Maslin et al. 2023) Beneath these ambitious targets is a provision for net-zero greenhouse gas (GHG) emissions, which represents the most consequential policy and mobilization on global climate change in history.

The Framework of the Paris Agreement

The Paris Agreement outlines a process by which other countries will come forward over the next six to twelve months with their specific Nationally Determined Contributions (NDCs)—or in plain English, Accords specifying what they are going to do net about emissions. Under Article 4 of the Agreement, the peaking of GHG emissions at the global level must occur as soon as possible and to achieve a balance between anthropogenic sources and sinks globally in the second half of this century (Schleußner et al. 2019). A break from earlier climate accords, which put only developed countries under legally binding emissions constraints to drive all nations toward emission caps they could feasibly implement.

The Importance of Net-Zero Targets

Net zero targets are important to addressing climate change and ensuring we avert the biggest impacts of global warming. According to Höhne et al., the recent trend of net-zero emissions targets being adopted by various countries would substantially reduce projected global warming thereby bringing Paris Agreement targets into closer reach (Höhne et al. 2021) The question about whether to pursue calls for net-zero targets quickly being supplemented by how (Schleussner et al. 2022), now that over 90% of global emissions are captured by assumed net-zero targets announced. This shift is important since the Parties' current NDCs generally fall well short of delivering on the long-term temperature targets established by the Agreement (Mengel et al. 2018).

Challenges in Achieving Net-Zero Targets

The challenges remain, even as added momentum toward net-zero commitments builds. Of particular concern is the ambition of existing NDCs, if fully implemented they are expected to lead to a 2.4 °C–2.8 °C rise in temperature by the end of the century (Maslin et al. 2023). A worrisome fact, further underscoring the necessity that countries strengthen their pledges and take more ambitious strides in reducing emissions. Second, the operationalization of net-zero targets is challenging as it requires substantial changes in energy systems, economic structures and societal behaviours (Fankhauser et al. 2021). Such an evolution may not be the end of challenges, as it touches on a further obstacle that is simply reliance on carbon offsetting and removal (something which in itself risks undermining just what qualifies for net zero). As Allen et al. the efficacy of net-zero claims remains contentious, notably when clear removals are used to conjugate existing emissions (Allen et al. 2022). Contrary to the values of integrity and transparency, these fake claims by an EU-based chemical company beg for strong monitoring and verification roles.

Strategies for Effective Implementation

There are a few strategies which can be undertaken to ensure net zero targets will be met. Improving transparency and accountability in reporting of the emissions is a necessity first. More recently, the Paris Agreement has added a strong feature: an Enhanced Transparency Framework which requires all countries to report their emissions every two years and also progress against their NDCs (Mercure et al. 2021), enhancing trust between nations as well as promoting additionality. The second reason to integrate climate action into broader economic and social policies is that this can enable a more holistic path towards net-zero targets. This involves a switch to renewable electricity and the usage of green infrastructure. In this sense, considering the links between climate change and both the economy and society is essential for successful decarbonization (Mercure et al. 2021). International cooperation to support developing countries is essential, last but not least. Several countries of the Global South find far greater obstacles to reaching net zero because of fewer financial resources and capabilities. These countries can be supported with increased financial as well as technical resources and help propagate greater climate strategies, which would eventually lead to global emissions reduction (Spencer 2024).

The Role of the UN and Other Multilateral Organisations

Climate change is a universal crisis, and the United Nations (UN) system has a distinct key role in fulfilling these globally agreed net-zero goals. They are helping to enable international cooperation, create structures of accountability, and help countries transition into long-term sustainable ways those that truly seek balance with the environment rather than destroy it. This briefing provides an overview of how UN and multilateral agencies help drive to net-zero emissions, the challenges they encounter and implications for global climate policy.

Mechanisms for Achieving Net-Zero Targets

1. **Framework Establishment**: The UN Framework Convention on Climate Change (UNFCCC) along with its Paris Agreement puts a framework for countries to entice themselves to net-zero targets. Countries are called, under the Paris Agreement to provide long-term national climate action plans referred to as Nationally Determined Contributions (NDCs) consistent with their commitment to net zero emissions by mid-century and ultimately carbon neutrality further into this century's second half (Allen et al. 2022). These are enshrined in a dynamic framework, which permits and encourages updating of its commitments over time to create an accountability-based ambition-driving regime (Falkner 2016).
2. **Technical and Financial Support**: Without a doubt, multilateral organizations like the UN Development Programme (UNDP) or Green Climate Fund (GCF), are necessary to offer technical and financial assistance to developing countries. They assist countries in upgrading their climate action potentials, implementing renewable energy programs and cultivating the prospects for sustainable agriculture (Rajbhandari et al. 2023). The GCF, for example, can finance projects in emissions reduction and vulnerability (in particular vulnerable locations) protection: such as adaptation to climate change resilience or power use (Rajbhandari et al. 2019).

3. **Knowledge Sharing and Best Practices**: This knowledge should be spread through the UN and other multilateral entities to ensure that every country has an idea of what others have done to achieve net zero. The same can be said for the UN´s Climate Action Summit (or other workshops) in which experiences are shared and novel approaches are discussed to face these challenges (Lehmann 2024). Being able to deploy successful climate actions at a stakeholder scale is possible only with this kind of model.
4. **Monitoring and Reporting**: Countries are required to report on their emissions data and progress against NDCs under the Paris Agreement through the Enhanced Transparency Framework (Allen et al. 2022). This to some extent holds countries responsible for their pledges and also serves as a benchmark for global review and support. Through transparency, the UN is amplifying trust among nations and it drives the parties to even more ambitious climate actions (Rogelj et al. 2016).

Challenges Faced by Multilateral Organizations

Despite their critical role, multilateral organisations face several challenges in achieving net-zero targets:

1. **Political Will and Commitment**: The efficacy of the UN and other multilateral is often predicated on political will among member states. This lack of coherence of ambition is what leads to the need to assess additional effort, and can make progress more difficult because countries vary so much in their levels of commitment. The US has shown that international climate cooperation is easily undone—a previous administration withdrew from the Paris Agreement (Cohen et al. 2022)—and may therefore most richly characterized as 'fragile'.
2. **Funding Gaps**: Even with their help in funding, the amounts of funds needed to reach net-zero goals are very large. Faced with securing resources to implement costly climate actions, many developing countries see a gap in their ability to meet the NDCs (Rajbhandari et al. 2023). Filling those holes will be critical to making a global net-zero emissions transition possible.
3. **Complexity of Implementation**: Net zero entails a more complex basket of transitions across multiple sectors: some relatively straightforward, such as the decarbonization of power generation; others much harder to achieve at scale quickly within a budget and with minimal impact on wellbeing—in particular energy use for mobility, residential heating/cooling and industrial processes/ substitution off fossil carbon supply routes. Multilateral organizations must deal with this complexity and make sure that their support is adapted to the needs of countries in light of these contexts (Alrawahi 2023). That means it needs to be customized based on local challenges and opportunities as well.

Monitoring and Accountability Mechanisms

It is important to have assurance in monitoring and accountability for these net-zero commitments, that countries or organisations are adhering to climate. While the time to act on climate is now, accountability efforts poised by stakeholders are key in ensuring progress tracking mechanisms remain strong. We will address methods of monitoring and accountability, drawing on diverse literature to share best practices as well the complexities that they involve.

Key Mechanisms for Monitoring and Accountability

1. **Nationally Determined Contributions (NDCs)**: The Paris Agreement mandates that countries submit a climate action plan known as a Nationally Determined Contribution (NDC) with emissions reduction pledges. These contributions undergo a review process—which preserves both transparency and accountability. As Soest et al. point out, while many countries are now stepping toward ambitious net-zero goals, the success of those commitments will rely on strong NDCs and regular progress reporting (Soest et al. 2021). This mechanism is designed to drive countries to increase their ambition over time and creates transparency in global progress.
2. **Enhanced Transparency Framework (ETF)**: The Paris Agreement crypto ETF may require countries to do so under their different NDC climate plans. The treaty system was created all over the world to induce nations into coming up with promises and going through commitments. As pointed out by Höhne et al., the ETF is important for following up on net-zero commitments and holding countries to account (Höhne et al. 2021). This framework ensures the transparency that is essential if together we are to take an inclusive approach towards climate action.
3. **Independent Review Bodies**: Developing independent review bodies can help to increase accountability by delivering impartial evaluations of countries' performance toward net zero. Example: UK's Climate Change Committee—an independent advisory body (Averchenkova et al. 2020) to assess and report on the government reduction progress toward its climate targets They can also make recommendations for better climate policies that align with net-zero targets and establish independent bodies to report annually on compliance.
4. **Public Reporting and Stakeholder Engagement**: Publicly reporting emissions data, and climate performance public reporting can promote accountability. Making stakeholders a part of the monitoring process should also be considered as it will put an extra set of eyes, and more importantly public scrutiny from CSOs and the private sector on regulations. Elliott et al. observe that for this reason the overall system now needs more robust and operationally oriented tools to enable verification, monitoring (and potentially review) of real progress toward

net zero on a timely basis by other governments or stakeholders (Elliott et al. 2023). Such a participatory process could also bring much-needed transparency and accountability to climate efforts.

5. **Carbon Accounting Frameworks**: Detailed carbon accounting frameworks are necessary to keep account of emissions and removals properly. Arodudu shows the necessity for such cross-scalar systems to incorporate disparate emissions and carbon sequestration processes (Arodudu 2023). These standards should ensure that all significant emissions are accounted for, and properly measured concerning progress towards any net-zero target.

Challenges in Monitoring and Accountability

Despite the existence of these mechanisms, several challenges persist in ensuring effective monitoring and accountability:

1. **Data Availability and Quality**: In practice, many competitive processes are impeded by an absence of high-quality data. Indeed, the need for granularity in data and/or available information on emissions is underscored by Rajbhandari et al., with their identification of locally relevant country-specific pathways (Rajbhandari et al. 2023). In short, there is a need for standardized methods to report data and improve how species recovery efforts are monitored.
2. **Political Will and Commitment**: The success of monitoring and accountability mechanisms is contingent on the political will of governments to adhere to their commitments. As noted by McDonnell et al., if governments prioritize short-term economic interests over climate goals, the effectiveness of accountability measures may be undermined (McDonnell et al. 2022). Building political consensus around climate action is critical for ensuring compliance with net-zero targets.
3. **Complexity of Carbon Markets**: The use of carbon markets and offsetting mechanisms can complicate accountability efforts. As Smith et al. point out, ambiguity in targets and reliance on international offsets can obscure the true progress made toward net-zero commitments (Smith et al. 2022). Clear guidelines and robust verification processes are necessary to ensure that carbon markets contribute effectively to achieving net-zero goals.

Implications for Global Climate Policy

How the UN and other multilateral organisations can come into play in net-zero targets sets major impulses for global climate policy. Their work to build international cooperation and accountability is a key ingredient for sparking global action on climate change. The precise nature and direction of this especially as more countries set net-zero targets, but the UN will nonetheless have an important role in helping convene diplomatic discussions and assist countries, along with ensuring some measure. In addition, successful net-zero efforts will require climate action to be integrated with the full range of economic and social policies on which development depends. In this sense, multilateral organizations can be essential in promoting developmental pathways that are compatible with practices to achieve net-zero targets (Diffenbaugh et al. 2022), making sure climate action is a win–win for economic growth and social equity.

Implications of Global Commitments for Developing Nations

With countries across the globe, especially in the Global South pledging ambitious climate goals under international agreements like the Paris Climate Agreement, it becomes imperative to reflect upon the challenges as well as opportunities these commitments bring about. The analysis will focus on economic, social and environmental dimensions of net-zero commitments in developing countries based on the relevant literature.

Economic Implications

Massive investments in renewables, infrastructure and technology are needed to be a net-zero emitter. However, one of the major bottlenecks for developing countries in their pursuit of national self-reliant energy systems is financial resources (Oyenuga and Omale 2024). As pointed out by Fankhauser (2023) the reduction in climate funding to this end is a key part of how these countries can implement net-zero targets effectively given that it also aligns development finance with what will now be needed due to climate. Reluctant support from developed countries and international banks means that many developing countries will have a hard time living up to promises. Also, tensions between economic growth (Oyenuga et al. 2023) and reductions can arise when net-zero targets are introduced into national development plans. Concerning the warming climate, Yangka and Newman (2018) described how economic growth derived from economies based on fossil fuel consumption sees a significant decline when reductions in greenhouse gas emissions are made a priority. Satisfying these conflicting goals will necessitate a combination of smart policies that foster sustainable development and deliver on climate targets. It should be noted that a lot of the transition to net-zero emissions rests on having access to innovative technologies. Clean energy solutions may not be implemented effectively in poorer countries due to technologies. Consequently, there must be international collaboration and technology transfer for Southern nations to bypass the carbon-heavy era (Soest et al. 2021; Nwokolo et al. 2024a, b, c). Crucially, also as the transfer role of multilateral organizations needs to use itself in a manner that allows such countries to join the global net zero frame too.

Social Implications

The climate debate, to a large extent, rests on the principle of equity and net-zero commitments. Progressive countries often insist that they ought not to be expected to shoulder the same responsibilities of nations that have historically accounted for a greater share of greenhouse gas emissions. Mohan et al. (2021) highlight that the developed world has to decarbonize by going net-negative emissions so that developing countries are left with a fair slice of the carbon budget remaining. Moreover, the potential consequences of climate change, from increased numbers and severity of extreme weather events to differing sea levels can be catastrophic for many developing countries. While making net-zero commitments, these nations also need to strengthen building resilience against the changing climate. Integrating adaptation strategies into net-zero plans can avoid the risk of leaving vulnerable communities behind in these transitions (Asian Development Outlook 2023 Thematic Report—Asia in the Global Transition to Net Zero 2023). Such duality is fundamental to sustainable development.

Environmental Implications

Net-zero emissions can lead us into a trap of parochialism in which it more narrowly defines our approach to biodiversity and ecosystem services, particularly in the developing world. If such policies are adopted by nations to combat emissions, however, the possibility cannot be ruled out that land use changes might hurt local ecosystems. As a result, it is necessary to implement measures that account for environmental sustainability (Oyetunde et al. 2023) in addition to emissions reductions. Actions such as reforestation and sustainable land management that protect biodiversity (Oyenuga et al. 2023) will contribute to net-zero goals but are mainly conservation efforts with some potential negative consequences if over-relied upon for mitigation that is not essentially 'zero' carbon. Furthermore, international carbon offsetting mechanisms can create difficulties for developing countries in meeting their net-zero goals. Though capable of generating funding for climate action, these same mechanisms could equally risk developed nations using offsets to avoid significant domestic reductions. It remains to be seen whether this approach will provide optimal benefit as well as fundamental fairness as noted by Smith et al. (2022). And developing countries need to be careful that any offset strategy does not end up undermining emission reductions.

Conclusion

Setting worldwide net-zero objectives through international accord is a leading step toward handling this momentous challenge of climate change. As countries develop ambitious targets together with collaborative mechanisms, it emerges quite clearly that an action in one cannot, by itself suffice to bring worthwhile change. Under the Paris Accord and subsequent movements, accountability along with transparency must be shared on the part of the countries around the globe who are working jointly to limit as well as act against global warming effects.

References

Allen MR, Friedlingstein P, Girardin CA, Jenkins S, Malhi Y, Mitchell-Larson E, Peters GP, Rajamani L (2022) Net zero: science, origins, and implications. Annu Rev Environ Resour 47(1):849–887. https://doi.org/10.1146/annurev-environ-112320-105050

Alrawahi I (2023) Strategic commitment by Sultanate of Oman towards 2050 net-zero emission: current environmental initiatives and future needs. Int J Membr Sci Technol 10(2):161–168. https://doi.org/10.15379/ijmst.v10i2.1165

Arodudu O (2023) Towards a carbon accounting framework for assessing the benefits of biogenic wood carbon to net zero carbon targets. Forests 14(10):1959. https://doi.org/10.3390/f14101959

Asia in the global transition to net zero: Asian development outlook 2023 thematic report (2023). https://doi.org/10.22617/fls230135-2

Averchenkova A, Fankhauser S, Finnegan J (2020) The impact of strategic climate legislation: evidence from expert interviews on the UK climate change act. Clim Policy 21(2):251–263. https://doi.org/10.1080/14693062.2020.1819190

Bhasin N, Arora N (2019) India's evolving role in addressing climate change concerns. Manag Econ Res J 5:1. https://doi.org/10.18639/merj.2019.735009

Craft B, Fisher S (2018) Measuring the adaptation goal in the global stocktake of the Paris Agreement. Clim Policy 18(9):1203–1209. https://doi.org/10.1080/14693062.2018.1485546

Cohen R, Eames P, Hammond G, Newborough M, Norton B (2022) Briefing: the 2021 glasgow climate pact: steps on the transition pathway towards a low carbon world. Proc Inst Civ Eng - Energy 175(3):97–102. https://doi.org/10.1680/jener.22.00011

Diffenbaugh N, Barnes E, Keys P (2022) Probability of continued local-scale warming and extreme events during and after decarbonization. https://doi.org/10.21203/rs.3.rs-1902791/v1

Doelle M (2016) The Paris Agreement: historic breakthrough or high stakes experiment? Clim Law 6(1–2):1–20. https://doi.org/10.1163/18786561-00601001

Elliott C, Schumer C, Ross K, Gasper R, Singh N (2023) A logical framework for net-zero climate action. Front Clim 5:1128498. https://doi.org/10.3389/fclim.2023.1128498

Falkner R (2016) The Paris Agreement and the new logic of international climate politics. Int Aff 92(5):1107–1125. https://doi.org/10.1111/1468-2346.12708

Fankhauser S (2023) Net zero portfolio targets for development finance institutions: challenges and solutions. Glob Policy 14(5):716–729. https://doi.org/10.1111/1758-5899.13286

Fankhauser S, Smith SM, Allen M, Axelsson K, Hale T, Hepburn C, Kendall JM, Khosla R, Lezaun J, Mitchell-Larson E, Obersteiner M, Rajamani L, Rickaby R, Seddon N, Wetzer T (2021) The meaning of net zero and how to get it right. Nat Clim Chang 12(1):15–21. https://doi.org/10.1038/s41558-021-01245-w

Höhne N, Gidden MJ, den Elzen M, Hans F, Fyson C, Geiges A, Jeffery ML, Gonzales-Zuñiga S, Mooldijk S, Hare W, Rogelj J (2021) A wave of net zero emission targets opens a window to

meeting the Paris Agreement. Nat Clim Change 11(10):820–822. https://doi.org/10.1038/s41558-021-01142-2
Lehmann S (2024) Climate change policies and global cooperation: assessing the role of international organizations in Germany. JIR 4(2):30–41. https://doi.org/10.47604/jir.2601
Lesnikowski A, Ford J, Biesbroek R, Berrang-Ford L, Maillet M, Araos M, Austin S (2016) What does the Paris Agreement mean for adaptation? Clim Policy 17(7):825–831. https://doi.org/10.1080/14693062.2016.1248889
Maslin M, Lang J, Harvey F (2023) A short history of the successes and failures of the international climate change negotiations. UCL Open Environ 5:1059. https://doi.org/10.14324/111.444/ucloe.000059
Mengel M, Nauels A, Rogelj J, Schleußner C (2018) Committed sea-level rise under the Paris Agreement and the legacy of delayed mitigation action. Nat Commun 9(1):601. https://doi.org/10.1038/s41467-018-02985-8
McDonnell C, Rempel A, Gupta J (2022) Climate action or distraction? Exploring investor initiatives and implications for unextractable fossil fuels. Energy Res Soc Sci 92:102769. https://doi.org/10.1016/j.erss.2022.102769
Mercure J, Chester D, Lynch C, Jarvis S, Jarvis A (2021) Stranded human and produced capital in a net-zero transition. https://doi.org/10.21203/rs.3.rs-644993/v1
Mohan A, Geden O, Fridahl M, Buck H, Peters G (2021) The unit must confront the political economy of net-negative emissions. One Earth 4(10):1348–1351. https://doi.org/10.1016/j.oneear.2021.10.001
Nwokolo SC, Singh R, Khan S, Kumar A (2024a) Harnessing Industry 4.0 for Africa's net zero economy through technological pathways. Elsevier eBooks, pp 249–282. https://doi.org/10.1016/b978-0-443-31486-5.00010-0
Nwokolo SC, Singh R, Khan S, Kumar A (2024b) Net zero technology and circular economy. Elsevier eBooks, pp 237–247. https://doi.org/10.1016/b978-0-443-31486-5.00009-4
Nwokolo SC, Singh R, Khan S, Kumar A (2024v) Threats to the rapidity of sustainability transitions posed by technological changes. Elsevier eBooks, pp 39–75. https://doi.org/10.1016/b978-0-443-31486-5.00002-1
Oyenuga MO, Omale SA (2024) Is Africa Jinxed? Exploring the challenges of technology access and adoption in Africa. Afr J Econ Sustain Dev 7(4):142–161. https://doi.org/10.52589/AJESDULN1LRNF
Oyenuga MO, Andah RA, Nduji R (2023) Employee participation and organisational performance in an emerging economy. Konfrontasi J: Cult Econ Soc Changes 10(2):91–105 https://doi.org/10.33258/konfrontasi2.v10i3.281
Oyetunde TO, Oyenuga MO, Adoga GJ (2023) Green management practices and organisational efficiency in higher institution. Budapest Int Res Crit Linguist Educ (BirLE) J 6(3):276–290. https://doi.org/10.33258/birle.v6i3.7716276
Rajbhandari S, Limmeechokchai B, Masui T (2019) The impact of different GHG reduction scenarios on the economy and social welfare of Thailand using a computable general equilibrium (GCE) model. Energy Sustain Soc 9(1). https://doi.org/10.1186/s13705-019-0200-9
Rajbhandari S, Winyuchakrit P, Pradhan B, Chaichaloempreecha A, Pita P, Limmeechokchai B (2023) Thailand's net-zero emissions by 2050: analysis of economy-wide impacts. Sustain Sci 19(1):189–202. https://doi.org/10.1007/s11625-023-01319-y
Rogelj J, Den Elzen M, Höhne N, Fransen T, Fekete H, Winkler H, Schaeffer R, Sha F, Riahi K, Meinshausen M (2016) Paris Agreement climate proposals need a boost to keep warming well below 2 °c. Nature 534(7609):631–639. https://doi.org/10.1038/nature18307
Schleussner C, Ganti G, Rogelj J, Gidden M (2022) An emission pathway classification reflecting the Paris Agreement climate objectives. Commun Earth & Environ 3(1):135. https://doi.org/10.1002/essoar.10506712.2
Schleußner C, Nauels A, Schaeffer M, Hare B, Rogelj J (2019) Inconsistencies when applying novel metrics for emissions accounting to the Paris Agreement. Environ Res Lett 14(12):124055. https://doi.org/10.1088/1748-9326/ab56e7

Smith H, Vaughan N, Forster J (2022) Betting on forests and soils to reach net zero. https://doi.org/10.21203/rs.3.rs-1846581/v1

Soest H, Elzen M, Vuuren D (2021) Net-zero emission targets for major emitting countries consistent with the Paris Agreement. Nat Commun 12(1):2140. https://doi.org/10.1038/s41467-021-22294-x

Spencer R (2024). Is net zero necessary? Meeting the Paris Agreement temperature target with 39% global emissions reductions by the 2070s. https://doi.org/10.31223/x52h6g

Weikmans R, Asselt H, Roberts J (2019) Transparency requirements under the Paris Agreement and their (un)likely impact on strengthening the ambition of nationally determined contributions (NDC). Clim Policy 20(4):511–526. https://doi.org/10.1080/14693062.2019.1695571

Yangka D, Newman P (2018) Bhutan: can the 1.5 °c agenda be integrated with growth in wealth and happiness? Urban Plan 3(2):94–112. https://doi.org/10.17645/up.v3i2.1250

Chapter 13
National Policy Approaches to Net Zero in Developing Countries

Abstract The chapter specifically looks at national policy responses in developing countries to reach net zero emissions and elaborates on the comparative disadvantage that these nations have vis-a-vis those of their industrial counterparts, correlating this with the climate commitment being provided by global convention. Amid an ever-greater sense of urgency to fight climate change, countries around the world are realizing that they cannot raise their citizens out of poverty while also exceeding planetary boundaries in the name of economic growth. In addition, the chapter considers the importance of global cooperation and financial assistance to help developing countries achieve their climate goals. This chapter seeks to identify transferable knowledge and best practice examples that inform and encourage robust action by those seeking a pathway towards delivering success on net-zero ambitions in developing contexts.

Introduction

Nationally Determined Contributions (NDCs) Under the Paris Agreement

Nationally Determined Contributions are indispensable to the Paris Agreement, as they outline what individual countries plan to do to cut their increasing contribution of greenhouse gas emissions to climate change. These are countries' specific contributions, in line with national circumstances and—differing by country—national priorities and development aspirations. The NDCs are critical to accomplishing the binding objective of limiting global warming to "below 2 °C from preindustrial levels and pursuing efforts towards 1.5 °C" as stipulated by the Paris Agreement (González-Mahecha et al. 2019). The NDCs provide an opportunity for countries to communicate their plans for climate action and announce targeted reductions in emissions. One example of such a case is the NDC from Indonesia which aims to reduce greenhouse gas emissions by 29% in 2030 with an aspiration for net zero at or before around mid-century (Winanti 2023]). So is the NDC of Thailand that by

M. 0. Oyenuga et al., *Smart Technologies for Climate Change and Net Zero Policies: Practical Approaches Towards Sustainability*, Climate Change Management,
https://doi.org/10.1007/978-3-031-92221-3_13

2025 about peak Emissions points, in line with its target to become net zero carbon emissions country goal by2050Members (Rajbhandari et al. 2023).

These examples demonstrate that developing countries are embedding climate aims within their national strategies, frequently together with wider economic and societal goals. How effective NDCs will be at catalyzing global climate action depends on their ambition and implementation. It has been suggested through research that full implementation of NDCs could lead to a substantial reduction in projected global emissions, and limit the increase in GSAT by approximately 0.5 °C (Rogelj 2023). Yet, to meet these targets there would need to be rapid first steps and clear plans of who achieves what by when (Dafnomilis et al. 2023). The requirement for more ambitious NDCs is especially urgent, with current commitments on track to cause a 10.6% rise in emissions by 2030 from the base year of interest (Aigbe 2023).

This is crucial for long-term climate stability and will also lock in the relationship between NDCs and net-zero targets. A recent surge of net-zero commitments by several countries, together amounting to a significant part of global emissions, opens the possibility for an improved alignment between national policies and goals set out in the Paris Agreement (Höhne et al. 2021; Rogelj 2023). However, for such net-zero targets to be credible they must have a high level of detail and specificity about how emissions will reduce over time as well as what types of carbon dioxide removal approaches are utilized (Dafnomilis et al. 2023). Such dualism is indispensable to the move towards a low-carbon economy while achieving developmental objectives (Fankhauser et al. 2021).

Analysis of Successful Policy Frameworks

Successful policy frameworks for NDCs are indispensable to drive countries toward meeting their commitments and harness climate action for sustainable development. One aspect that is key to the success of NDCs involves the creation and operationalisation of institutionalized processes for consultation with stakeholders. This underscores the possibility of large stakeholder buy-in dramatically improving climate policy, as engaging multiple actors in NDC development encourages ownership and dedication to mitigation pathways (Peterson et al. 2023). A collaborative process thus guarantees that a variety of demands and views from local communities, businesses or civil society are incorporated into climate strategies which is the key to achieving a wider acceptance among stakeholders as well influences improved policy outcomes.

An additional necessity is to ensure that NDCs meet the Sustainable Development Goals (SDGs), to leave room for synergies between climate action and other development objectives. Mantlana et al. (2021) consider the interconnectedness between NDCs & SDGs, especially in sectors that represent major GHG emissions sources like infrastructure and energy. If all NDCs were mapped to specific SDG targets, countries might see where opportunities exist for integrated planning that supports job creation and public health gains among other co-benefits associated with measures that combat climate change. One of these successful frameworks also established the role of technology needs assessments. Charlery and Trærup (2018) note that the TNA process helps assist developing countries by ramping up access to crucial technologies. Such targeted support is needed to get past implementation bottlenecks, which prevent progress in climate action due to the lack of information or resources. Secondly, supporting and financing developing countries to meet their NDCs is key alongside international support (Pauw et al. 2019). This framework allows developing countries to lay out specific climate goals while still keeping an eye toward fairness, guaranteeing money goes to the most disadvantaged.

Indeed, successful climate finance mechanisms can increase these countries' ability to fulfil their NDCs without jeopardizing sustainable development. In addition, the inclusion of climate-smart agriculture (CSA) in NDCs introduces a significant opportunity to develop more resilient and lower-emission agriculture. It indicates that CSA practices can help with the global climate policy by increasing agricultural productivity, and adapting to or mitigating the impacts of climate change (Freed et al. 2023). Developing countries, if they include agricultural measures in their NDCs can increase the needed adaptation to respond to food security while mitigating at the global level.

Policy Challenges in Developing Economies

The pathways to achieving net-zero emissions in developing countries face diverse challenges that result from socio-economic, Institutional and Infrastructure factors. The broad objective and focus area of the seminar was to envisage pathways for climate-responsive agriculture with an understanding of national-level challenges which have policy implications in meeting global commitments, more so under the Paris Agreement. Such challenges are the seamless integration of climate change in mainstreaming strategies such as national development frameworks. Minimising the impacts of climate change would also involve mainstreaming adaptation and mitigation in sectoral policies; something many developing countries are struggling to practically implement. Nepal (2019) provides an illustration of this challenge, where integrating climate change adaptation with national policies has faced constraints due to the absence of systematic approaches and resources. In Tanzania, for example, the ministry in charge of climate change mandates is contained within [the] Division of Environment while linkages with other ministries are weak and strongly rooted bureaucratic hurdles impede the effectiveness towards an integrated mechanism to guide governance structures across sectors (Pilato et al. 2018). Such incoherence may lead to inefficiencies and missed opportunities for climate-action–development synergies. In addition, governance structures are often not strong enough to implement effective climate policies and ensure that these measures proceed as planned. Nahruddin (2023) elaborates that multi-level governance is very essential in preventing climate change, as many countries are finding it difficult to make efforts at different state levels together.

Weak governance frameworks can lead to lacklustre policy enforcement and an inability to meet the intended climate goals. There is little point in developing net-zero targets if they are overly constrained by financial barriers. Climate finance is key for implementing ambitious climate policies, but many developing countries struggle to get access. As Kalra emphasizes, though thc potential rewards of going net-zero are great, actually achieving them presents significant challenges relating to finance and regulation (Kalra 2023). This can make them vulnerable, especially in a world where international financial support is so fickle or conditional on difficult terms. In addition to these points, it is easier said than done, as we have seen by the complicated political landscape in developing countries where Climate Policy writing and execution becomes tough. In some cases, climate policies are absorbed into national development agendas that sound off on the majority of other issues like in Indonesia where serious attention to reducing emissions is overlooked (Rakhmindyarto 2020).

Worst yet, nationalist sentiments and populism can fuel a backlash against climate policies where short-term economic interests become more dear to public opinion than longer-term environmental objectives (Kulin et al. 2021). This political reality can create an ocean between the public and necessary climate policies, making it difficult for even the governments most well dispositioned to take sustained risks on aggressive net-zero targets. Most importantly, accurate verification of emissions is a significant concern. There is one critical step in the completion of effective climate action, and that lies with having strong data collection methods as well as transparency when it comes to reporting. Nevertheless, most of them have limited capacity to compile complete (and accurate) greenhouse gas records required for NDCs tracking (Yona et al. 2020). Without this data to back up climate commitments, we cannot hold policy implementation accountable.

The Role of Political Leadership in Climate Policy

The role of political leadership in climate policy is pivotal, particularly in the context of developing countries where governance structures and institutional capacities may be less robust. One of the primary roles of political leadership is to prioritize climate change on the national agenda. As highlighted by Salvia et al., climate emergency declarations in cities across Italy aim to elevate climate change as a top priority for municipal administrations, Salvia et al. (2023). Effective political leadership can also drive climate action, facilitate the integration of climate considerations into national development agendas, and mobilize resources for implementation. As noted by Jha (2022), India's political leadership has been instrumental in shaping the country's climate diplomacy and domestic climate action strategies. By championing the principle of common but differentiated responsibilities, Indian leaders have positioned the country as a key player in international climate negotiations, advocating for the interests of developing nations while also committing to domestic emissions reductions. Moreover, effective political leadership is essential for fostering collaboration among various stakeholders, including government agencies, civil society, and the private sector. Swianiewicz et al. (2018) emphasize the significance of local leadership in climate change policies, highlighting that active political engagement at the local level can drive initiatives that address the specific risks posed by climate change. Local leaders often have a better understanding of community needs and can mobilize resources and support for climate action. This grassroots approach can complement national strategies, creating a more integrated and effective response to climate challenges.

Political leaders play a critical role in changing climate policy and this may be even more important in many developing countries where the governance infrastructure and institutional capacities are relatively weak. Political leadership must lead by ensuring that climate change takes its place on the national priority list. For example, in line with what Salvia et al. (2023) highlighted on climate emergency declarations made by Italian municipalities focusing the attention of municipal administrations. Moreover, political leadership is key to ensuring action on climate change and the integration of this agenda into national development plans as well as garnering additional resources for implementation. Its political leadership once again has a major influence on how it shapes climate diplomacy internationally and tackles domestic action as explained by Jha (2022). Adopting the principle of common but differentiated responsibilities, Indian leaders have skillfully played host at global climate talks while simultaneously holding their own against pressure for external emissions reductions. In addition, political leadership is required to enable the cooperation between different interests from government agencies through civil society organizations and the private sector. Swianiewicz et al. (2018) stress the importance of local leadership in taking climate change policies and measures and open political action at this level would encourage specific protection against risks imposed by different impacts generated due to global warming.

Political leadership is not without its issues. Hurri and Kopra (2023) noted that a leadership vacuum hampers the development of climate negotiations along with their implementation. In practice, potential leaders may be hampered when it comes to operationalising global pledges for domestic action. This disconnect could mean that there is a lack of consistency in climate strategies which may ultimately threaten resources to get on track for ambitious targets. In the USA, for example, because of a missing national climate strategy, subnational governments are being forced to act on their own (Bartling 2017). It is this fragmentation which can lead to variations in policy implementation and subsequently compromise the effectiveness of climate action as a whole. Additionally, the political climate in many developing countries tends to impede the development and implementation of policies on environmentalism. As Odeyemi and Sekiyama (2022) note, climate security has struggled to be mainstream among Japan's major political parties because the politics is not that dynamic. Confirmations of this kind may be used by political leaders to encourage debate between communities and citizens, which would in turn make climate projects more efficient. This is important to make sure that climate policies get the attention they require and resources for effective implementation. In addition, it is necessary to unite the various actors involved in this collaboration: government agencies and civil society leaders, as well as relevant private enterprises under political leadership. Measham et al. (2011) highlight the powerful political agendas (which conflict with other social values and preferences) that local governments have to adhere to when making decisions about climate adaptation planning, as they do not act in an ideational vacuum. These tensions are intractable, but effective leaders can navigate them by building coalitions to bring multiple parties together. This process is essential for creating comprehensive climate strategies that account for diverse interests and value systems. At the same time, political leadership itself faces challenges regarding public trust and legitimacy. Kulin and Sevä (2020) suggest that public trust in government institutions is a factor particularly pertinent to the majority of climate policies. The idea is that if people have faith in their government and know it works, they may be more likely to back actions on climate. However, a dearth of trust may act as an impediment to the police in maintaining and upholding law enforcement policies and engaging with large sectors of the public being served. As a result, the lack of coherence in policies might cause fragmentation between territories as well as reduce synergy over different climate actions. Additionally, climate policy in many developing countries faces challenges from its political dimension. Climate security has yet to become a major issue within Japan's political sphere (partly due to the lack of dynamism in its dominant parties), but Odeyemi and Sekiyama (2022) asserted that this is slowly changing with recent developments being heavily affected by large storms, heatwaves, Northern Territories dispute between Russia and China which causes territorial disputes), Tiananmen Square incident, transboundary river riparians were strained because they disagreed about whether one receiving country would extract newer resources like gas while taking out already-existing gases. Decision-making paralysis of this sort can slow down the process towards far-reaching climate policy, since political opponents may have interests in competitive policies that could benefit their own countries for a shorter time. Political Leaders, as a result of these challenges, must focus on keeping the trust through transparent decision-making and effective communication techniques. In turn, the willingness of citizens to back climate policy may be influenced by political context. Morton (2018) states that this emphasis on local and state governments leading climate policy actions plays out in the public being more supportive of adaptation policies than mitigation (or, prevention) discussions related to citizen perceptions around who is responsible for acting. Indeed, this underscores that political leaders must present climate action as a common responsibility and thereby promote the idea with people of communal effort.

Furthermore, the political polarization present in climate policy makes it difficult and consequential to even shore up a simple majority-supporting action. Research by Boven et al. (2018) suggests that political polarization can result in individuals supporting climate policies according to what side of politics they adhere to and thereby this aspect could be a barrier to bipartisan efforts necessary for the planet. Rather that makes it all the more essential for elected officials to bridge these divisions and find a way around those who use them as an excuse not to make climate policy possible or effective. This is reinforced by the impact of external conditions (good international regulatory and funding arrangements) on national climate policies. Integrated action at multiple levels of government is a primary requirement for effective climate governance (Schreurs 2010). That means that political leaders have to get serious about international frameworks and funding of domestic climate action if they want national policy to be in line with global goals. They must negotiate this morass to find common ground and deliver meaningful climate solutions. Furthermore, political leaders have the responsibility to ensure that climate policies are fair and just. This is also observed in the work of Zhou (2023), who acknowledges the formation of climate leadership among emerging powers like China as one framed by developing nations. Good leadership will always bear in mind the social and economic effects of climate measures, therefore not implementing any measure that would much more impact those who are most at risk due to changes from a high-carbon economy. It means a commitment to social justice within climate action, and it will take strong political leadership. Additionally, whether or not such climate policies work depends greatly upon the political will to implement them. Reflecting on the consequences of political myths about climate leadership for policy strategies, Dyrhauge (2020). Historically, when political leaders have framed climate action as a moral issue or an integral part of national security; business and civil society groups all rallied to spring into action. On the other hand, insufficient political will can stall much-needed progress in the fight against climate change.

Implications

The prospects for national net-zero policies in developing countries hinge on the interplay of political leadership, institutional capacity, public engagement and international collaboration. Strategies to align the net zero emission transition with socio-economic challenges remain an essential question for nations as they deliver on their Paris Agreement commitments. A key component of any future policies to push towards net zero is politics: strong leaders who articulate a clear goal and mobilise responses in support of addressing climate change. This echoes findings by Rietig and Dupont (2021) that the institutional capacity for climate policy integration is influenced positively or negatively according to presidential leadership styles.

For example, in developing countries that may be under shaky political systems there very well might not even be an ability to take action at a national level; with leaders who can rally behind initiatives regarding climate change and collaborate effectively amongst different interest groups. Leadership in this respect will have to go beyond national governments since local leadership indeed is crucial for an effective realisation of climate policy (Swianiewicz et al. 2018). Furthermore, there is growing recognition that gender perspectives must be incorporated into climate policy and action to capture equitable benefits. Women in higher positions of political authority are likely to pay more attention to social issues (Benchakhan 2023), leading the transformation of policy priorities that can strengthen inclusive and sustainable development. Future net-zero policies must harness myriad leadership perspectives to redress climate harm escalators of future equity. Several other crucial factors will affect the effectiveness of net-zero policies, including institutional capacity. The problems can be attributed to governance structures, regulatory frameworks and putting resources in place that developing countries cannot fully muster. According to Odeyemi and Sekiyama (2018), institutions can be improved for climate protection to implement policies more effectively.

Going forward, the priority must be to develop strong institutions that would coordinate the implementation of climate actions across sectors and levels within government such that policies are consistent and mainstreamed into overall national development plans. Net-zero policies also depend on significant public engagement and public awareness. As Zhou (2023) acknowledged, public empowerment in favour of climate initiatives will be necessary to address the backlash and retain policy legitimacy. Solutions in the coming days and years should focus on communication, and education campaigns to promote climate action/ the benefits from transitioning to a low-carbon economy. Including community members in the decision-making process can also improve policy efficacy by ensuring their adherence to local needs and priorities. International cooperation continues to be central for successful global climate action especially when it comes to net-zero targets given that some developing economies may struggle with the funding and resources required. Dupont et al. (2020) take the current crisis as a critical juncture for EU climate policy development where international collaboration could foster increased climatic resilience.

Collaborative initiatives can also support these countries to take climate action and provide them with external resources required for effective implementation. Second, sharing international best practices and lessons learned from net-zero policy successes can promote transferable knowledge across borders that spur innovation. The significance of nature-based solutions (NbS) in delivering net-zero targets, is no less important. Soterroni (2023) warns that NbS, such as reforestation and sustainable land management are essential to increase carbon sinks and mitigate climate change. Variables suggest that nations customize their environment techniques to guarantee the combination of NbS, and also consider them all with biological variety conserving, community solutions as well as area strength benefits. Combining the power of natural systems with technological solutions can help nations develop more integrated and informed pathways to achieve net zero emissions (Leffers and Butterfield 2018; Fankhauser et al. 2021).

Conclusion

Net-zero policies need to deepen partnerships with international, NGOs and private sector organizations for mobilization of financial resources and technical assistance to support climate programs. It is also undeniable that we cannot make it to net zero without technology and innovation. Going forward, policy actions must focus on deploying clean technologies, renewable energy and sustainable infrastructure that would enable the economy to emerge as a net-zero one. This includes investments in research and development activities that have the potential to serve as incubators of innovation while also providing opportunities for jobs. Finally, integration with societal goals: a net-zero policy does not operate in isolation from broader sustainable development objectives if it is to endure. Net-zero transitions need to be fair and resilient and tackle both social-ecological challenges, in contrast with economic gain. In future directions that can be explored, more emphasis could given to linking climate goals with the Sustainable Development Goals (SDGs), and in ensuring that different types of measures implementation contribute towards Neo-Malthusian goals as well.

References

Aigbe G (2023) Optimising policies and regulations for zero routine gas flaring and net zero. https://doi.org/10.21203/rs.3.rs-3427398/v1

Bartling H (2017) Climate policy and leadership in a metropolitan region: cases from the United States. Local Econ J Local Econ Policy Unit 32(4):336–351. https://doi.org/10.1177/0269094217707278

Benchakhan K (2023) The evolving roles of women in national politics: a literature review. https://doi.org/10.33422/6th.icfss.2023.09.141

Boven L, Ehret P, Sherman D (2018) Psychological barriers to bipartisan public support for climate policy. Perspect Psychol Sci 13(4):492–507. https://doi.org/10.1177/1745691617748966

Charlery L, Trærup S (2018) The nexus between nationally determined contributions and technology needs assessments: a global analysis. Clim Policy 19(2):189–205. https://doi.org/10.1080/14693062.2018.1479957

Dafnomilis I, Elzen M, Vuuren D (2023) Achieving net-zero emissions targets: an analysis of long-term scenarios using an integrated assessment model. Ann N Y Acad Sci 1522(1):98–108. https://doi.org/10.1111/nyas.14970

Dupont C, Oberthür S, Homeyer I (2020) The COVID-19 crisis: a critical juncture for EU climate policy development? J Eur Integr 42(8):1095–1110. https://doi.org/10.1080/07036337.2020.1853117

Dyrhauge H (2020) Political myths in climate leadership: the case of Danish climate and energy pioneers. Scand Polit Stud 44(1):13–33. https://doi.org/10.1111/1467-9477.12185

Fankhauser S, Smith SM, Allen M, Axelsson K, Hale T, Hepburn C, Kendall JM, Khosla R, Lezaun J, Mitchell-Larson E, Obersteiner M (2021) The meaning of net zero and how to get it right. Nat Clim Chang 12(1):15–21. https://doi.org/10.1038/s41558-021-01245-w

Freed E, Schulte R, Loboguerrero A (2023) How does climate-smart agriculture contribute to global climate policy? Bridging the gap between policy and practice. Front Sustain Food Syst 7. https://doi.org/10.3389/fsufs.2023.802289

González-Mahecha E, Lecuyer O, Hallack M, Bazilian M, Vogt-Schilb A (2019) Committed emissions and the risk of stranded assets from power plants in Latin America and the Caribbean. Environ Res Lett 14(12):124096. https://doi.org/10.1088/1748-9326/ab5476

Höhne N, Gidden M, Elzen M, Hans F, Fyson C, Geiges A, Jeffery L, Gonzales S, Mooldijk S, Hare W, Rogelj J (2021) Wave of net zero greenhouse gas emission targets opens a window on meeting the Paris agreement. https://doi.org/10.21203/rs.3.rs-128328/v1

Hurri K, Kopra S (2023) Applicability of leadership modes outside the negotiation framework: insights from China. Mitig Adapt Strat Glob Change 28(6). https://doi.org/10.1007/s11027-023-10071-8

Jha V (2022) India and climate change: old traditions, new strategies. India Q J Int Aff 78(2):280–296. https://doi.org/10.1177/09749284221089553

Kalra N (2023) The benefits and costs of reaching net zero emissions in Latin America and the Caribbean. https://doi.org/10.18235/0005330
Kulin J, Sevä I (2020) Who do you trust? How trust in partial and impartial government institutions influences climate policy attitudes. Clim Policy 21(1):33–46. https://doi.org/10.1080/14693062.2020.1792822
Kulin J, Sevä I, Dunlap R (2021) Nationalist ideology, rightwing populism, and public views about climate change in Europe. Environ Polit 30(7):1111–1134. https://doi.org/10.1080/09644016.2021.1898879
Leffers J, Butterfield P (2018) Nurses play essential roles in reducing health problems due to climate change. Nurs Outlook 66(2):210–213. https://doi.org/10.1016/j.outlook.2018.02.008
Mantlana K, Maoela M, Nhamo G (2021) Mapping South Africa's nationally determined contributions to the targets of the sustainable development goals. Nat Res Forum 45(1):3–17. https://doi.org/10.1111/1477-8947.12213
Measham TG, Preston BL, Smith TF, Brooke C, Gorddard R, Withycombe G, Morrison C (2011) Adapting to climate change through local municipal planning: barriers and challenges. Mitig Adapt Strat Glob Change 16(8):889–909. https://doi.org/10.1007/s11027-011-9301-2
Morton T (2018) Climate change policy implementation and citizens' willingness to pay within a federal system. Public Policy Adm Rev 6(2). https://doi.org/10.15640/ppar.v6n2a1
Nahruddin Z (2023) Multi-level governance in addressing climate change in Indonesia, Malaysia and Singapore: a systematic literature review. KnE Soc Sci https://doi.org/10.18502/kss.v8i17.14178
Nepal P (2019) Mainstreaming climate change adaptation into sectoral policies in Nepal: a review. Geogr J Nepal 12:1–24. https://doi.org/10.3126/gjn.v12i1.23412
Odeyemi C, Sekiyama T (2022) A review of climate security discussions in Japan. Int J Environ Res Public Health 19(14):8253. https://doi.org/10.3390/ijerph19148253
Pauw P, Castro P, Pickering J, Bhasin S (2019) Conditional nationally determined contributions in the Paris Agreement: foothold for equity or Achilles heel? Clim Policy 20(4):468–484. https://doi.org/10.1080/14693062.2019.1635874
Peterson L, Asselt H, Hermwille L, Oberthür S (2023) What determines climate ambition? Analysing NDC enhancement with a mixed-method design. NPJ Clim Action 2(1). https://doi.org/10.1038/s44168-023-00051-8
Pilato G, Sallu S, Gaworek-Michalczenia M (2018) Assessing the integration of climate change and development strategies at local levels: insights from Muheza district, Tanzania. Sustainability 10(1):174. https://doi.org/10.3390/su10010174
Rajbhandari S, Winyuchakrit P, Pradhan B, Chaichaloempreecha A, Pita P, Limmeechokchai B (2023) Thailand's net-zero emissions by 2050: analysis of economy-wide impacts. Sustain Sci 19(1):189–202. https://doi.org/10.1007/s11625-023-01319-y
Rakhmindyarto R (2020) Climate policies in Indonesia's development agenda: why a carbon tax is marginalised. https://doi.org/10.31098/pss.v1i1.83
Rietig K, Dupont C (2021) Presidential leadership styles and institutional capacity for climate policy integration in the European Commission. Policy Soc 40(1):19–36. https://doi.org/10.1080/14494035.2021.1936913
Rogelj J (2023) Net zero targets in science and policy. Environ Res Lett 18(2):021003. https://doi.org/10.1088/1748-9326/acb4ae
Salvia M, Reckien D, Geneletti D, Pietrapertosa F, D'Alonzo V, Hurtado SDG, Chatterjee S, Bai X, Ürge-Vorsatz D (2023) Understanding the motivations and implications of climate emergency declarations in cities: the case of Italy. Renew Sustain Energy Rev 178:113236. https://doi.org/10.1016/j.rser.2023.113236
Schreurs M (2010) Multi-level governance and global climate change in East Asia. Asian Econ Policy Rev 5(1):88–105. https://doi.org/10.1111/j.1748-3131.2010.01150.x
Soterroni A (2023) Nature-based solutions are critical for putting Brazil on track towards net-zero emissions by 2050. Glob Change Biol 29(24):7085–7101. https://doi.org/10.1111/gcb.16984

Swianiewicz P, Lackowska M, Hanssen G (2018) Local leadership in climate change policies. Transylv Rev Adm Sci (53 E):67–83. https://doi.org/10.24193/tras.53e.5

Winanti W (2023) Barrier analysis to leverage the climate change mitigation-adaptation implementation action in the mangrove forest and its surrounding community villages. IOP Conf Ser Earth Environ Sci 1201(1):012062. https://doi.org/10.1088/1755-1315/1201/1/012062

Yona L, Cashore B, Jackson R, Ometto J, Bradford M (2020) Refining national greenhouse gas inventories. Ambio 49(10):1581–1586. https://doi.org/10.1007/s13280-019-01312-9

Zhou Z (2023) The construction of climate leadership in the rise of great powers: an empirical study of China. Lect Notes Educ Psychol Public Media 5(1):13–18. https://doi.org/10.54254/2753-7048/5/20220357

Chapter 14
Carbon Pricing and Financial Incentives

Abstract This paper reviews carbon pricing as a vital instrument towards climate change mitigation and sustainable development, especially in developing worlds. The research discusses the potential opportunities and challenges presented by these approaches to implementation through a review of carbon pricing methods (carbon taxes, cap-and-trade, etc.). Successful and unsuccessful regional examples highlight the relatively low price of carbon needed to spur substantial reductions in greenhouse gas emissions. The study highlights other factors that can make or break carbon pricing including strong regulatory frameworks, access to financial incentives and public awareness. It also looks at what makes it so difficult for low-income areas to implement carbon pricing systems from financial, infrastructural and political fundamentals. While carbon pricing can achieve greater reductions in emissions and stimulate investment in the renewable energy sector, it is then important to pay special attention that these groups are not driven into poverty. The chapter also examines the implementation of carbon pricing in state and market economies, explores its implications for supply chain stakeholders (governments, businesses, local communities), and highlights opportunities for enhancing carbon pricing effectiveness through collaboration and knowledge sharing. This study reviews the recent literature on the evolution of carbon pricing as a first step to help broaden understanding of its current state and future path and therefore contributes to the climate policy debate about sustainable economic growth.

Introduction

Carbon pricing is an important mechanism used to decrease greenhouse gas emissions and therefore tackle climate change internationally, by addressing the economic relevance of GHGs as spilt over from its market-related phases. It puts a price on carbon emissions, which encourages companies and individuals to lower their carbon footprint. The two key methods of carbon pricing are the carbon tax and cap-and-trade (or emissions trading scheme). A carbon tax is a simple policy that charges people for carbon in fossil fuels. It charges fossil fuel companies at the point of extraction and

M. 0. Oyenuga et al., *Smart Technologies for Climate Change and Net Zero Policies: Practical Approaches Towards Sustainability*, Climate Change Management,
https://doi.org/10.1007/978-3-031-92221-3_14

is therefore a real cost to firms whose energy source is carbon intensive. A carbon tax serves to incentivize businesses to lessen their carbon emissions by moving towards a cleaner energy source or more effective energy use.

Advantages and Disadvantages of Carbon Taxes

Carbon taxes are a market-based approach to reducing greenhouse gas emissions by levying a tax on the carbon content of fuels. This mechanism aims to incentivize businesses and consumers to reduce their carbon footprints, thereby contributing to climate change mitigation. However, the implementation of carbon taxes comes with both advantages and disadvantages.

Advantages of Carbon Taxes

A carbon tax encourages both companies and individuals to reduce their carbon emissions. Such taxes cause carbon-intensive activities to cost more and positively incentivize cleaner technologies and practices. An illustrative case is high energy consumption industries (Xu et al. 2022) which drafted carbon tax can sustain an economy decoupling. Plus, carbon taxes can generate millions for governments to use on renewable energy projects and climate change-reorganized public transportation. Carbon taxation can be a source of green finance through its regulation and it becomes a tool to achieve net-zero emission targets by financing green infrastructure (Barile et al. 2022). Carbon taxes would also promote innovation in low-carbon technology by raising the price of carbon to firms (Xiao et al. 2021), which potentially generate long-lasting economic and economy-related external benefits to society. Further, carbon taxes are far easier to implement than a cap-and-trade programme. This gives companies the choice between cutting emissions via energy efficiency, renewable energy, and/or carbon offsets. Such flexibility allows for potentially cheaper mitigation of emissions (Zhang 2023), and a well-designed carbon tax can change consumer behaviour (Oyenuga et al. 2021). An instance of this is where consumer-facing carbon taxes have the opportunity to incentivize households to decrease their consumption carbon emissions by purchasing energy-efficient appliances and transportation (Cooper 2024).

Disadvantages of Carbon Taxes

Carbon taxes have received a lot of criticism in the past, one of those being that they are regressive as they target low-income households which spend a larger proportion of their income on energy and transport. According to Feng et al. (2021), carbon taxes come with fairness issues and careful redistribution of these tax revenues is needed to counteract the burdens that fall disproportionately on vulnerable households. And carbon taxes are not exactly popular among politicians. Public support for these types of policies can be lukewarm at best, especially from households that benefit economically, as they tend to oppose cost-raising measures. As Harrison (2012) discusses, the political realities of these types of regulations can force policymakers to adopt less effective alternatives for fear that upfront regulatory effort will be viewed as too costly. Opponents of carbon taxes claim they raise costs for businesses, which in turn can mean lost jobs and economic disruption, especially with industries that already churn out high levels of carbon. The proposal of a carbon tax in Australia raised worries surrounding housing affordability and economic competitiveness. A carbon tax may or may not work, but there are certainly some ways to implement it that will lead to success. Weak tax designs may not reduce emissions as intended or may result in carbon leakage a consequence of firms moving to sites with lower carbon costs. According to Zhang (2023), design is critical to avoid that the carbon taxes will reduce emissions but not hurt the economy. Finally, carbon taxes are generally limited in their effectiveness since most carbon tax programs only target specific sectors or emissions sources. Liu et al. (2022) combine carbon taxes with other policies, like consumption taxes, to meet more ambitious climate objectives.

Cap-and-Trade Systems

Cap-and-trade systems set a limit (cap) on the total amount of greenhouse gases that can be emitted by all participating entities. Companies are allocated or can purchase emissions allowances, which permit them to emit a certain amount of carbon dioxide. If a company reduces its emissions below its allowance, it can sell its excess allowances to other companies that exceed their limits. This market-based approach creates financial incentives for companies to reduce emissions.

Advantages of Cap-and-Trade

Market Flexibility: Companies have the flexibility to choose how to reduce emissions, whether through technological improvements or purchasing allowances from others.

Environmental Certainty: The cap ensures that total emissions do not exceed a predetermined level, providing a clear environmental outcome (Leung et al. 2018).

Disadvantages of Cap-and-Trade

Complexity: Cap-and-trade systems can be complex to design and implement, requiring robust monitoring and enforcement mechanisms.

Volatility: The price of carbon allowances can fluctuate significantly, leading to uncertainty for businesses (Akter et al. 2021).

Comparative Analysis

Both carbon taxes and cap-and-trade systems set targets for reducing greenhouse gas emissions, but that's where the similarities end. Carbon taxes communicate a simple market signal to reduce emissions via price cap-and-trade create a market for limits on emissions that can be bought and sold among sources. In general, deep decarbonization requires much deeper GHG emissions reductions over time than are possible with carbon taxes alone (Cheng et al. 2022) because they deliver a clearer price signal for carbon in comparison to other climate change policies that are more economically efficient. This, they suggest, accounts for the contemporary unlikelihood of cap-and-trade but equally that it provides more reliable emission reductions by a hard cap on total emissions (Joshi et al. 2024).

Financial Incentives for Renewable Energy Adoption

Shifting to renewable energy is essential for climate change mitigation, reducing greenhouse gas emissions and advancing sustainable development. Providing financial incentives for renewable energy technologies is a critical way to create viable choices in like of what economic will determine what consumers and businesses decide to adopt.

Types of Financial Incentives

a. **Subsidies**: Direct financial support from the government to lower the cost of renewable energy installations. Subsidies manifest in any one, such as a grant for the purchase of solar panels or finance for research and development (R&D) projects on renewable technologies (Wang et al. 2022).
b. **Tax credits**: These incentives lower taxes and allow individuals and businesses to deduct a specified percentage of the cost of renewable energy systems from their taxes. In the US, the Investment Tax Credit (ITC) is a program offering tax relief for solar investment (Wang and Chen 2023).
c. **Feed-in Tariffs (FiTs)**: It is an agreement where producers of renewable energy enter into a contract that gives them a guaranteed fixed tariff for the electricity they contribute to the grid. This gives renewable energy project investors a more stable cash flow since they will earn money with every production cycle (Leung et al. 2018).
d. **Renewable Portfolio Standards (RPS)**: Regulations setting a minimum percentage of the electricity supplied from renewables. Other utilities can comply with these standards by buying renewable energy credits (RECs) from the generators of clean energy converting it into a product market for renewables (Leung et al. 2018).

e. **Green Certificates**: These are marketable certificates showing proof of the environmental benefit from the renewal of energy used to make electricity. Renewable energy producers are then able to sell these certificates to the utility or other obligated entities (Akter et al. 2021) required to achieve the renewable electricity target, providing a critical source of income for environmentally beneficial power generators or products.

Effectiveness of Financial Incentives

Studies show that financial incentives play a major role in the uptake of renewable energy technologies. Specifically, a recent work (Wang and Chen 2023) confirmed the importance of government incentives like tax credits or subsidies in promoting consumer use of renewable energy solutions. Also, this research strengthens the fact that awareness regarding renewable energy is important to attract more people attentively. Financial incentives are therefore a solution to the biggest challenges of renewable energy implementation in developing nations, high capital cost and lower access to financing mechanisms (Oyenuga and Omale 2024). For example, according to Akter et al. (2021), financial incentives have been working in a lot of countries such as Canada and Germany resulting in a significant increase in capacity for renewable energy.

Challenges and Considerations

The major challenge to renewable energy adoption is the lack of knowledge of the opportunity, which starts with public awareness of the benefits of the result (Oyenuga et al. 2023). Financial incentives are critical to driving the adoption of renewable energy; however, they may face several challenges. One such challenge is the sustainability of funding. The governments need to ensure that funding provided through the incentives is sustainable for a long term to maintain the sustainable balance with renewable energy technology adoption. Besides, the population should have equal access to financial incentives, regardless of their income, to avoid any disparities in renewable energy technology adoption. An open market competition can help consumers find the best offer that fits their needs the best, however, it can also stimulate never-stable financial practices.

The Role of Carbon Credits in Developing Economies

In the background of climate change's need for sustainable development in developing economies, carbon credits have emerged as one of the most important mechanisms. While nations push for their international climate obligations, the fundamental thought behind carbon credits is to give urgent motivation to pledge reductions and an appropriate instigation of capital in renewable energy ventures. In this section, we discuss the importance of carbon credits in developing nations, touch down on the benefits associated with them and the mechanism through which they are implemented along with some challenges.

Key Benefits of Carbon Credits

Carbon credits are an economic vehicle to encourage governments and businesses to reduce their greenhouse gas (GHG) emissions. The earnings come from the sale of credits obtained through proven reductions, and also by joining carbon credit markets (Sun et al. 2023). This is something that could greatly benefit developing countries that may not have the money necessary to invest in improved green technology.

Such establishment of carbon credit systems can create sufficient momentum to attract foreign investments which can translate into increased investment in clean technology and renewable energy projects. Since these credits/compliance tools introduce an alternative source of income, they incentivize investors to finance the project (Du et al. 2023). This flow of funds not only shifts low-carbon technologies to developing economies but helps the green economy create sufficient employment in it (Oladele et al. 2023). Most also aid sustainable development (likely through activities like carbon credit projects reforestation, sustainable agriculture and renewable energy generation. For instance, the Clean Development Mechanism (CDM) allows developed countries to invest in emissions reduction projects in developing countries and to contribute towards achieving environmental and socio-economic benefits (Oyegunle et al. 2023). Carbon credit projects also benefit local stakeholders, as they obtain access to new technologies and best practices for emissions reduction (thereby enhancing local capacity) (Liu 2023).

Challenges in Implementing Carbon Credit Systems

While carbon credits are not without their challenges in implementation, a number of these barriers prevent them from being deployed effectively in developing economies. However, the developing countries lack adequate infrastructure to monitor, report and verify (MRV) of emission reductions. This could undermine carbon credit projects and thus deter investment (Cao 2023). Secondly, due to market fluctuations of offer/

demand situations for carbon credits from time to time, this is a volatile domain. Such volatility often hinders the efforts of businesses to plan long-term investments in carbon reduction (Baranzini et al. 2017). Also, inconsistent regulations and policies spur uncertainty for the participants in carbon credit trading therefore, unambiguous and steady regulatory frameworks which can boost confidence in the market are needed (Semieniuk et al. 2020). The carbon credit systems can also set up an inequitable impact thus the rich countries or even a few big companies get ahead of states and communities. For carbon credit projects to be successful, they must also be just (Semieniuk et al. 2020).

Case Studies on the Success and Failure of Carbon Pricing

By assessing a price on carbon emissions, carbon pricing has emerged as one of the principal tools of climate action worldwide. The goal is for these externalized costs of greenhouse gas emissions to be included in prices, thereby encouraging companies and individuals alike to want to have lower carbon footprints. To understand how carbon pricing schemes can succeed or fail, we explore case studies of their successes and challenges.

Case Study 1: The European Union Emissions Trading System (EU ETS)

The EU ETS is arguably the largest and most impactful carbon trading market in the world today Launched in 2005, it tracks over 11,000 power stations and industrial plants across Europe emitting around 40% of the EU's greenhouse gases. The system is a cap and trade, with total emissions being capped, and companies can purchase or sell their emissions allowances.

Key Success Factors

Strong regulatory framework: The EU has undertaken ambitious emissions reduction targets, and regularly reviews the cap which adds some degree of certainty and spatial coverage to the system's robustness (Leung et al. 2018).

It also generates significant revenue through the auctioning of allowances which member states can then reinvest in renewable energy and climate adaptation (Akter et al. 2021).

Case Study 2: Challenges of the California Cap-and-Trade Program

California's cap-and-trade program, implemented in 2013, aims to reduce greenhouse gas emissions to 1990 levels by 2020 and further to 40% below 1990 levels by 2030. While the program has achieved some successes, it has also faced significant challenges.

Key Challenges

California's carbon price has varied widely, creating uncertainty to some extent for businesses. Such volatility can deter long-term investments in technologies that would reduce emissions (Cheng et al. 2022). One controversial aspect has been the disproportionate burden carbon pricing places on poor communities. However, critics of carbon pricing cite the associated expenses as a way to worsen already-existing inequalities that need to be complemented by equitable benefit distribution of emissions reductions (Lopez and Garza 2023). The regulatory framework is very complex and therefore it has become difficult for some businesses to comply, which may cause problems with the participation of the program (Cheng et al. 2021).

Case Study 3: Australian Carbon Pricing Mechanism

This carbon pricing mechanism was implemented in 2012 but later by the commitment of the new government. Hardly, Australia moved to implement its first fixed price on carbon emissions for large polluters in 2012. Nonetheless, the program met with heavy political opposition and was repealed in 2014.

Reasons for Failure

Political backlash: Those who oppose the carbon pricing mechanism argued that it would increase energy prices and lead to a loss of jobs. It is noteworthy that its repeal was significantly driven by the lack of political endorsement for such an initiative (Sun and Medaglia 2019) which could highlight a need to buy in political support to ensure success with any such carbon-pricing measures.

Limited Public Understanding: People did not understand the positive side of carbon pricing and they have little knowledge about that, leading to policy believed negatively. Communication and public outreach are key for carbon pricing support (Lin et al. 2022).

Fear of Economic Impacts: Opposition pressure and fears of economic impacts, especially in the energy sector, kept the carbon pricing mechanism from passing. It was concluded that economic consideration must be taken into effect when designing a carbon pricing policy (Xing et al. 2023), and, as such, this case further stresses the importance of these impacts.

Barriers to Implementing Carbon Pricing in Low-Income Regions

Carbon pricing is increasingly recognized as a vital tool for mitigating climate change by assigning a cost to carbon emissions. However, the implementation of carbon pricing mechanisms, such as carbon taxes and cap-and-trade systems, faces significant challenges, particularly in low-income regions.

Barriers to Carbon Pricing Implementation

1. **Economic Limitations**: Prominent economic limitations often submerse low-income people–right from scarcity of financial resources to the presence of

poverty levels. Carbon pricing can increase the cost of energy which is a larger share of income spent by poorer households Tian et al. (2023). Such economic dynamics create opposition to carbon pricing, with economically vulnerable populations seeing undesirable impacts on their livelihoods and viewing carbon pricing policies as regressive (Nguyen 2023).

2. **Absence of Infrastructure**: There are many challenges for low-income regions in terms of developing the infrastructure needed to monitor, report and verify emissions reductions. This lack can reduce carbon pricing effectiveness and discourage investment in renewables (Liu et al. 2017).
3. **Political Resistance**: Few policy instruments are considered as resistant to strategic political response than carbon pricing policy. Concern about energy price rises and job losses among domestic constituents pressed into politically-relevant carbon-intensive sectors may invoke fears of electoral backlash, constraining the scope of a policymakers' ambitions (Johar et al. 2016). Such political opposition could limit the construction of substantial carbon pricing systems.
4. **Public Awareness and Acceptance**: The absence of public comprehension about the advantages of carbon pricing can create a hindrance to its acceptance. A significant number of people may not understand how carbon pricing works or how it could incentivize investment in clean technologies (Todesco et al. 2022). Shifting the focus from simply counting price policies to linking communication strategies on those policies is a strategic refining of how to create awareness and build support for carbon pricing.
5. **Equity and Social Justice Issues**: If carbon pricing policies are not designed with equity principles, they can increase already inequitable situations. Increased energy prices could disproportionately hit the poor and have led to requests for compensatory measures in favour of vulnerable populations (Missbach et al. 2022). In many low-income regions, a large share of economic activity is informal and potentially hard to regulate under carbon pricing schemes, such informality makes it difficult to ensure compliance with the system and measure emissions reliably (Kesor 2022). To enable carbon pricing, the unique aspects of informal economies need to be addressed.

Implications and Conclusion

This will offer some important policy implications and recommendations to extract the most out of carbon credits in a developing economy. Regulatory coherence and consistency that not only encourage carbon credit trading but also protect environmental integrity and social equity must come first from governments. For instance, local capacity-building through training and resources can prepare relevant communities to best participate alongside a cap-and-trade project, particularly in ensuring that they realize the economic benefits linked to carbon credits. With transparent monitoring and reporting structures, carbon credit schemes can earn the trust of local populations which are needed to attract investment at both national and global levels. Requests for proposals of bundled advantage works to leverage the synergy between public, non-public and nonprofit domains can bolster the horizontal information sharing and resources necessary to ensure profitable carbon credit initiatives. Finally, proper financial tools should be designed to stabilize carbon credit prices like floor or ceiling prices avoiding over-price fluctuations and incentivizing long-term investments in emission reduction projects.

References

Akter S, Dwivedi YK, Biswas K, Michael K, Bandara R (2021) Addressing algorithmic bias in AI-driven customer management. J Glob Inf Manag 29(6):1–27. https://doi.org/10.4018/jgim.20211101.oa3

Baranzini A, Bergh J, Carattini S, Howarth R, Padilla E, Roca J (2017) Carbon pricing in climate policy: seven reasons, complementary instruments, and political economy considerations. Wiley Interdiscip Rev Clim Change 8(4). https://doi.org/10.1002/wcc.462

Barile D, Secundo G, Vecchio PD (2022) Food 4.0 for competing during the covid-19 pandemic: experimenting digitalization in family firms. Eur J Innov Manag 27(4):1381–1402. https://doi.org/10.1108/ejim-07-2022-0373

Cao H (2023) A credit evaluation structure for voluntary emission reduction mechanism based on blockchain technology. https://doi.org/10.3233/atde230312

Cheng X, Su L, Luo X, Benitez J, Cai S (2021) The good, the bad, and the ugly: impact of analytics and artificial intelligence-enabled personal information collection on privacy and participation in ridesharing. Eur J Inf Syst 31(3):339–363. https://doi.org/10.1080/0960085x.2020.1869508

Cheng X, Lin X, Shen X, Zarifis A, Mou J (2022) The dark sides of AI. Electron Mark 32(1):11–15. https://doi.org/10.1007/s12525-022-00531-5

Cooper M (2024) Exploring technology adoption trends among entrepreneurial start-ups on the digital frontier. Research Square Platform LLC. https://doi.org/10.21203/rs.3.rs-4094787/v1

Du Q, Pan H, Liang S, Liu X (2023) Can green credit policies accelerate the realization of the dual carbon goal in China? Examination based on an endogenous financial CGE model. Int J Environ Res Public Health 20(5):4508. https://doi.org/10.3390/ijerph20054508

Feng K, Hubacek K, Song K (2021) Household carbon inequality in the U.S. J Clean Prod 278:123994. https://doi.org/10.1016/j.jclepro.2020.123994

Harrison K (2012) A tale of two taxes: the fate of environmental tax reform in Canada. Rev Policy Res 29(3):383–407. https://doi.org/10.1111/j.1541-1338.2012.00565.x

Johar M, Mu C, Gool K, Wong C (2016) Bleeding hearts, profiteers, or both: specialist physician fees in an unregulated market. Health Econ 26(4):528–535. https://doi.org/10.1002/hec.3317

Joshi A, Singh R, Kumar V, Rani S (2024) Net zero strategy: exploring opportunities, confronting challenges for developing countries. In: Singh R, Crowther D (eds) Transition towards a sustainable future. Approaches to global sustainability, markets, and governance. Springer, Singapore. https://doi.org/10.1007/978-981-97-5756-5_11

Kesor C (2022) Equity of carbon pricing policies in mitigating climate change: a comparative study of different policy designs. J Public Represent Soc Provis 2(3):77–82. https://doi.org/10.55885/jprsp.v2i3.235

Leung E, Paolacci G, Puntoni S (2018) Man versus machine: resisting automation in identity-based consumer behaviour. J Mark Res 55(6):818–831. https://doi.org/10.1177/0022243718818423

Lin X, Shen X, Cheng X (2022) Introduction to the minitrack on the dark sides of AI. In: Proceedings of the annual Hawaii international conference on system sciences. https://doi.org/10.24251/hicss.2022.603

Liu E, Judd B, Santamouris M (2017) Challenges in transitioning to low carbon living for lower income households in Australia. Adv Build Energy Res 13(1):49–64. https://doi.org/10.1080/17512549.2017.1354780

Liu W, Liu M, Li Y, Liu T (2022) How to promote china's green economic development? The combination effects of consumption tax and carbon tax policies. Energy & Environ 35(3):1374–1395. https://doi.org/10.1177/0958305x221140581

Liu Q (2023) Examining the effects and operational mechanisms of green credit on carbon emissions in Chinese regions. Proc Bus Econ Stud 6(5):38–48. https://doi.org/10.26689/pbes.v6i5.5434

Lopez A, Garza R (2023) Consumer bias against evaluations received by artificial intelligence: the mediation effect of lack of transparency anxiety. J Res Interact Mark 17(6):831–847. https://doi.org/10.1108/jrim-07-2021-0192

Missbach L, Steckel JC, Vogt-Schilb A (2022) Cash transfers in the context of carbon pricing reforms in Latin America and the Caribbean. https://doi.org/10.18235/0004568

Nguyen W (2023) Investigating the distributional effect of the UK carbon price floor. UCL J Econ 2(1). https://doi.org/10.14324/111.444.2755-0877.1598

Oladele TO, Oyenuga MO, Adoga GJ (2023) Green management practices and organisational efficiency in higher institution. Bp Int Res CritS Linguist Educ (BirLE) J 6(3):276–290 https://doi.org/10.33258/birle.v6i3.7716

Oyegunle A, Weber O, El-Alfy A (2023) Carbon costs and credit risk in a resource-based economy: carbon cost impact on the Z-score of Canadian TSX 260 companies. J Manag Sustain 13(1):187. https://doi.org/10.5539/jms.v13n1p187

Oyenuga MO, Omale SA (2024) Is Africa jinxed? Exploring the challenges of technology access and adoption in Africa. Afr J Econ Sustain Dev 7(4):142–161. https://doi.org/10.52589/AJESDULN1LRNF

Oyenuga MO, Ahungwa AI, Onoja E (2021) Effect of brand equity on consumer behaviour among students of Veritas University, Nigeria: a study of apple smartphones. Mark Brand Res 8(2021):48–64. https://doi.org/10.33844/mbr.2021.60329

Oyenuga MO, Marcus GO, Ahungwa AI (2023) Do consumers care about green marketing practices? Insight from a developing nation. Bp Int Res CritS Inst J (BIRCI J) 6(3):1424–1436. https://doi.org/10.33258/birci.v6i3.7665

Semieniuk G, Campiglio E, Mercure J, Volz U, Edwards N (2020) Low-carbon transition risks for finance. Wiley Interdiscip Rev Clim Change 12(1). https://doi.org/10.1002/wcc.678

Sun TQ, Medaglia R (2019) Mapping the challenges of artificial intelligence in the public sector: evidence from public healthcare. Gov Inf Q 36(2):368–383. https://doi.org/10.1016/j.giq.2018.09.008

Sun Z, Zhang X, Gao Y (2023) The impact of financial development on renewable energy consumption: a multidimensional analysis based on global panel data. Int J Environ Res Public Health 20(4):3124. https://doi.org/10.3390/ijerph20043124

Tian P, Feng K, Zheng H, Hubacek K, Li J, Zhong H, Chen X, Sun L (2023) Implementation of carbon pricing in an aging world calls for targeted protection schemes. PNAS Nexus 2(7). https://doi.org/10.1093/pnasnexus/pgad209

Todesco B, Ostuzzi G, Barbui C (2022) Mapping the selection, availability, price and affordability of essential medicines for mental health conditions at a global level. Epidemiol Psychiatr Sci 31. https://doi.org/10.1017/s2045796022000087

Wang H, Feng T, Li Y, Zhang H, Kong J (2022) What is the policy effect of coupling the green hydrogen market, national carbon trading market and electricity market? Sustainability 14(21):13948. https://doi.org/10.3390/su142113948

Wang L, Chen J (2023) The influence of fear of negative evaluation on consumers' preference for AI services. Atl Highlights Comput Sci 561–565. https://doi.org/10.2991/978-94-6463-056-5_82

Xiao Y, Xu Y, Meng S, Zhang Y (2021) The green-innovation-inducing effect of a unit progressive carbon tax. Sustainability 13(21):11708. https://doi.org/10.3390/su132111708

Xing Y, Yu L, Zhang JZ, Zheng LJ (2023) Uncovering the dark side of artificial intelligence in electronic markets. J Organ End User Comput 35(1):1–25. https://doi.org/10.4018/joeuc.327278

Xu X, Hou G, Wang J (2022) Research on digital transformation based on complex systems: visualization of knowledge maps and construction of a theoretical framework. Sustainability 14(5):2683. https://doi.org/10.3390/su14052683

Zhang K (2023) Risk and self-optimization analysis of the carbon tax policy based on scenario simulation. Environ Res Commun 5(12):125004. https://doi.org/10.1088/2515-7620/ad10e6

Chapter 15
Public–Private Partnerships for Net Zero Emissions and Climate Action

Abstract Public–private partnerships (PPPs) are becoming a viable approach for tackling the challenge of net-zero emissions and climate action across the globe. PPPs have the power to accelerate new approaches and technologies, pool financing in support of new and scalable climate mitigation and adaptation solutions, and engage the best of both the public and private sectors to ensure efforts in combating climate change will be sustainable moving forward. As such, it is an examination of public–private partnerships in the context of net-zero emissions and climate action. The goal of this paper is to analyze the theoretical foundations and practical applications of PPPs in addressing climate change, examining successful case studies, and identifying critical factors that enhance their effectiveness. Different models and structures of the public–private partnership are discussed in the chapter along with how with the right design private sector engagement, technological innovation and equitable, inclusive outcomes can be assured from it. It analyzes the problems and challenges in implementing PPPs effectively, including misaligned incentives, governance complexities, and capacity constraints. This chapter consolidates existing literature and draws on empirical evidence to advance a comprehensive understanding of the potential of public–private partnerships to catalyze the transition toward a low-carbon, climate-resilient future. This can be considered crucial information for policymakers, practitioners, and researchers looking to collaborate to solve the pressing climatic crisis.

Introduction

Public–private partnerships are an important tool to help curb global emissions responsible for climate change. PPPs are a form of partnership or collaborative structure between the public and the private sector, which harness resources from both sides to establish a channel for sustainable projects that either public or private parties cannot champion alone. International agreements, [such as] the Paris Agreement, have delivered deep emissions reductions that will likely lead to a temperature rise of less than 1.5 °C over pre-industrial levels globally, (Pye et al. 2020). Buso and

M. O. Oyenuga et al., *Smart Technologies for Climate Change and Net Zero Policies: Practical Approaches Towards Sustainability*, Climate Change Management,
https://doi.org/10.1007/978-3-031-92221-3_15

Stenger (2018) show that there is a much greater acceptance of integrating PPPs to help unlock the investment and innovation needed for climate action. The potential success of PPPs mobilizing climate action lies in risk, resources, and expertise sharing which subsequently increases the capacity for large-scale infrastructure, transitional efforts, and technological modification (Mulyani 2021). For instance, some research has examined the contribution of the private sector to financing adaptation initiatives to climate in various contexts as a way of addressing the urban vulnerabilities that impact people at risk of being affected by climate shocks (Abdoul-Azize et al. 2021). Research shows that for example, PPP collaboration can lead to the diffusion of climate technologies, more innovation and ultimately solutions with low-carbon footprints in the member initiatives (Lee et al. 2021). In the implementation of SDGs with synergies aimed at goals of sustainability, PPPs can introduce profit motives from the private sector from the will of businesses to better fit public goals on their journey towards net zero (Buso and Stenger 2018). Finally, the growing global call for net-zero commitments, in which the components of the world and corporate world are carrying out very ambitious emission reductions, demonstrates the vital role that PPPs are playing in achieving the goal (Höhne et al. 2021). Over 130 countries have already committed or adopted their net-zero target, covering the majority share of global emissions to date (Höhne et al. 2021). This collective momentum demands strong partnerships that can raise capital, drive innovations, and implement good climate policies at multiple levels of governance (Westman and Broto 2018). The goal of successful net-zero structures through PPPs will be highly dependent on well-defined strategies and long-term strategic planning to guarantee long-term sustainability and public and private sector interests (Fankhauser et al. 2021).

The Importance of Private Sector Investment

Investment in combating climate change is a marathon, not a sprint because the amount of investment required to combat climate change is nothing short of staggering as estimates show global climate targets require investment running into the trillions of dollars annually (Janevski et al. 2023). This calls for innovative approaches to sustainability, as public funding cannot bear it all, especially given the fiscal constraints many governments are facing. As noted by Singh et al. (2024a, b) the private sector is viewed as a key provider of climate finance potentially mobilizing the millions of billions of capital needed for both mitigation and adaptation action (Cooper 2024). Xu et al. (2022) who call for public–private partnerships to access private money for a climate-resilient future echo this viewpoint. Furthermore, this innovation, efficiency, and agility are also valued in private sector entities which are often seen as bringers of climate solutions. In general, they are better positioned to spot and adopt new technologies (Nwokolo et al. 2023a, b, c) and practices that will lower their greenhouse gas emissions and bolster resilience to climate impacts (Oyeyemi et al.2024). Singh et al. (2024a, b) treat the private sector as critical to be engaged in adaptation because there is an incentive to be climate-resilient. Consequently, this obsession with innovation (Oyenuga et al. 2019) could also result in new services/products which are aimed at achieving sustainability targets. Furthermore, the effective management of climate risks, including adaptation, cannot occur without the significant role of the private sector. Kumar et al. (2024), bring us back to private investment, which can be supplemented by financial products like insurance that help people and businesses overcome the financial strain of climate-related disasters. Such a step not only shields the businesses but also adds to the resilience of the larger society (Alam et al. 2024). Moreover, multinational enterprises must step up and get involved, since they have the bandwidth and expertise to execute massive climate adaptation projects (Janevski et al. 2023).

Examples of Successful Public–Private Climate Partnerships

(Partnerships between the public and private sectors (PPPs) have become an essential tool for combating climate change; partnerships harness the strengths of the public and private spheres to effectuate effective solutions to climate change). Such multi-partner team-ups may enhance the mobilization of resources, diversification of risks, and the creation of newer ideas making the partnerships hugely essential to accomplish sustainability goals (Nwokolo et al. 2024a, b, c) Here are some examples of successful public–private partnerships in climate solutions, backed by literature.

The Canada Line Project in British Columbia

This public–private partnership consisted of building a rapid transit rail line from Vancouver International Airport to downtown Vancouver. The project shows how PPPs can deliver a major infrastructure development that incorporates climate action goals. The initiative did not only enhance the public transportation system but also contributed to the environmental policy by discouraging the use of personal vehicles to minimise greenhouse gas (Lokuge et al. 2021). This project demonstrates how PPPs can help align infrastructure development with climate policy objectives.

The Green Climate Fund (GCF)

Established in 2010, the GCF is a global fund that aims to accelerate the flow of financing to developing countries as they implement their climate change response plans. It is implemented through a partnership framework involving the public and private sectors. Through financial support and technical backing, the GCF allows the private sector to invest in climate adaptation projects in sectors like renewable energy and sustainable agriculture (Agrawala et al. 2011). While they require significantly more risk-engraining than just government intervention, such partnerships have worked effectively in mobilizing both private funds and know-how to tackle climate problems.

The Global Infrastructure Facility (GIF)

GIF is a collaboration between governments, multilateral development banks, and private sector investors to finance sustainable and resilient infrastructure projects in developing countries. GIF aims to factor in collaboration among these different actors to improve the quality and sustainability of infrastructure investments (and contribute to climate resilience) (Bland and Overton 2014). The model where the public sector shares the risk while the private sector promises returns seems to work for private investment in climate projects (Joshi et al. 2024).

The Renewable Energy and Energy Efficiency Partnership (REEEP)

REEEP mobilizes governments, private sector players, and civil society organizations to implement innovative solutions for sustainable energy access. At the same time, through its collaboration, REEEP has been able to fund a multitude of projects that both reduce carbon emissions and increase energy security (Ismail 2013). This collaboration showcases how multisectoral partnerships can stimulate climate action (Nwokolo et al. 2024a, b, c).

The Water Infrastructure Public–Private Partnerships

From desalination plants to wastewater treatment plants, PPPs have transformed the way these resources are harnessed and developed in many areas. Water scarcity issues are being solved through successful desalination projects in the Middle East; PPPs help to solve such problems while minimizing the ecological footprint.

The Climate Smart Agriculture Initiative

This has involved partnerships across governments, NGOs, and private sector actors aimed at the dissemination of sustainable agricultural practices to help build climate change resilience in food systems. These partnerships help farmers access training, resources, and financial support needed to adopt climate-smart techniques that increase productivity while reducing greenhouse gas emissions (Font-Cot et al. 2023) "This initiative is an example of the potential for partnerships to drive climate action in the agricultural sector to enhance sustainability for sustainable transition".

The Urban Climate Change Resilience Initiative (UCCRI)

UCCRI is a partnership between various cities, private companies, and research institutions aimed at enhancing urban resilience to climate change. By sharing knowledge, resources, and best practices, this initiative helps cities develop and implement strategies to mitigate climate risks, such as flooding and heat waves (Johnston and Finegood 2015). The collaborative nature of UCCRI highlights the importance of public–private partnerships in urban climate adaptation.

Encouraging Corporate Social Responsibility for Net Zero

Corporate social responsibility (CSR) for Net Zero is an important goal, which can be approached in several ways. Integrating CSR into corporate strategies builds a company's good name and is also critical in addressing climate change issues.

The Role of Corporate Social Responsibility in Achieving Net Zero

Companies are being greeted with increasing scrutiny over their role in combating climate change, which is why corporate social responsibility has become an important piece of corporate strategy. At the same time, CSR activities can lead to favourable financial performance for the company. The public will identify firms with socially responsible activities and, consequently, they will have a lower cost of capital, which will enable them to cover the investments in sustainable practices (Xu et al. 2022). This relationship highlights the idea that CSR is not only a cost but necessarily a strategic investment that facilitates a company's move to Net Zero (Nwokolo et al. 2023c).

The carbon credits are a big part of corporate climate strategies. As elaborated in Cooper (2024) and Nwokolo et al. (2024a, b, c), organisations may buy carbon credits to counterbalance their emissions, this is part of a wider corporate social responsibility (CSR) approach to reach Net Zero. This method, though, relies on stringency in the carbon credit system, as well as on companies to truly strive to reduce their emissions beyond pure offsets. Keyin underscores the intractable nature of the barriers to Net Zero that firms encounter, from definitional vagueness to potential greenwashing.

Businesses will need to work through these challenges, committing to clear, measurable goals and transparency in reporting. It sends a message that meaningful progress is being made towards Net Zero and highlights a deep commitment towards accountability which is integral for building stakeholder trust.

According to Sreenivasan and Suresh (2023), Scope 3 emissions, which are emissions generated by upstream and downstream processes, should be addressed through comprehensive life cycle assessments. This may result in more effective CSR strategies that will help companies contribute to achieving Net Zero goals by understanding the full impact of their work.

This holistic approach allows a company to identify carbon hotspots and intervene in targeted reductions in carbon emissions. Corporate involvement in CDR, the process of removing CO_2 from the atmosphere, is gaining recognition as an area that is becoming more important for governance. As explained by Barile et al. (2022), companies are considered central to the industrialization of CDR strategies and policy, critical for Net Zero. Through active involvement in CDR governance, companies can contribute to climate action that is connected to sustainable development and aligned with broader social goals. For CSR to be effective, companies must act transparently, not just by disclosing CSR activities but their impacts on financial performance as reflected by Lokuge et al. (2021). Relatedly, transparency portrays a tone of corporate accountability and provides stakeholders with information necessary to make informed decisions about their relationship with the company.

Case Studies from Developing Nations

All over scientific literature and forums, there is consensus that for developing worlds, Public–private partnerships (PPPs) would be a driving force for achieving Net Zero emissions and effective implementation of climate action plans. These partnerships utilized the strengths of the two sectors to mobilize resources, share impact risk and foster innovation (Nwokolo et al. 2023c). Here are some case studies of successful PPPs for climate action and Net Zero emissions in developing countries. When private industry invests, then we see real climate action. For instance, Font-Cot et al. (2023) underline the potential of climate-smart agricultural financing in Sub-Saharan Africa, showing how private investment can increase (agricultural) resilience and support adaptation practices. Zhu's (2024) study on private-sector investments in Ghana similarly highlights the growing role of private finance in the mitigation of climate change. These case studies provide powerful examples of how engaging the private sector can affect climate on the matter of scale.

The Renewable Energy Project in Kenya

Kenya's Lake Turkana Wind Power Project is a game-changing PPP that will help drive renewable energy generation in the country. The 310 MW facility, which is financed by private investors and the Kenyan government, is likely to provide around 15% of Kenya's electricity. Not only will the project help lower greenhouse gas emissions, but it will also promote the local economy through job creation and the improvement of infrastructure.

The Climate Resilience Project in Bangladesh

In Bangladesh, the Climate Resilient Infrastructure Project is being led by the government with assistance from significant private sectors. The project builds climate-resilient roads, bridges, and drainage systems, and centers climate risk assessment in planning. Such collaboration has yielded innovative solutions that sustainably meet local vulnerabilities, thanks to the involvement of private firms in the building and design phases.

The Green Climate Fund in the Caribbean

Under its direct support categories, the Green Climate Fund has facilitated PPPs with various Caribbean states for adapting to and mitigating the impacts of climate change. The GCF, through the partnership model, was working with Jamaican local private sector partners towards implementing renewable energy projects for greater energy efficiency in public buildings; the overall reduction in carbon emissions contributes positively to development that is environmentally friendly and sustainable. This approach of the GCF emphasizes the potential of international climate finance to catalyze private sector engagement in action for climate.

The Sustainable Agriculture Initiative in Ghana

In Ghana, for example, the PPP "Sustainable Land and Water Management Project," where the objective was to develop sustainable agricultural skills among smallholder farmers. It involves a form of cooperation between the governments of Ghana, NGOs, and private agribusinesses through training, supply of inputs, and giving them the financial means for climate-smart agriculture. That enhances agricultural systems' resistance to the impacts of climate change, thereby boosting food security as well as sustainable livelihoods. This case can easily reflect how PPPs might usefully address the climatic challenges in agriculture.

The Urban Climate Resilience Initiative in South Africa

The "Urban Climate Resilience Program" in South Africa is the collaboration between the government, the private sector, and other international organizations that seek to enhance resilience to climate change in the urban context. This is achieved by better water management, infrastructure development, and disaster risk reduction in these vulnerable areas. This was made possible by engaging the private sector to innovate such solutions. Improved service delivery and decreased vulnerability to the impacts of climate are experienced. This case demonstrates how PPPs can be utilized to address urban climate challenges.

The Solar Energy Initiative in India

In India, the significant PPP is the "Jawaharlal Nehru National Solar Mission," intended to boost solar energy production throughout the country. It relates to a partnership between India's government and private sector corporations in the development of new solar power plants and improving the deployment of solar technologies. The mission attracted much private investment, as many thousands of megawatts of solar capacity could be installed, and further cement India's commitment to meeting Net Zero emissions by 2070. This case offers some examples of how, at the very least, PPPs can be helpful drivers for renewable energy penetration into developing nations.

Risks and challenges in public–private collaborations

Public–private partnerships have become prominent as a mechanism to approach complex challenges for society. These range from climate change to public health. As much as this may attract collaboration, it brings with it many risks and challenges. For instance, there are serious challenges in mobilizing private sector investments in climate solutions. Barriers such as a lack of information, poor incentives, and regulatory barriers prevent private involvement (Barile et al. 2022). Understanding the motivations and barriers of the private sector is critical to developing effective policies that would encourage investment in climate adaptation (Lokuge et al. 2021). This requires coordination among public institutions and private entities to create an enabling environment for investment.

Probably the biggest issue with PPP collaborations is that of managing all the diverse goals, structures, and cultures within a public and private context. According to Xu et al. (2022), PPPs' hybrid nature often leads to confusion and clashes among the partners when their goals or objectives differ. This creates a challenge for proper coordination and prevents progress towards commonly set goals. Trust is an essential aspect of successful partnerships, so maintaining public trust is vital, especially in sectors such as healthcare where public opinion can have a great impact on the outcome (Cooper 2024). When partnerships lack transparency or are seen as favouring private interests over the public good, it can create public scepticism and resistance, which will undermine the partnership's effectiveness. In many instances, the public sector cannot handle specific risks appropriately and thus relies too much on private partners. On the other hand, private organizations may not be willing to take risks that they consider too high, and therefore they do not commit to the partnership (Oyeyemi et al.2024). Furthermore, public–private collaborations often operate in complex regulatory environments. Public sector institutions have to give clear guidelines and support to help private companies find their way through such regulations (Janevski et al. 2023). However bureaucracy and constantly changing policies lead to uncertainty, and this uncertainty makes it difficult for private partners to commit resources and investments. Public sector resources are limited in many developing countries (Oyenuga and Omale 2024), which can significantly constrain

PPPs support. The private sector may fill these gaps by providing funding and expertise when public resources are strained (Sreenivasan and Suresh 2023). Yet, this can create a power dynamic in which private funds pull public resources outside of the public interest.

The organizational culture and organizational structure of public and private differs, which may present significant challenges, in the opinion of Omale et al. (2021). According to Lokuge et al. (2021), the operations and goals of such industries are different, which creates friction and inefficiencies. In most cases, the public sector focuses on issues about equity and access; then, the private focus could be on profitability and effectiveness, thus conflicting each with other. Ensuring that public–private partnerships are sustainable over the long term is yet another challenge. As such, partnerships should be structured in a manner that will sustain public trust and in ways that reflect principles of universality and equality (Cooper 2024). If there is no commitment to long-term collaboration and shared benefits, the outcome may not be achieved in the partnership. It is worth noting that the inclusion of private entities in public services presents ethical issues, especially when it concerns making profits and exploitation. When sensitive data, like health information, is included in the partnerships, then ethical issues may arise with public mistrust (Barile et al. 2022). Ethical consideration in partnerships is important to avoid public distrust.

Implications

Public–Private Partnerships (PPPs) Public–Private Partnerships (PPPs) are a critical enabler of the transition to net-zero GHG emissions and effective climate action. All together, these changes indicate a fundamental reconfiguration of how we do climate action, beyond greater efficiency and effectiveness through market transactions. This type of investment enables the regular financing of sustainable moods and infrastructures. However, studies have shown that economic PPPs can help mobilize private capital, but it is also needed to monetize large projects which will offset the levels of greenhouse gases (Ayangbah 2024). These partnerships can spur the growth of innovative solutions that would not be viable through public funding (Khan et al. 2020). Also, the risk and know-how sharing that come along with the PPP's collaborative manner can help advance climate action further. This is a significant driver of environmentally sustainable economic growth, all the more so when investing in PPP and technological innovation (Khan et al. 2020). It leverages such synergies that, rather than lead to the development of any one renewable energy source, enhances and magnifies the overall rate of energy consumption and therefore carbon reduction (Pattberg 2010).

Conclusion

The role of Public–private partnership (PPP) is a crucial tool in net-zero emissions and climate action. Furthermore, with the spiralling fears of climate change, public entities and private stakeholders must join forces to achieve the objectives. This chapter covered different roles that PPPs can play in mobilizing resources, innovation and actual sustainable practices that contribute to climate resilience. It can, therefore, be complemented with public sector oversight and accountability toward the creation of robust infrastructures and services to make progress on climate goals and integrated with private sector efficiency and innovation. Strong PPPs can thus ease the transition to renewable sources of energy, to energy efficiency and to sustainable land-use practices, which all together would greatly dial down greenhouse gas emissions. This is backed up by a handful of different case studies demonstrating that aligning public and private stakeholders is several earned benefits, both environmentally and economically (Khan et al. 2022; Klaaßen and Steffen 2023) but there is an essential package of elements upon which success in PPPs around climate action depend.

A further issue to note in this regard is to address the challenge of current issues caused by misaligned incentives, risk-sharing mechanisms, and capacity building (Höhne et al. 2021; Mulyani 2021). From the literature, the things that are significantly responsible for the pattern of carbon emissions are technological innovations and PPP investments (Adebayo et al. 2021; Owusu-Manu et al. 2020). In this era, both the government and the private sector will have to create a partnership culture that transcends classical engagement models. Accelerating transformative changes in PPP may be possible through discussion that includes the full breadth of different stakeholders, as well as outreach that aims to mobilize all collective expertise and resources toward this end. Transparency and accountability will be key to rebuilding public trust around an equitable and effective climate action agenda. The public–private partnerships have an unmatched opportunity to catalyse the transition to a future net zero. Besides leveraging the best of both worlds, we shall strive for creative solutions that serve not just to mitigate climate change, but to speeden its social equity and sustainable developments, both of which have urgent imperatives. It will be all the more a path of the kind of complex game that needs to be played from a perspective of collaboration, investment, yet also environmental concern to create a world that is about resilience and sustainability.

References

Abdoul-Azize H, Salé A, Soulé M (2021) Private telecommunication companies and climate finance: evidence from the Niger Republic. Int J Environ Clim Change 374–385. https://doi.org/10.9734/ijecc/2021/v11i1230588

Adebayo T, Genç S, Castanho R, Kırıkkaleli D (2021) Do public-private partnership investments in energy and technological innovation matter for environmental sustainability in the East Asia and

Pacific region? An application of a frequency domain causality test. Sustainability 13(6):3039. https://doi.org/10.3390/su13063039

Agrawala S, Carraro M, Kingsmill N, Lanzi E, Mullan M, Prudent-Richard G (2011) Private sector engagement in adaptation to climate change: approaches to managing climate risks. https://doi.org/10.1787/5kg221jkf1g7-en

Alam MN, Turi JA, Bhuiyan AB, Kharusi SA, Oyenuga M, Zulkifli N, Iqbal J (2024) Factors influencing intention for reusing virtual reality (VR) at theme parks: the mediating role of visitors satisfaction. Cogent Soc Sci 10(1). https://doi.org/10.1080/23311886.2023.2298898

Ayangbah S (2024) Economic policies for climate change mitigation in the United States. Int J Res Publ Rev 5(2):1450–1456. https://doi.org/10.55248/gengpi.5.0224.0443

Barile D, Secundo G, Vecchio PD (2022) Food 4.0 for competing during the COVID-19 pandemic: experimenting digitalization in family firms. Eur J Innov Manag 27(4):1381–1402. https://doi.org/10.1108/ejim-07-2022-0373

Bland R, Overton M (2014) Assessing the contributions of collaborators in public-private partnerships. Am Rev Public Adm 46(4):418–435. https://doi.org/10.1177/0275074014559246

Buso M, Stenger A (2018) Public-private partnerships as a policy response to climate change. Energy Policy 119:487–494. https://doi.org/10.1016/j.enpol.2018.04.063

Cooper M (2024) Exploring technology adoption trends among entrepreneurial start-ups on the digital frontier. https://doi.org/10.21203/rs.3.rs-4094787/v1

Fankhauser S, Smith SM, Allen M, Axelsson K, Hale T, Hepburn C, Kendall JM, Khosla R, Lezaun J, Mitchell-Larson E, Obersteiner M, Wetzer T (2021) The meaning of net zero and how to get it right. Nat Clim Change 12(1):15–21. https://doi.org/10.1038/s41558-021-01245-w

Font-Cot F, Navarra PL, Serradell-López E (2023) Digital transformation policies to develop an effective startup ecosystem: the case of Barcelona. Transform Gov People, Process Policy 17(3):344–355. https://doi.org/10.1108/tg-01-2023-0006

Höhne N, Gidden MJ, den Elzen M, Hans F, Fyson C, Geiges A, Jeffery ML, Gonzales-Zuñiga S, Mooldijk S, Hare W, Rogelj J (2021) Wave of net zero emission targets opens window to meeting the Paris agreement. Nat Clim Chang 11(10):820–822. https://doi.org/10.1038/s41558-021-01142-2

Ismail S (2013) Critical success factors of public-private partnership (PPP) implementation in Malaysia. Asia-Pac J Bus Adm 5(1):6–19. https://doi.org/10.1108/17574321311304503

Janevski Z, Imeri B, Boshkoska-Klisaroski M (2023) Importance of digital transformation for business growth: looking over today's world business trends. Econ Dev 25(3):21–34. https://doi.org/10.55302/ed23253021j

Johnston L, Finegood D (2015) Cross-sector partnerships and public health: challenges and opportunities for addressing obesity and noncommunicable diseases through engagement with the private sector. Annu Rev Public Health 36(1):255–271. https://doi.org/10.1146/annurev-publhealth-031914-122802

Joshi A, Singh R, Kumar V, Rani S (2024) Net zero strategy: exploring opportunities, confronting challenges for developing countries. In: Singh R, Crowther D (eds) Transition towards a sustainable future. Approaches to global sustainability, markets, and governance. Springer, Singapore. https://doi.org/10.1007/978-981-97-5756-5_11

Khan Z, Muhsin A, Kırıkkaleli D, Wahab S, Jiao Z (2020) The impact of technological innovation and public-private partnership investment on sustainable environment in China: consumption-based carbon emissions analysis. Sustain Dev 28(5):1317–1330. https://doi.org/10.1002/sd.2086

Khan S, Akbar A, Nasim I, Hedvičáková M, Bashir F (2022) Green finance development and environmental sustainability: a panel data analysis. Front Environ Sci 10. https://doi.org/10.3389/fenvs.2022.1039705

Klaaßen L, Steffen B (2023) Meta-analysis on necessary investment shifts to reach net zero pathways in Europe. Nat Clim Change 13(1):58–66. https://doi.org/10.1038/s41558-022-01549-5

Kumar K, Singh R, Sharma A, Lehri BPS (2024) Climate change adaptation strategies for achieving net-zero economy. In: Approaches to global sustainability, markets, and governance, pp 163–183. https://doi.org/10.1007/978-981-97-5756-5_8

Lee W, Juskenaite I, Mwebaza R (2021) Public-private partnerships for climate technology transfer and innovation: lessons from the climate technology centre and network. Sustainability 13(6):3185. https://doi.org/10.3390/su13063185
Lokuge S, Sedera D, Cooper V, Burstein F (2021) Digital transformation: environmental friend or foe? Panel discussion at the Australasian conference on information systems 2019. Commun Assoc Inf Syst 48(1):616–634. https://doi.org/10.17705/1cais.04846
Mulyani S (2021) Critical success factors in public-private partnership. J Account Audit Bus 4(1):81–86. https://doi.org/10.24198/jaab.v4i1.31953
Nwokolo SC, Singh R, Khan S, Kumar A, Luthra S (2023a) Remedies to the challenges of renewable energy deployment in Africa. In: CSR, sustainability, ethics & governance, pp 59–74. https://doi.org/10.1007/978-3-031-44514-9_3
Nwokolo SC, Singh R, Khan S, Kumar A, Luthra S (2023b) Global investment and development in Africa. In: CSR, sustainability, ethics & governance, pp 15–58. https://doi.org/10.1007/978-3-031-44514-9_2
Nwokolo SC, Singh R, Khan S, Kumar A, Luthra S (2023c) Technological pathways to net-zero goals in Africa. In: CSR, sustainability, ethics & governance, pp 93–210. https://doi.org/10.1007/978-3-031-44514-9_5
Nwokolo SC, Singh R, Khan S, Kumar A (2024a) Key players in net zero technology. In: Elsevier eBooks, pp 193–215. https://doi.org/10.1016/b978-0-443-31486-5.00007-0
Nwokolo SC, Singh R, Khan S, Kumar A (2024b) Net zero technology and circular economy. In: Elsevier eBooks, pp 237–247. https://doi.org/10.1016/b978-0-443-31486-5.00009-4
Nwokolo SC, Singh R, Khan S, Kumar A (2024c) Technological pathways for Africa' s net-zero economy. In: Elsevier eBooks. https://doi.org/10.1016/c2023-0-52499-1
Omale SA, Oyenuga MO, Oriaku C (2021) Effect of organizational culture on the performance of hospitality industry in Covid-19 era. Abhigyan Manag J 39(3):24–32. https://doi.org/10.56401/abhigyan/39.3.2021.24-32
Owusu-Manu D, Adjei T, Sackey D, Edwards D, Hosseini R (2020) Mainstreaming sustainable development goals in Ghana's energy sector within the framework of public-private partnerships: challenges, opportunities and strategies. J Eng Des Technol 19(3):605–624. https://doi.org/10.1108/jedt-06-2020-0255
Oyenuga MO, Omale SA (2024) Is Africa jinxed? Exploring the challenges of technology access and adoption in Africa. Afr J Econ Sustain Dev 7(4):142–161. https://doi.org/10.52589/AJESDULN1LRNF
Oyenuga MO, Andah RA, Marcus GO, Agabi AU (2019) Effects of customer relationship management on product innovation in Nigeria. A case study of Xerox HS Nigeria Limited. Am J Theor Appl Bus 5(4):113–126. https://doi.org/10.11648/j.ajtab.20190504.15
Oyeyemi NOP, Kess-Momoh NAJ, Omotoye NGB, Bello NBG, Tula NST, Daraojimba NAI (2024) Entrepreneurship in the digital age: a comprehensive review of start-up success factors and technological impact. Int J Sci Res Arch 11(1):182–191. https://doi.org/10.30574/ijsra.2024.11.1.0030
Pattberg P (2010) Public-private partnerships in global climate governance. Wiley Interdiscip Rev Clim Change 1(2):279–287. https://doi.org/10.1002/wcc.38
Pye S, Broad O, Bataille C, Brockway P, Daly HE, Freeman R, Gambhir A, Geden O, Rogan F, Sanghvi S, Tomei J, Watson J (2020) Modelling net-zero emissions energy systems requires a change in approach. Clim Policy 21(2):222–231. https://doi.org/10.1080/14693062.2020.1824891
Singh R, Crowther D, Ravesangar K (2024a) Harnessing human capital for a sustainable future: HR's pivotal role toward net zero policies. In: Approaches to global sustainability, markets, and governance, pp 27–47. https://doi.org/10.1007/978-981-97-5756-5_2
Singh R, Gupta A, Kandpal V (2024b) ESG Excellence in the digital world. Circ Econ E-Bus Solut 193–204. https://doi.org/10.1002/9781394234035.ch10
Sreenivasan A, Suresh M (2023) Digital transformation in start-ups: a bibliometric analysis. Digit Transform Soc 2(3):276–292. https://doi.org/10.1108/dts-12-2022-0072

Westman L, Broto V (2018) Climate governance through partnerships: a study of 150 urban initiatives in China. Glob Environ Change 50:212–221. https://doi.org/10.1016/j.gloenvcha.2018.04.008

Xu X, Hou G, Wang J (2022) Research on digital transformation based on complex systems: visualization of knowledge maps and construction of a theoretical framework. Sustainability 14(5):2683. https://doi.org/10.3390/su14052683

Zhu X (2024) Information technology capability, digital transformation strategy and digital innovation performance: basis for digital transformation development framework. Int J Res Stud Manag 12(1). https://doi.org/10.5861/ijrsm.2024.1014

Chapter 16
Strengthening Legal and Institutional Frameworks for Net Zero Economy

Abstract A net-zero economy is the critical imperative that transitions to mitigating climate change and sustainable development. Strengthening legal and institutional frameworks is necessary, as they will form the basis of proper climate policy implementation and hold those involved in the process accountable. This paper discusses the building blocks of a strong legal framework statutory targets, regulatory mechanisms, and enforcing strategies needed to achieve net zero. In addition, it reflects on how institutions can enhance coordination for climate action mobilisation and public participation. It discusses the challenges that developing nations have faced while adopting international legal frameworks such as the Paris Agreement and, thereby calls for the need to implement tailored approaches that would cater to local contexts and capacities. In addition, the paper discusses the added value of integrating multidisciplinary perspectives to enhance the feasibility of net-zero transitions from the lessons learned from various sectors, energy, transportation, and industry. This analysis aims to provide a complete framework to policymakers and practitioners seeking the strengthening of legal and institutional capacities for a net-zero economy. Ultimately, the results affirm the need for mutual support between governments, private sectors, and civil society to bring about resilient and equitable systems supporting the transition to a sustainable future.

Introduction

The global transition to a net-zero economy is an urgent necessity (Singh et al. 2024): it has been mandated by the increasing impacts of climate change, alongside sustainable development. Effecting net-zero emission requires an adequate legal and institutional framework that isn't only to be established in terms of grandiose ambitions but also by its effective implementations with accountability; it is also essential in ensuring that national as well as subnational efforts follow the right tracks toward decarbonization, support economic growth with social equity, and gives foundations to climate actions by making obvious the obligations among governments, industries, and peoples. They define the boundaries in which climate policies are formulated

M. O. Oyenuga et al., *Smart Technologies for Climate Change and Net Zero Policies: Practical Approaches Towards Sustainability*, Climate Change Management,
https://doi.org/10.1007/978-3-031-92221-3_16

and implemented so that all parties work towards a common goal: reducing greenhouse gas emissions. Furthermore, strong legal systems enable climate considerations to be integrated into different sectors like energy, transportation, and agriculture, promoting a more comprehensive approach towards sustainability. These systems are complemented by institutional frameworks by aligning action at different levels of government and sectors. Good institutions can bring in the needful resources to create public involvement and meet legality aspects. They monitor the surveillance of progress through the enforcement of various regulations and adjust policies based on emerging challenges and scientific frontiers (Elliott et al. 2023). However, many developing countries face constraints in establishing and strengthening those frameworks, such as shortage of financial resources, restrictions in capacity, and high political instability (Costa 2012).

Building Legal Frameworks for Climate Action

The establishment of robust legal frameworks is essential for effective climate action and achieving net-zero emissions. Legal frameworks not only provide the necessary structure for implementing climate policies but also ensure accountability, transparency, and public participation in decision-making processes.

Key Elements of Effective Legal Frameworks

A few core elements should therefore be included in effective legal frameworks for climate action.

a. **Statutory Targets and Accountability**: The very basis of climate action is setting legally binding targets to reduce emissions. For example, the UK Climate Change Act establishes short-term and long-term targets as well as an independent advisory body monitoring progress and making recommendations (Averchenkova et al. 2020). Such frameworks improve accountability and ensure that governments maintain their commitment to climate goals.
b. **Adaptive Governance**: Legal frameworks need to be adaptive and changeable in response to the dynamic nature of climate change. This includes the principles of resilience theory and adaptive management, where laws could change according to new scientific evidence and policy experiments (Garmestani et al. 2013). In this way, the legal frameworks can respond more effectively to the uncertainties and complexities of climate change.
c. **Public Participation and Equity**: This calls for public involvement in making climate policies to ensure that policymakers engage with diverse perspectives for equitability.

Legal frameworks should ensure public participation in decision-making. For instance, the Aarhus Convention requires that the right to access information, public participation, and access to justice be implemented in environmental matters (Perlaviciute and Squintani 2020). This will increase the legitimacy of climate policies and contribute to community ownership and support.

The Role of Judiciary Systems in Climate Governance

The judiciary has been highly instrumental in climate governance for reasons of being the defender of the rule of law and the catalyst of accountability regarding climate change. The ever-increasing severity of climate change impacts renders the judiciary a more relevant agency in enforcing the environmental law of rights to have governments and companies held accountable in climate-related litigations.

Judiciaries form necessary bases for enforcing whatever rules have been put forward for environmental legal enactments. Courts should ensure that neither public nor private respondents violate rules enacted regarding environmental protection due to legal standards. The Urgenda Climate Case in the Netherlands and the Neubauer case in Germany are two landmark cases, showing how the judiciary can make governments act on climate change through a ruling mandating emission cuts and adherence to international climate commitments Kotzé (2021). This demonstrates how the judiciary can intervene to fill the gap left by the legislative branch if it fails to take a decisive move on climate governance. The judiciary is essential in protecting the rights of individuals and communities that are affected by climate change.

Courts can find a place where climate-related damages are violating the fundamental right to health, clean air, and a stable climate. This includes, as in the context of climate litigation, focusing much more on children and indigenous populations who are disproportionally affected by climate impacts (Fambasayi and Addaney 2021).

In addition, judicial systems enhance accountability and transparency in climate governance through public interest litigation. Public interest litigation can be filed by citizens, non-governmental organizations, and advocacy groups against governments and corporations for failing to live up to their climate responsibilities. Public interest litigation also makes entities accountable for their actions while raising the awareness of the public towards climate issues and adherence to environmental laws (Setzer and Vanhala 2019). This is an increasingly real situation in climate litigation that symbolizes recognition of the role of the judiciary to be a check on the executive and legislative powers on making climate commitments seriously. Indeed, the judiciary has to be able to tread through complex landscapes of climate change. They have to deal with issues of scientific uncertainties, changing climate science, and interplays of national and international legal frameworks. The ability of the judiciary to go through scientific evidence and expert testimonies is therefore crucial for determining climate cases, as Augustinis (2023) points out.

Besides, comparative judicial approaches between the jurisdictions have highlighted diverse strategies of climate litigation and the need for adaptive legal frameworks to address unique climate change challenges that differ from conventional cases (Xu 2018; Lange 2023).

Strengthening Institutions for Climate Monitoring and Enforcement

Climate governance effectiveness strongly depends on strong and robust monitoring and enforcement institutions for climate policies (Kumar et al. 2024). With the exacerbating impacts of climate change, it is also necessary to have well-functioning institutional frameworks that can enforce and implement climate action. Strong and adequate institutional frameworks provide adequate structure for action coordination involving cross-sectoral as well as intergovernmental levels with the integration of climate objectives into broader policy agendas (Ranabhat et al. 2018). Effective institutions facilitate stakeholders' collaboration through government agencies, civil society, as well as private sectors leading to an improved overall climate action capacity (Lu et al. 2021). Apart from that, institutions form the backbone of developing accountability and promoting transparency in climate governance. This is so because when defined institutions develop their different roles that assist in regulating them, the system will have an opportunity to monitor compliance based on climate change regulations and prosecute the defaulters. When the system is part of international agreements, sometimes, it assists in developing a feeling of whether countries have achieved their provisions agreed upon or not (Mohammad and Husted 2019).

Developing robust mechanisms of monitoring and reporting would be quite important for strengthened institutions of climate monitoring. Designed in a specific way, it should collect as well as make analyses regarding various data concerning the emission of greenhouse gases, climate-related impacts, or the effectiveness of mitigation and adaptation measures (Kim and Kim 2022; Nwokolo et al. 2023a, b, c, d, e). For instance, developing national climate finance institutions will mobilize and manage resources dedicated to climate action while giving an assurance of effective channelling of funds. Moreover, strengthening an institution's capacity to make technological provisions for monitoring purposes would go a long way in enhancing the accuracy and access to data (Sheriffdeen et al. 2020). Technologies such as remote sensing, satellite imagery, and data analytics can unveil critical environmental changes and the impacts of climate policies (Eurich 2023; Apata et al. 2024, Alam et al. 2024). Broadly, technology may play a role in enhancing institutions that support the implementation of compliance as well as monitor the implications of climate measures (Nwokolo et al. 2023e, 2024a). Appropriately adapted legal instruments may therefore establish a basis for monitoring and enforcement in that what needs to be done and what is to be undertaken in case it does not happen is codified legally (Averchenkova et al. 2020). The UK Climate Change Act stands as one of the best examples of statutory frameworks mandating legally binding limits on emission reductions and distinct methods for enforcing them (Averchenkova et al. 2020). This will also render legal frameworks to allow public participation in climate governance. That is, the interest groups affected by the decision possess a share in the decision-making and can hold the institutions accountable (McDonald 2011).

Such a participatory approach, apart from enhancing transparency, enhances ownership of communities, which is important for the success of the implementation of climate policy. Capacities for personnel who work in the area of climate governance also make a significant impact on strengthening climate monitoring and law enforcement. These capacities provide capability to the employees in terms of training programs or capacity building so they would do well in formulating and enacting policies into practice on matters related to the climate (Scotford et al. 2019). Besides, the institutions need to be climate change resilient (Bernstein and Hoffmann 2018). This means developing adaptive management strategies that would allow institutions to respond to emerging climate risks and uncertainties (Naeku 2020). Thus, governments will be in a position to ensure that climate governance remains effective under the changing environmental conditions if resilience is developed within the institutions.

Legal Challenges in Holding Industries Accountable

The accountability of industries to environmental degradation and climate change calls for robust legal frameworks and mechanisms of enforcement (Nwokolo et al. 2024b). Though the recognition of corporate responsibility in environmental matters is growing, several legal challenges impede the effective holding of industries accountable (Oyedele et al. 2023). The complexity of jurisdictional issues is one of the significant challenges in holding multinational enterprises accountable for environmental harm. This results in MNEs facing challenges in the application of uniform legal standards and parent company accountability since subsidiaries often operate in various countries. Zhang et al. (2023) argue that with the principle of separate legal personality, parent companies can avoid liability by arguing that their subsidiaries are independent entities. This legal loophole complicates the pursuit of accountability by having to navigate the diverse legal systems of different jurisdictions, each with its very own regulatory frameworks and mechanism of enforcement. Moreover, most countries do not have any all-inclusive legal frameworks that speak directly to corporate environmental responsibility. Existing laws may fail to define the scope of liability for environmental damage adequately or provide sufficient penalties for violations (Roy and Ghosh 2018). For example, in many countries, civil liability mechanisms of environmental harm have been legislated, but these could hardly be implemented or do not contain provisions to legally bring corporations accountable for climate change contributions (Downs 2024).

Lack of clear legal standards leads to inconsistent enforcement and a weaker deterrent against environmentally harmful activities. In addition, the environmental damage caused by corporate operations is often not directly linkable to the latter through a causal chain (Oyetunde et al. 2023). Environmental science is complex, and the causes of climate change are multifaceted, making it difficult for specific harms to be laid at the door of specified corporate actions (Li 2024). In cases like this, courts demand great scientific evidence to show causation, which can be hard to obtain and would lead to lengthy and expensive lawsuits. This is worsened by the fact that corporations often have more resources to fight against legal cases, thereby leaving the judicial process unevenly balanced. Another challenge is that of access to justice, a very critical issue in environmental litigation, especially for the affected communities and individuals. Most harmed by environmental damage in developing countries are deprived of the capital to pursue lawsuits against large corporations (Nyberg and Wright 2012). The law may also be slanted in favour of corporate interests, which may act as a hindrance to having people and communities pursue a course of redress. Unclear standards in the law will lead to inconsistent practice in its enforcement and a lack of deterrence towards practices that can harm the environment. In addition, the existence of a direct causal relationship between corporate activities and environmental damage is considered the biggest hurdle in lawsuits in most cases.

The scientific aspects of the environment are complex; climate change issues are multifactorial, hence difficult to specifically attribute harm towards certain corporate actions, according to Li (2024). The court demands heavy scientific proof that has been relatively hard to find so far. Mostly, it translates into long-drawn and very costly litigation cases. This is because corporations in many cases have more finances to invest in legal procedures, which may create judicial imbalance. Another challenge is access to justice, an issue that is very key in environmental litigation, especially among the affected communities and persons. Most victims of environmental harm, especially in the developing world, cannot afford legal action against powerful firms (Nyberg and Wright 2012). Additionally, legal systems may be biased in favour of corporate interests, further hindering the ability of individuals and communities to seek redress.

International Legal Frameworks and Their Applicability to Developing Nations

International legal frameworks therefore play an important role by setting up standards and obligations that should be met and followed in countries to abate the problem. Nonetheless, such frameworks tend to face immense challenges, both in applicability and effectiveness within developing nations.

Among all the significant international legal instruments to fight climate change is this one, known as the United Nations Framework Convention on Climate Change.

This organization was created in 1994 and gives binding obligations toward developed countries regarding a limitation of greenhouse gas emissions alongside facilitation to developing nations while fighting the effects of climatic change (Tari and Diah 2024). Such treaties as the Paris Agreement from 2015 stress once more global cooperation and framework conditions in which countries independently develop their NDC to mitigate global warming (Du and Zhang 2022). Other key agreements are the Kyoto Protocol—specifying concrete emission reduction targets among the developed countries, or the series of regional-level climate change impacts and adaptation techniques agreements. Collectively, these frameworks are observed to develop a coherent response to climate change. But, their effectiveness continues to remain highly varies from country to country.

For developing nations, the task of applying international legal frameworks may not be easy due to scarce resources, capacity, and development priorities. In international agreements, the principle of "common but differentiated responsibilities" is recognized in terms of the fact that developed countries should take a leading role in emissions reduction, which makes it very challenging for developing nations to meet their commitments without financial and technical support (Perkumienė et al. 2020). For instance, the Climate Change Act of Nigeria and other domestic measures are an indication of fulfilling the country's international agreement obligations. In such settings, however, such jurisprudence systems are not effective in practice because of weak enforcement and scarcity of funds for climate schemes. Developing countries also lack the institutional capacity to monitor compliance with international obligations, thereby having gaps in accountability and transparency (Oyenuga and Omale 2024).

Barriers to Effective Implementation

Several challenges deny international legal frameworks an efficient implementation in developing countries. Most developing countries do not have enough funds to implement effective climate action plans. Funding mechanisms exist internationally, for example, the Green Climate Fund, but the access process is complex and bureaucratic, which often delays key projects (Scotford and Minas 2018; Joshi et al. 2024). Additionally, developing countries, no thanks to COVID-19 (Omale et al. 2023) also might lack the technical understanding and institutional capability to construct and enforce effective climate policy. This might encompass issues that are as elementary as the collection of data, and monitoring and reporting requirements seen as being the bare-bone basics of international commitment (Setzer and Vanhala 2019). Political instability and governance can hinder the effective implementation of climate policies. In a few instances, competing interests and corruption might shift the focus and resources from climate actions (Naeku 2020). A major limitation of a legal framework is the poor public awareness of climate change issues. Climate governance cannot be implemented effectively without community and stakeholder engagement for support as well as the relevance of policies to the needs at the local level (Toussaint 2020).

However, there are prospects for international legal frameworks' improvement to be more applicable in developing countries. Domestic legal frameworks of developing countries can be made more robust to correspond with the obligations of international legal frameworks. Clear rules, enforcement mechanisms, and accountability mechanisms should be put in place (Scotford et al. 2019). International cooperation will support capacity building and technical support for developing countries. Developed nations will help the developing nations to build capacity to effectively implement climate policies through training and resource allocation (Ogunbode 2021). Public participation in climate governance will result in increased transparency and accountability. This is because the local communities are engaged in the decision-making process to ensure that their needs and priorities are represented in policies, thereby encouraging greater support for climate initiatives (Yiwen 2023).

Implications

A net-zero economy requires strengthening legal and institutional frameworks. Such strengthening has huge implications for the achievement of climate goals and sustainable development. Strong legal frameworks give a clear mandate for action. Binding targets and accountability mechanisms force governments and industries to reduce greenhouse gas emissions. This transparency further goes on to strengthen compliance while creating trust in climate policies by the public as they get to see a visible commitment in the stakeholders' book (Smith and Christie 2021; Nwokolo et al. 2023b). On the other hand, effective institutional frameworks have to facilitate coordination among all sectors and levels of government to ensure that climate action is integrated into broader economic and social policies. In terms of fostering cross-sector collaboration, these frameworks can draw on resources and expertise to promote more innovative and effective decarbonization solutions (Elliott et al. 2023). For instance, by taking into account climate in urban planning and transportation policies, net-zero strategies will be ensured to be effective overall, as presented in different case studies all over the world (Elliott et al. 2023). Strengthening the legal and institutional frameworks could be of great benefit to increase resilience against climate impacts, particularly for developing countries, as they are more susceptible to climate change impacts. Empowering the communities and their people raises their interests; hence, social equity and justice can be boosted during this transition into a net-zero economy (Booth 2021). It has become increasingly important to make the voices of the marginalized heard and to make sure their needs are reflected in the climate policies. Strong legal frameworks can be used to drive investment in green technologies and practices for a better, more sustainable future. Clear regulations and incentives will attract private sector investments and innovation and economic development in emerging industries: renewable energy, energy efficiency, and sustainable agriculture (Sakrabani et al. 2023; Nwokolo et al. 2023c, d). All this will contribute to fulfilling the net-zero goals by opening new job opportunities and improving energy security.

Conclusion

Key among these is building a net-zero economy, the need for strengthening legal and institutional frameworks in light of the acute challenges thrown up by climate change. Indeed, as countries seek to meet their commitment toward international agreements that they signed onto, solid legal frameworks can provide part of the scaffolding needed so accountability and transparency can be assured in implementing climate policy. This paper discussed the importance of statutory targets, regulatory mechanisms, and enforcement strategies in attaining net-zero emissions. The findings confirm that one-size-fits-all policies should be flexible and responsive to the specific situations and capabilities of countries, more particularly in developing states. Such countries are typically hindered by numerous challenges: they have small financial capacities, inadequate institutional ability, and political impediments towards the effective realization of international law. These barriers can be addressed through a collective effort to strengthen domestic legal frameworks, build institutional capacity, and promote public participation in climate governance. Integration of multiple perspectives, in different areas, will help integrate more comprehensive strategies in aid of the transition to net-zero economies. Experience at national and regional levels across different countries has proven that governmental and private sector cooperation or interaction with civil society are important in successful implementation mechanisms that lead to innovative approaches in promoting sustainability and equity. Thus, while the journey toward a net-zero economy seems tough, it provides opportunities for reform and innovation all the more. Strengthening the legal and institutional frameworks of countries can help build systems that are resilient but more effective in mitigating climate change while promoting development and social justice. It, therefore, focuses on collaborative effort and an adaptive legal framework that is capable of answering the challenges climate poses in all its new forms; it opens the way to a sustainable and fair future.

References

Alam MN, Turi JA, Bhuiyan AB, Kharusi SA, Oyenuga M, Zulkifli N, Iqbal J (2024) Factors influencing intention for reusing virtual reality (VR) at theme parks: the mediating role of visitors satisfaction. Cogent Soc Sci 10(1). https://doi.org/10.1080/23311886.2023.2298898

Apata SB, Oyenuga MO, Abeoba MI, Ugom MK, Abiodun AO (2024) Internet of Things (IoT) solutions for smart transportation infrastructure and fleet management. J Propuls Technol 45(4):1492–1505 https://propulsiontechjournal.com/index.php/journal/article/view/8325

Augustinis J (2023) Judicial approaches to science and the procedural legitimacy of climate rulings: comparative insights from the Netherlands and Germany. Eur Law J 29(3–6):378–392. https://doi.org/10.1111/eulj.12483

Averchenkova A, Fankhauser S, Finnegan J (2020) The impact of strategic climate legislation: evidence from expert interviews on the UK climate change act. Clim Policy 21(2):251–263. https://doi.org/10.1080/14693062.2020.1819190

Bernstein S, Hoffmann M (2018) The politics of decarbonization and the catalytic impact of subnational climate experiments. Policy Sci 51(2):189–211. https://doi.org/10.1007/s11077-018-9314-8

Booth R (2021) Pathways, targets and temporalities: analysing English agriculture's net zero futures. Environ Plan E Nat Space 6(1):617–637. https://doi.org/10.1177/25148486211064962

Costa L (2012) Sustainable development in emerging markets & CSR codes of conduct: oil and gas industry in Brazil. J Oper Supply Chain Manag 44–66. https://doi.org/10.12660/joscmv1n1p44-66

Downs S (2024) Civil liability for climate change? The proposed tort in Smith v Fonterra with reference to France and the Netherlands. Rev Eur Comp & Int Environ Law 33(1):31–44. https://doi.org/10.1111/reel.12532

Du H, Zhang H (2022) Climate neutrality in the EU and China: an analysis of the stringency of targets and the adaptiveness of the relevant legal frameworks. Rev Eur Comp & Int Environ Law 31(3):495–509. https://doi.org/10.1111/reel.12453

Elliott C, Schumer C, Gasper R, Ross K, Singh N (2023) Realizing net-zero emissions: good practices in countries. WRIPUB. https://doi.org/10.46830/wriwp.22.00134

Eurich J (2023) Diverse pathways for climate resilience in marine fishery systems. Fish Fish 25(1):38–59. https://doi.org/10.1111/faf.12790

Fambasayi R, Addaney M (2021) Cascading impacts of climate change and the rights of children in Africa: a reflection on the principle of intergenerational equity. Afr Hum Rights Law J 21(1). https://doi.org/10.17159/1996-2096/2021/v21n1a3

Garmestani A, Allen C, Benson M (2013) Can law foster social-ecological resilience? Ecol Soc 18(2). https://doi.org/10.5751/es-05927-180237

Joshi A, Singh R, Kumar V, Rani S (2024) Net zero strategy: exploring opportunities, confronting challenges for developing countries. In: Singh R, Crowther D (eds) Transition towards a sustainable future. Approaches to global sustainability, markets, and governance. Springer, Singapore. https://doi.org/10.1007/978-981-97-5756-5_11

Kim J, Kim S (2022) International agreements and global initiatives for low-carbon cooling. https://doi.org/10.56506/rpae4386

Kotzé L (2021) Neubauer et al. versus Germany: planetary climate litigation for the Anthropocene? Ger Law J 22(8):1423–1444. https://doi.org/10.1017/glj.2021.87

Kumar K, Singh R, Sharma A, Lehri BPS (2024) Climate change adaptation strategies for achieving net-zero economy. In: Approaches to global sustainability, markets, and governance, pp 163–183. https://doi.org/10.1007/978-981-97-5756-5_8

Lange D (2023) What climate litigation reveals about judicial competence. Eur Law J 29(3–6):312–325. https://doi.org/10.1111/eulj.12492

Li Z (2024) Corporate environmental infringement, legal regulation, and sustainable development: punitive damages as a perspective. Front Environ Sci 12. https://doi.org/10.3389/fenvs.2024.1325375

Lu X, Zhu E, Campbell L, Hafner M, Noussan M, Raimondi P (2021) Comparison between China, the EU and US climate and energy governance: how policies are made and implemented at different levels. SSRN Electron J. https://doi.org/10.2139/ssrn.4017427

McDonald J (2011) The role of law in adapting to climate change. Wiley Interdiscip Rev Clim Change 2(2):283–295. https://doi.org/10.1002/wcc.96

Mohammad S, Husted B (2019) Law-abiding organizational climates in developing countries: the role of institutional factors and socially responsible organizational practices. Bus Ethics Eur Rev 28(4):476–493. https://doi.org/10.1111/beer.12228

Naeku M (2020) Climate change governance: an analysis of the climate change legal regime in Kenya. Environ Law Rev 22(3):170–183. https://doi.org/10.1177/1461452920958398

Nwokolo SC, Singh R, Khan S, Kumar A, Luthra S (2023a) Remedies to the challenges of renewable energy deployment in Africa. In: CSR, sustainability, ethics & governance, pp 59–74. https://doi.org/10.1007/978-3-031-44514-9_3

Nwokolo SC, Singh R, Khan S, Kumar A, Luthra S (2023b) Decarbonizing hard-to-abate sectors in Africa. In: CSR, sustainability, ethics & governance, pp 211–236. https://doi.org/10.1007/978-3-031-44514-9_6

Nwokolo SC, Singh R, Khan S, Kumar A, Luthra S (2023c) Global investment and development in Africa. In: CSR, sustainability, ethics & governance, pp 15–58. https://doi.org/10.1007/978-3-031-44514-9_2

Nwokolo SC, Singh R, Khan S, Kumar A, Luthra S (2023d) Africa's awakening to climate action. In: CSR, sustainability, ethics & governance, pp 299–310. https://doi.org/10.1007/978-3-031-44514-9_9

Nwokolo SC, Singh R, Khan S, Kumar A, Luthra S (2023e) Technological pathways to net-zero goals in Africa. In: CSR, sustainability, ethics & governance, pp 93–210. https://doi.org/10.1007/978-3-031-44514-9_5

Nwokolo SC, Singh R, Khan S, Kumar A (2024a) Threats to the rapidity of sustainability transitions posed by technological changes. In: Elsevier eBooks, pp 39–75. https://doi.org/10.1016/b978-0-443-31486-5.00002-1

Nwokolo SC, Singh R, Khan S, Kumar A (2024b) Key players in net zero technology. In: Elsevier eBooks, pp 193–215. https://doi.org/10.1016/b978-0-443-31486-5.00007-0

Nyberg D, Wright C (2012) Justifying business responses to climate change: discursive strategies of similarity and difference. Environ Plan Econ Space 44(8):1819–1835. https://doi.org/10.1068/a44565

Ogunbode T (2021) Climate change scenario in Nigeria: local perceptions and the way forward. Int J Hydrol 5(2):84–85. https://doi.org/10.15406/ijh.2021.05.00269

Omale SA, Oyenuga M, Gurin IM (2023) Effects of COVID-19 pandemic on organizational transformation and employees' performance. J Manag Econ Ind Organ 7(1):16–33. https://doi.org/10.31039/jomeino.2023.7.1.2

Oyedele MO, Marcus GO, Ahungwa AI (2023) Do consumers care about green marketing practices? Insight from a developing nation. Bp Int Res CritS Inst J (BIRCIJournal) 6(3):1424–1436. https://doi.org/10.33258/birci.v6i3.7665

Oyenuga MO, Omale SA (2024) Is Africa jinxed? Exploring the challenges of technology access and adoption in Africa. Afr J Econ Sustain Dev 7(4):142–161. https://doi.org/10.52589/AJESDULN1LRNF

Oyetunde TO, Oyenuga MO, Adoga GJ (2023) Green management practices and organisational efficiency in higher institution. Bp Int Res Crit Linguist Educ (BirLE) J 6(3):276–290. https://bircu-journal.com/index.php/birle/article/view/7716

Perkumienė D, Pranskūnienė R, Vienažindienė M, Grigienė J (2020) The right to a clean environment: considering green logistics and sustainable tourism. Int J Environ Res Public Health 17(9):3254. https://doi.org/10.3390/ijerph17093254

Perlaviciute G, Squintani L (2020) Public participation in climate policy-making: toward reconciling public preferences and legal frameworks. One Earth 2(4):341–348. https://doi.org/10.1016/j.oneear.2020.03.009

Ranabhat S, Ghate R, Bhatta L, Agrawal N, Tankha S (2018) Policy coherence and interplay between climate change adaptation policies and the forestry sector in Nepal. Environ Manage 61(6):968–980. https://doi.org/10.1007/s00267-018-1027-4

Roy A, Ghosh S (2018) Determinants of corporate environmental disclosure from an Asian perspective. Iim Kozhikode Soc & Manag Rev 8(2):171–189. https://doi.org/10.1177/2277975218784796

Sakrabani R, Garnett K, Knox JW, Rickson J, Pawlett M, Falagan N, Girkin NT, Cain M, Alamar MC, Burgess PJ, Harris J, Simmons R (2023) Towards net zero in agriculture: future challenges and opportunities for arable, livestock and protected cropping systems in the UK. Outlook Agric 52(2):116–125. https://doi.org/10.1177/00307270231178889

Scotford E, Peeters M, Vos E (2019) Is legal adjudication essential for enforcing ambitious climate change policies? pp 191–206. https://doi.org/10.4324/9780429446252-14

Scotford E, Minas S (2018) Probing the hidden depths of climate law: analysing national climate change legislation. Rev Eur Comp & Int Environ Law 28(1):67–81. https://doi.org/10.1111/reel.12259

Setzer J, Vanhala L (2019) Climate change litigation: a review of research on courts and litigants in climate governance. Wiley Interdiscip Rev Clim Change 10(3). https://doi.org/10.1002/wcc.580

Sheriffdeen M, Nurrochmat D, Perdinan P, Gregorio M (2020) Indicators to evaluate the institutional effectiveness of national climate financing mechanisms. For Soc 358–378. https://doi.org/10.24259/fs.v4i2.10309

Singh R, Crowther D, Ravesangar K (2024) Harnessing human capital for a sustainable future: HR's pivotal role toward net zero policies. In: Approaches to global sustainability, markets, and governance, pp 27–47. https://doi.org/10.1007/978-981-97-5756-5_2

Smith S, Christie I (2021) Knowledge integration in the politics and policy of rapid transitions to net zero carbon: a typology and mapping method for climate actors in the UK. Sustainability 13(2):662. https://doi.org/10.3390/su13020662

Tari VA, Diah EC (2024) Challenges and prospects of the legal frameworks for combatting climate change in Nigeria. AKSU J Adm Corp Gov 4(1):109–121. https://doi.org/10.61090/aksujacog.2024.009

Toussaint P (2020) Loss and damage and climate litigation: the case for greater interlinkage. Rev Eur Comp & Int Environ Law 30(1):16–33. https://doi.org/10.1111/reel.12335

Xu H (2018) Legal and policy pathways of climate change adaptation: a comparative analysis of the adaptation practices in the United States, Australia and China. Transnatl Environ Law 7(2):347–373. https://doi.org/10.1017/s2047102518000092

Yiwen X (2023) A comparative analysis of the Urgenda climate case and climate case Ireland. Sci Law J 2(12). https://doi.org/10.23977/law.2023.021218

Zhang Y, Zhao H, Zhu M (2023) Multinational enterprises and environmental degradation–what makes multinational enterprises hard to be held accountable and solutions. Highlights Bus Econ Manag 7:127–134. https://doi.org/10.54097/hbem.v7i.6933

Part IV
Barriers and Opportunities in Net-Zero Implementation

Chapter 17
Barriers to Technology Adoption in Developing Countries

Abstract In developing countries, adopting technology is constrained by many barriers that block progress and benefit from technological advancement. There are many hurdles these countries face in adopting new technologies, from energy grids and internet access to cost and capacity. Barriers relate to high technology acquisition costs, low educational systems leading to low technology literacy, and institutional weakness leading to ineffective policymakers. The digital divide further compounds these issues, leading to inequality in technology access among various socio-economic groups. This chapter addresses these barriers in the context of technology adoption in developing countries, contributing to the literature that aims to define pathways to overcome the major obstacles to technology that promotes sustainable development.

Introduction

While the prospects of technology adoption lead to economic growth, many developing nations, are subjected to technology barriers that do not allow them to fully tap into it (Nwokolo et al. 2023a). Hence, recognizing these barriers is of paramount importance for policymakers, stakeholders, and researchers striving to enhance technology adoption and foster sustainable development in these regions. The high cost of acquiring and implementing new technologies is one of the main barriers to technology adoption in developing countries. As stressed by Ghansah et al. (2021), the financial barrier that smart building technologies impose within the context of Ghana; the economic challenges present a relevant point in how the lack of financial access inhibits exploration of more optimized building technologies and ultimately holds back fruitful possibilities. Furthermore, what is equally mentioned by Ligon et al. (2019) is that this is also due to the unavailability of credit, meaning that businesses and individuals cannot get financing easily to invest in new technologies.

The major underlying factor for the low adoption of technology other than economic considerations is the lack of infrastructure (Oyenuga and Omale 2024;

M. 0. Oyenuga et al., *Smart Technologies for Climate Change and Net Zero Policies: Practical Approaches Towards Sustainability*, Climate Change Management,
https://doi.org/10.1007/978-3-031-92221-3_17

Nwokolo et al. 2023b). Many developing nations suffer from inadequate infrastructure, unreliable electricity supply, little or no access to the internet, and poor transportation networks. Such infrastructural constraints impede the successful deployment and use of technology. (Zhou et al. 2022) discussed how access to technology impacted users' engagement, especially for older adults. Such communities are largely disconnected from the benefits of technology as they lack a solid infrastructure. Additionally, sociocultural influences drive the acceptance of new technology. New technologies face resistance, fear of change, and fear of the unknown. For instance, evidence from MOOCs in developing nations shows that when the technologies are perceived as being foreign impositions, there is a reluctance to adopt them (Ma and Lee 2018; Oyedele and Iember 2021).

Moreover, the digital divide has aggravated these difficulties, given that unequal access to technology commonly reflects socioeconomic standing and thus further consolidates facets of disadvantage (Sujarwoto 2023). Technology adoption is also heavily impacted by the education system. New technologies can be ineffective if the workforce lacks the technological literacy and the skills to use the technology, hence the need to work with governments to help empower citizens by training them in technology (Oyedele et al. 2024; Nwokolo et al. 2024a, 2023c; Singh et al. 2024a, b). Fuentes and Mies (2020) claimed that a country's absorptive capacity, influenced by its educational quality and accessibility, is critical for effective technology adoption. The potential gains will go unrealised without a workforce that is sufficiently educated to use and get the most out of new technologies; by using and working with computers/automated machinery. These barriers to technology adoption in developing countries are multifaceted and shaped by economic, infrastructural, sociocultural and educational factors. Confronting these challenges demands a holistic approach that includes engagement from all sectors: governments, businesses, and society.

Infrastructure Challenges: Energy Grids and Internet Access

Strong infrastructure in developing countries is an important base for economic development and societal progress. Two of the most important infrastructure challenges are energy grids and internet access, both of which are critical to powering modern economic activity and increasing quality of life.

Energy Grids

Energy grids are the foundation of an energy supply for a given country, enabling it to generate, transmit and distribute its electricity. However, there are huge obstacles to building and running optimal energy grids in many developing countries.

One of the main problems is the failed investment in elements of energy infrastructure that cause constant power outages and supply of electricity based on high failure rates. According to Seidu et al. (2022), with its potential to improve the performance of energy infrastructures, disruptive technologies are often disrupted by inadequate funding and policy inconsistencies. Besides, energy sectors in most developing countries depend on non-renewable exploitable energy resources, which leads to environmental degradation and energy insecurity. Infrastructural deficits and regulatory uncertainties challenge the transition into renewable energy sources (Adepoju 2023).

The implementation of Industry 4.0 technologies like the Internet of Things (IoT) and artificial intelligence has the potential to improve energy generation and consumption efficiency (Apata et al. 2024; Nwokolo et al. 2024b). But these advances will be untapped unless there is a strong energy infrastructure in place. The consequences of feeble energy grids aren't limited to economic productivity; they're also a scourge to health and education. In the first instance, irregular electricity disrupts health delivery as well as educational access, particularly in rural areas, which have insufficient energy access. The erratic electricity supply can make it impossible for a clinic to stay open and a school to run, keeping people trapped in poverty and underdevelopment (Joshi et al. 2024).

Internet Access

Bridging the digital divide alongside energy infrastructure, internet access is a vital facet of contemporary development. In the world today, in many developing nations, the internet is still an unaffordable luxury due to infrastructural roadblocks, high costs and socio-economic factors, creating a noticeable digital divide. According to a review by Clark and Gómez (2011), only 26% of the population in the world have internet access at home, with internet users mostly living in developed countries. These numbers in this case show a clear and present need for targeted interventions to improve internet connectivity in these unserved areas.

Limited internet access has far-reaching implications for multiple sectors, from education and health care to economic development (Nwokolo et al. 2024c). Another good example can be seen in the COVID-19 pandemic where infrastructure and high internet costs made online learning a non-real scenario for many students in developing countries (Baticulon et al. 2020). This is why developing nations need to invest in digital infrastructure so that everyone has equal access to education and information (Oyenuga et al. 2024). Even more so, this vacuum can choke innovations and entrepreneurs. In most developing countries, SMEs have limited access to the internet, to global or even regional markets to achieve economies of scale and to use digital tools to increase their efficiency. The inclusion of ICT into economic activities (Nduji et al. 2023) contributes to the quality of growth as well as to help bridge gaps in the region.

High Upfront Costs of Green Technologies

Green technologies are critical to sustainable development and climate action (Singh 2013). Still, high upfront costs for these types of technologies are some of the biggest barriers to their implementation. This financial barrier is especially acute in developing nations with scarce financial resources and competing economic needs that prevent them from investing in green solutions.

Financial Barriers to Green Technology Adoption

The upfront cost of green technologies can be high, which discourages businesses, particularly small and medium-sized enterprises (SMEs), from transitioning. Reynolds (2024) further cites the high implementation costs of green supply chain practices as a major cause for concern among industry leaders. The long payback periods of most green investments add to the financial burden and make it difficult for most companies to justify economically such expenditures in a short-term oriented world. In developing economies, however, access to financing options is usually limited and the risk of investing in unproven technologies can put potential adopters off completely. Yangle (2023) reports that for green buildings, the high cost of investment in sustainable construction practices discourages many property developers from pursuing them. The long-term advantages of green buildings, the reduction of operational costs and increased energy efficiency, among others, are widely cited in literature but come with an upfront cost that can be a barrier.

Similarly, Marotta et al. (2023) highlight that the additional expenses of developing and constructing environmentally-friendly buildings can be compensated through operational savings, yet this initial reluctance to invest still exists, especially in less developed areas.

Renewable Energy Technologies

The high upfront costs stifle the use of renewable energy technologies in particular. According to Usman (2024), high upfront capital investment remains one of the most significant obstacles to the adoption of renewable energy solutions in the manufacturing sector. Even though renewable energy has the long-term potential of drastically lowering costs and also helping the environment, the upfront cost of implementation and infrastructure is often too much. Štreimikienė et al. (2022) identified that high upfront costs and long payback periods are among the most typical barriers of renewable energy uptake by households in Lithuania. Moreover, Kiprop et al. (2019) emphasize that the large cost of renewable energy equipment is still one of the major barriers to investment in these technologies. This presents a significant challenge in developing countries as many households and businesses do not have the financial support to invest in renewable energy solutions that could lower energy costs and carbon footprints in the long run (Oyetunde et al. 2023).

Addressing the Financial Barrier

Targeted policy interventions can help to address financial barriers linked to large initial investments. The policy framers must therefore focus on providing renewable energy incentives, which include tax credits, subsidies, and low-interest loans (Kamau 2024). Such measures can help ease costs which will be beneficial from a business point of view as well as make green technologies more affordable for households. New financing models such as community solar projects and power purchase agreements also can help broaden participation in renewable energy projects without needing substantial initial investment. Moreover, education and awareness campaigns about the sustainable advantage offered green unfold relatively the cost of deployment of the related technologies as emphasized by Khalid et al. (2021) but should focus on increasing awareness about renewable energy benefits, as this is a key driver for adoption (Oyedele et al. 2023).

Technical Skills Gap and Capacity Limitations

The technical skills gap is the gap between the skills needed to adopt technology effectively, and the skills the workforce currently possesses. Many organizations in developing countries, where skilled workers trained in the latest technologies are in short supply, struggle to implement and leverage such technologies. Kyobe (2011) surmises that the ability to adopt and use information and communication technology (ICT) is often hindered by a lack of exposure and training, meaning that less effective deployment of technology is a main factor (Omale et al. 2024). And for certain sectors like agriculture, there is a clear gap between the potential of innovative practice and the actual practice of innovation critically driving productivity and sustainability.

With new technologies springing up more quickly than educational institutions can keep pace, the skills gap continues to widen as universities and colleges face pressure to revise curricula and training programs to match an accelerating rate of change. Insufficient knowledge and skills relating to new technologies are making it harder for farmers to adopt agricultural technologies (Awuor 2023). Therefore, there is a need for continuous training and capacity building that would produce a skilled workforce that will be competitive in an advanced economy.

Capacity Limitations in Organizations

Apart from individual skills, organisational capability is essential for technology adoption. According to Gitonga (2018) the ability of an organization to deploy turnaround strategies and adopt new technologies greatly determines its competitiveness. Struggling to keep up with rapid advancements in new technologies, organizations with limited capacity may find themselves unable to integrate these technologies into existing models, thus fostering inefficiencies and reducing their competitiveness in the market (Alam et al. 2024). Small and medium-sized enterprises (SMEs) are especially affected by this since they often do not have the capacity or know-how to invest in advanced technologies to the same level that larger companies can (Sendstad et al. 2023).

Addressing the technical skills gap and capacity limitations requires a concerted effort to enhance education and training programs. Wulandari and Djufry (2021) and Oyenuga et al. (2023) note that providing targeted training and capacity-building initiatives is critical for empowering individuals and organizations to adopt new technologies effectively. This approach should involve collaboration between educational institutions, industry stakeholders, and government agencies to ensure that training programs align with the labour market's needs (Nwokolo et al. 2024d). Moreover, fostering a culture (Omale et al. 2021) of continuous learning and professional development is essential for keeping pace with technological advancements. As highlighted by Meena and Singh, enhancing the knowledge base of scientists and practitioners is crucial for addressing the challenges associated with technology adoption in agriculture.

Political and Bureaucratic Roadblocks

Political and bureaucratic impediments act as considerable hindrances to governance and these roadblocks get in the way of any expedient policy decision-making, but more so in the context of sustainable development and the adoption of innovative technologies. Such obstacles can include inadequate regulatory systems, insufficient political will, or entrenched bureaucratic practices that are all too familiar in our country.

Nature of Political Roadblocks

Another driver of political roadblocks is a lack of commitment from political leaders to make sustainable development initiatives a priority. As Gadekar et al. (2023) have noted, innovative practices especially lean-green and Industry 4.0 practices that are critical for manufacturing can be hindered by political inertia, which realizes considerable cost savings due to a lack of thorough output of control. If people do not unite and advocate fiercely for these initiatives then there will not be enough funding, policies or business incentives in place to tackle the issues we are facing. Political connections also play a role in retaining resources and prioritizing development projects. Zhang et al. (2019) show that firms with strong political ties are more likely to produce green technology innovation, noting that political connections facilitate uneven access to resources and opportunities. As a result of this dynamic, some politically connected organizations could become monopolistic in the benefits that they accrued, and other organizations might find it more difficult to access the support they need for sustainable activities.

Bureaucratic Challenges

Bureaucratic roadblocks are generally defined as inflexible administrative procedures, fragmentation of government agencies, and inertia to change public infrastructure. Yolles (2019) stresses that the success of policy implementation is heavily reliant on existing bureaucratic structures. Moreover, the wasteful bureaucratic process tends to cause a slowdown of decisions, higher expenses and finally the collapse of good leaning policies. Many developing countries suffer from bureaucratic inefficiencies caused by poor training and a lack of manpower. As observed by Mosly (2015) the lack of skilled personnel in the public sector may not provide an adequate approach to green building as there may be a bureaucracy without specific knowledge of the regulatory issues. As a consequence of that skills gap, resources might be mismanaged and policies poorly executed. Moreover, the interaction of political and bureaucratic obstacles can form a vicious circle of maladministration. There might be insufficient bureaucratic support behind political leaders to pursue their agenda to the limits, just as bureaucrats might sometimes believe political directions are ill-advised or do not fit bureaucratic operational realities. Such implication may lead to an aversion to accountability and transparency, thereby eroding public trust in government institutions.

Overcoming Political and Bureaucratic Roadblocks

The political and bureaucratic obstacles can only be subdivided and addressed holistically. Positive political commitment to sustainable development initiatives, among others, is of first-order importance. Advocacy campaigns emphasizing the long-term economic growth and social well-being costs of sustainability will help make this a reality. Collaborating with stakeholders, including those in civil society and the private sector, can create a coalition of support for sustainable policies. Innovation is critical to policy implementation because improving bureaucratic efficiency is a prerequisite for policy implementation. Improving the efficiency of being able to manage the services provided is an important function necessary to be able to focus resources on the mission of the work at hand to be done. Providing the necessary skills and knowledge helps bureaucrats to be more equipped to address changing challenges, which enhances policy implementation at the hands of governments. Lastly, to regain public trust, improved transparency and accountability mechanisms are required in the political and bureaucratic domains. Consultation with these stakeholders early on serves to increase policy effectiveness and create a sense of ownership by stakeholders.

Overcoming Barriers Through Innovation and Policy

Several barriers make it difficult for developing countries to adopt technology: economic barriers, lack of infrastructure, and insufficient technical skills. But with innovation and better policy frameworks, nations can leverage technology to overcome such negatives and create solutions for them to utilise the benefits of ingenuity towards sustainable development.

The Role of Innovation

Innovation plays a decisive role in using technology when traditional methods have lost their steam. According to Hwabamungu and Williams (2010), there is a range of innovative technologies that have the potential to drastically improve healthcare delivery, especially in resource-constrained settings, with one of these technologies being mobile health (m-Health) applications. Innovation is highly contextual, which means that countries are better suited to develop localised solutions that respond to domestic needs and exigencies, making it all the more likely that technology can be adapted to benefit an entire population. Carayannis et al. (2021) affirm that this peril has aroused the attention of practitioners and policy-makers and that of people outside the field, tapping into their collective efforts to support this cause, and highlights the necessity of collaboration between various sectors, including state, academic, private and civil society. It also enables the co-creation of a knowledge base and resources that lead to better solutions, through the inclusion of important stakeholders in the process of designing technology.

Policy Interventions

Such an enabling environment would be best guided by good policy frameworks that can lead to technology adoption. Developing countries require specific strategies to navigate these trails, and that is something they can not do alone. Innovation policy has to develop beyond what Mingjie and Xiong (2020) termed a systematic transformation in response to socio-economic challenges arising from technological advancements. This may involve their sets of regulation jobs, financial incentives or research and development initiatives. Another example is Mandala (2024) which showcases how technology can improve efficiency and transparency in public services. Digital technologies can improve efficiency, effectiveness, and transparency, and reduce corruption, in public procurement processes; increasing trust and further encouraging the uptake of new forms of technology. Policies like these strengthen governance systems but also create an enabling environment for private sector firms to invest in technology.

Addressing the Skills Gap

According to Sikder (2024), human capital needs to be improved through education and training programs to allow people and organizations to adopt disruptive technologies into actions within organizations. In this regard, both governments and educational institutions should collaborate to develop curricula that would enable students to fulfil the labour market demand and provide relevant skills to the workforce to stay up-to-date with the demands of a technology-led economy.

Also developing a culture of continuous learning and professional development can help close the skills gap. In doing so, countries will help prepare their citizens for the new world and innovations required to combat and resolve the various changes that will arise. Not only does this approach build up individual abilities, but it also improves the economy as a whole.

Implications

These obstacles to technology adoption in developing nations create major issues, leading to decreased economic growth, hampered social advancement, and stalled progress in general. Understanding these impediments is important for policymakers, businesses, and stakeholders who intend to ensure a climate for technological progress. These barriers have varied ramifications across sectors such as agriculture, healthcare, and manufacturing and require sector-specific interventions to address them effectively.

In developing countries, this stifles economic growth due to the inability to adopt modern technologies. As noted by Javaid et al. (2022), the lack of infrastructure and the high cost of technology restrain the way businesses innovate and compete in the international market. Barriers to the adoption of technology also have deep consequences for social development. This is formalized in the health sector well by Petersen et al. (2019) who demonstrated how access to digital health technologies impacts the quality of care delivered. When paired with the existing inequalities in health and care access, this unmet need can further widen health gaps and impede patients from receiving appropriate care from providers. To overcome environmental challenges, we need to take advantage of sustainable technologies. However, as Ghansah et al. As (2021) noted, the high costs of smart building technologies can discourage investment in sustainability. Green technologies will be the only way for developing countries to uphold their commitments to the environment and to combat climate change. Any barriers to technology adoption stifle innovation and drive down competitiveness. As Raj et al. (2020) noted without structured national strategies, organizations may end up with inconsistent adoption of technology at the organizational level. Such an ad hoc approach may cause the country to fail to develop a strong and connected innovation ecosystem that builds collaboration and knowledge.

Conclusion

Overcoming these barriers to technology adoption in developing countries is an essential catalyst for unlocking their potential for economic growth, social development, and environmental sustainability. Such countries will only gain from comprehensive foreign policies adapted to their interests. Some concrete ways of doing this would include investment in infrastructure, offering financial incentives, and encouraging innovative practice through education and training. Developing an enabling environment for technology uptake also requires partnerships across governments, private sector actors and civil society. Understanding the positive aspects found in the existing systems and through cross-type collaboration among non-homogeneous actors will be significant to making robust systems the final clients as well as the beneficiaries of transformation across economies and societies using technology. So in short, solving the barriers to technology adoption is not just about improving access to technology; it is about creating an ecosystem for innovation, inclusivity, and sustainability. However, by addressing these obstacles, they can pave the way for success in a digital, interconnected world.

References

Adepoju O (2023) Integrating human-computer interactions in the Nigerian energy system: a skills requirement analysis. J Digit Food Energy & Water Syst 4(2). https://doi.org/10.36615/digital_food_energy_water_systems.v4i2.2933

Alam MN, Turi JA, Bhuiyan AB, Kharusi SA, Oyenuga M, Zulkifli N, Iqbal J (2024) Factors influencing intention for reusing virtual reality (VR) at theme parks: the mediating role of visitors satisfaction. Cogent Soc Sci 10(1). https://doi.org/10.1080/23311886.2023.2298898

Apata SB, Oyenuga MO, Abeoba MI, Ugom MK, Abiodun AO (2024) Internet of Things (IoT) solutions for smart transportation infrastructure and fleet management. J Propuls Technol 45(4):1492–1505. https://propulsiontechjournal.com/index.php/journal/article/view/8325

Awuor F (2023) Leveraging mobile phones in crop pests and disease surveillance-perspectives of farmers in Kenya. Arch Agri Res Technol 4(2):1–8. https://doi.org/10.54026/aart/1051

Baticulon RE, Sy JJ, Alberto NRI, Baron MBC, Mabulay REC, Rizada LGT, Tiu CJS, Clarion CA, Reyes JCB (2020) Barriers to online learning in the time of COVID-19: a national survey of medical students in the Philippines. https://doi.org/10.1101/2020.07.16.20155747

Carayannis E, Dezi L, Gregori G, Calò E (2021) Smart environments and techno-centric and human-centric innovations for industry and society 5.0: a quintuple helix innovation system view towards smart, sustainable, and inclusive solutions. J Knowl Econ 13(2):926–955. https://doi.org/10.1007/s13132-021-00763-4

Clark M, Gómez R (2011) The negligible role of fees as a barrier to public access computing in developing countries. Electron J Inf Syst Dev Ctries 46(1):1–14. https://doi.org/10.1002/j.1681-4835.2011.tb00323.x

Fuentes J, Mies V (2020) Technological absorptive capacity and development stage: disentangling barriers to riches. Macroecon Dyn 25(6):1589–1624. https://doi.org/10.1017/s1365100519000804

Gadekar R, Sarkar B, Gadekar A (2023) Analysis and evaluation of roadblocks hindering lean-green and Industry 4.0 practices in Indian manufacturing industries. Int J Decis Support Syst Technol 15(1):1–36. https://doi.org/10.4018/ijdsst.325350

Ghansah F, Owusu-Manu D, Ayarkwa J, Edwards D, Hosseini M (2021) Exploration of latent barriers inhibiting project management processes in adopting smart building technologies (SBTS) in developing countries. Constr Innov 21(4):685–707. https://doi.org/10.1108/ci-07-2020-0116

Gitonga M (2018) Determinants of turnaround strategy implementation on the competitiveness of the insurance industry in Kenya. strategicjournals.com 5(4). https://doi.org/10.61426/sjbcm.v5i4.893

Hwabamungu B, Williams Q (2010) m-Health adoption and sustainability prognosis from a caregivers' and patients' perspective, pp 123–131. https://doi.org/10.1145/1899503.1899517

Javaid M, Khan S, Haleem A, Rab S (2022) Adoption of modern technologies for implementing Industry 4.0: an integrated MCDM approach. Benchmarking Int J 30(10):3753–3790. https://doi.org/10.1108/bij-01-2021-0017

Joshi A, Singh R, Kumar V, Rani S (2024) Net zero strategy: exploring opportunities, confronting challenges for developing countries. In: Approaches to global sustainability, markets, and governance, pp 233–252. https://doi.org/10.1007/978-981-97-5756-5_11

Kamau J (2024) Efficiency of renewable energy sources in reducing carbon footprint in developing countries. Int J Environ Sci 7(2):1–12. https://doi.org/10.47604/ijes.2637

Khalid B, Urbański M, Kowalska-Sudyka M, Wysłocka E, Piontek B (2021) Evaluating consumers' adoption of renewable energy. Energies 14(21):7138. https://doi.org/10.3390/en14217138

Kiprop E, Matsui K, Maundu N (2019) The role of household consumers in adopting renewable energy technologies in Kenya. Environments 6(8):95. https://doi.org/10.3390/environments6080095

Kyobe M (2011) Investigating the key factors influencing ICT adoption in South Africa. J Syst Inf Technol 13(3):255–267. https://doi.org/10.1108/13287261111164844

Ligon E, Malick B, Sheth K, Trachtman C (2019) What explains low adoption of digital payment technologies? Evidence from small-scale merchants in Jaipur, India. PLoS One 14(7):e0219450. https://doi.org/10.1371/journal.pone.0219450

Ma L, Lee C (2018) Understanding the barriers to the use of MOOCs in a developing country: an innovation resistance perspective. J Educ Comput Res 57(3):571–590. https://doi.org/10.1177/0735633118757732

Mandala N (2024) The impact of technology adoption on efficiency and transparency in public procurement processes in Kenya. Eur Sci J ESJ 25. https://doi.org/10.19044/esipreprint.1.2024.p483

Marotta A, Porras-Amores C, Sánchez A (2023) Are green buildings an indicator of sustainable development? Appl Sci 13(5):3005. https://doi.org/10.3390/app13053005

Mingjie M, Xiong H (2020) Upgrading of technological capability and innovation policy transformation in China under a new situation. J Adv Health 2(2):83–98. https://doi.org/10.3724/sp.j.2096-5141.2020.0005

Mosly I (2015) Barriers to the diffusion and adoption of green buildings in Saudi Arabia. J Manag Sustain 5(4):104. https://doi.org/10.5539/jms.v5n4p104

Nduji R, Orji M, Oyenuga M, Oriaku C (2023) Assessing e-business and organisational performance in Nigeria today: evidence from Jumia Ltd., Lagos. Br Int HumIties Soc Sci J 5(2):81–92 https://doi.org/10.33258/biohs.v5i2.897

Nwokolo SC, Singh R, Khan S, Kumar A, Luthra S (2023a) Remedies to the challenges of renewable energy deployment in Africa. In: CSR, sustainability, ethics & governance, pp 59–74. https://doi.org/10.1007/978-3-031-44514-9_3

Nwokolo SC, Singh R, Khan S, Kumar A, Luthra S (2023b) Technological pathways to net-zero goals in Africa. In: CSR, sustainability, ethics & governance, pp 93–210. https://doi.org/10.1007/978-3-031-44514-9_5

Nwokolo SC, Singh R, Khan S, Kumar A, Luthra S (2023c) Global investment and development in Africa. In: CSR, sustainability, ethics & governance, pp 15–58. https://doi.org/10.1007/978-3-031-44514-9_2

Nwokolo SC, Singh R, Khan S, Kumar A (2024a) Harnessing Industry 4.0 for Africa's net zero economy through technological pathways. In: Elsevier eBooks, pp 249–282. https://doi.org/10.1016/b978-0-443-31486-5.00010-0

Nwokolo SC, Singh R, Khan S, Kumar A (2024b) Role of digitalization and connectivity for achieving a net zero economy in Africa. In: Elsevier eBooks, pp 175–192. https://doi.org/10.1016/b978-0-443-31486-5.00006-9

Nwokolo SC, Singh R, Khan S, Kumar A (2024c) Key players in net zero technology. In: Elsevier eBooks, pp 193–215. https://doi.org/10.1016/b978-0-443-31486-5.00007-0

Nwokolo SC, Singh R, Khan S, Kumar A (2024d) Technological pathways for Africa' s net-zero economy. In: Elsevier eBooks. https://doi.org/10.1016/c2023-0-52499-1

Omale SA, Oyenuga MO, Oriaku C (2021) Effect of organizational culture on the performance of hospitality industry in Covid-19 era. Abhigyan Manag J 39(3):24–32. https://doi.org/10.56401/abhigyan/39.3.2021.24-32

Omale SA, Oyenuga MO, Abdullahi D, Madu I (2024) Moderating role of entrepreneurial education on innovative work behaviour and resilience of SMEs. Innovations 77(2):1166–1185. www.journals-innovations.com

Oyedele OM, Iember AA (2021) Covid-19 and the future of higher education. IEEE Technol Policy Ethics 6(4):1–3. https://doi.org/10.1109/ntpe.2021.9778140

Oyedele MO, Marcus GO, Ahungwa AI (2023) Do consumers care about green marketing practices? Insight from a developing nation. Bp Int Res CritS Inst J (BIRCI J) 6(3):1424–1436. https://doi.org/10.33258/birci.v6i3.7665

Oyedele MO, Sunday AO, Abuh AI (2024) Fostering technological-enhanced training and development for business survival and performance in the new normal. J Propuls Technol 45(3):1858–1869. https://propulsiontechjournal.com/index.php/journal/article/view/7479

Oyenuga MO, Omale SA (2024) Is Africa jinxed? Exploring the challenges of technology access and adoption in Africa. Afr J Econ Sustain Dev 7(4):142–161. https://doi.org/10.52589/AJESDULN1LRNF

Oyenuga MO, Andah RA, Nduji R (2023) Employee participation and organisational performance in an emerging economy. Konfrontasi J: Cult, Econ Soc Chang 10(2):91–105. https://doi.org/10.33258/konfrontasi2.v10i3.281

Oyenuga MO, Apata SB, Sulaiman MM (2024) Pre-service teachers' perceptions and utilisation of ChatGPT in higher institutions using UTAUT theory: evidence from an emerging economy. Br J Contemp Educ 4(2):28–42 https://doi.org/10.52589/BJCE-DBIT7S4T

Oyetunde TO, Oyenuga MO, Adoga GJ (2023) Green management practices and organisational efficiency in higher institution. Bp Int Res CritS Linguist Educ (BirLE) J 6(3):276–290. https://doi.org/10.33258/birle.v6i3.7716 276

Petersen F, Brown A, Pather S, Tucker W (2019) Challenges for the adoption of ICT for diabetes self-management in South Africa. Electron J Inf Syst Dev Ctries 86(5). https://doi.org/10.1002/isd2.12113

Raj A, Dwivedi G, Sharma A, Jabbour A, Rajak S (2020) Barriers to the adoption of Industry 4.0 technologies in the manufacturing sector: an inter-country comparative perspective. Int J Prod Econ 224:107546. https://doi.org/10.1016/j.ijpe.2019.107546

Reynolds S (2024) Barriers to adopting green supply chain practices: perspectives from industry leaders. https://doi.org/10.20944/preprints202406.1313.v1

Seidu S, Owusu-Manu D, Kukah A, Adesi M, Oduro-Ofori E, Edwards D (2022) An assessment of the implications of disruptive technologies on the performance of energy infrastructure projects in Ghana. Int J Energy Sect Manage 17(5):887–903. https://doi.org/10.1108/ijesm-09-2021-0007

Sendstad L, Chronopoulos M, Hagspiel V (2023) Optimal risk adoption and capacity investment in technological innovations. IEEE Trans Eng Manage 70(2):576–589. https://doi.org/10.1109/tem.2021.3056142

Sikder A (2024) Technology adoption and leadership strategies for enhancing human capital development in Bangladeshi NPOS. World J Adv Res Rev 21(3):1185–1190. https://doi.org/10.30574/wjarr.2024.21.3.0747

Singh K (2013) Conservation agriculture: innovations, constraints and strategies for adoption. SSRN Electron J. https://doi.org/10.2139/ssrn.2318985

Singh R, Gupta A, Kandpal V (2024a) ESG excellence in the digital world. E-Bus Solut 193–204. https://doi.org/10.1002/9781394234035.ch10

Singh R, Crowther D, Ravesangar K (2024b) Harnessing human capital for a sustainable future: HR's pivotal role toward net zero policies. In: Approaches to global sustainability, markets, and governance, pp 27–47. https://doi.org/10.1007/978-981-97-5756-5_2

Štreimikienė D, Lekavičius V, Stankūnienė G, Pažėraitė A (2022) Renewable energy acceptance by households: evidence from Lithuania. Sustainability 14(14):8370. https://doi.org/10.3390/su14148370

Sujarwoto S (2023) Facilitators and barriers to the adoption of mhealth apps for covid-19 contact tracing: a systematic review of the literature. Front Public Health 11. https://doi.org/10.3389/fpubh.2023.1222600

Usman F (2024) Integrating renewable energy solutions in the manufacturing industry: challenges and opportunities: a review. Eng Sci & Technol J 5(3):674–703. https://doi.org/10.51594/estj.v5i3.865

Wulandari S, Djufry F (2021) Technology transfer strategy for women in coffee livestock integration as climate-smart agriculture practice. E3s Web Conf 316:04010. https://doi.org/10.1051/e3sconf/202131604010

Yangle S (2023) The efficacy of building information model in cost control of green buildings in Shanghai. J Adv Res Appl Sci Eng Technol 33(1):471–489. https://doi.org/10.37934/araset.33.1.471489

Yolles M (2019) Governance through political bureaucracy: an agency approach. Kybernetes 48(1):7–34. https://doi.org/10.1108/k-09-2017-0329

Zhang L, Ye F, Yang L, Zhou G (2019) Impact of political connections on corporate environmental performance: from a green development perspective. Sustainability 11(5):1317. https://doi.org/10.3390/su11051317

Zhou T, Mao J, Xie L, Zhao Q, Chen Y (2022) Mobile medical applications' state quo and underlying barriers: a review of literature†. Front Nurs 9(3):241–246. https://doi.org/10.2478/fon-2022-0029

Chapter 18
Political and Social Challenges to Net Zero Policies Action

Abstract Net zero emissions is now a paramount goal for the nations of the world but the path to achieving it throws up political and social roadblocks. We turn now to the several interacting barriers to the realization of net zero policies one of which is the political context, and the other the policies that we all strive to implement. Politically, there are also major barriers in the form of a lack of integrated policy frameworks, inadequate political will, and vested interests that may counter transformative change. Simultaneously, social issues, including public perception, equitability, and social exclusion, complicate net zero uptake. Based on empirical evidence, this chapter argues that overcoming these critical challenges requires linking innovative policy solutions to inclusive social strategies. Mitigating climate change and social injustice can indeed be achieved together, firstly through collaboration among the stakeholders to achieve net zero regulation and through facilitating the conducive space for engagement of social capital with the regulatory body. This chapter ultimately suggests the need for an integrated focus that balances political and social considerations in the pursuit of climate action sustainability and equity.

Introduction

Governments and organizations around the world are now setting net zero emission targets in line with international efforts to tackle climate change. But the path to realizing this ambitious vision is strewn with political and social challenges that threaten to interrupt progress. Oftentimes, it is political will that drives success when it comes to climate policy, with leaders favouring short-term conveniences over sustainability during most times. Social resistance to changing consumption habits and ingrained cultural norms can also block acceptance of the lifestyle changes that will be essential for reducing emissions. At the same time, an understanding of these issues, with properly targeted interventions, can increase the efficacy of community engagement with stakeholder groups. According to Moleka, the way SMEs conceive the concept of net zero goals determines how they respond to engaging sustainability initiatives. On land, the measure will also be ultimately dependent on political will

M. O. Oyenuga et al., *Smart Technologies for Climate Change and Net Zero Policies: Practical Approaches Towards Sustainability*, Climate Change Management,
https://doi.org/10.1007/978-3-031-92221-3_18

and public acceptance, as Meijerink and Stiller also emphasise. The political and social challenge to net zero scales down adverse impacts or diverts good towards the post-industrial world to find out the pathways leading us to sustainability is what we need first, which we do not see in practice today for some major sectors. Removed from policy and limiting to existing approaches, overcoming these challenges is vital to the high ambition, equity and deliverability of net zero policies for all those who have an interest in, or are affected by, the outcomes.

Lack of Political Will in Certain Regions

Political will is the determination of political leaders and institutions to design, mobilize the resources and focus the organization of policy-makers on the outcomes many would understand to be achieved when sustainable development goals are put at the front of the line for resources. This commitment is frequently undermined by nationalist politics, weak governance systems and socio-economic problems that plague many developing nations.

Competing Political Interests

Where political stability is fragile, the interests of a politician to win an election may outweigh long-term environmental goals. According to Qasim (2023), new technologies can generate political losers that make it difficult for incumbent elites to accept transformative change, as this can be damaging to their political or economic interests. Such pushback might be felt in terms of slowness to implement new policies, a failure to allocate budget for means to go green or inadequate laws that don't provide incentives for eco-friendly options.

Governance and Institutional Challenges

There is also the lack of politics which is exacerbated by weak governance structures. According to Mazhambe (2023), lack of capacity, poor communication, and coordination between stakeholders can greatly influence the ability of strategic plans to achieve their targets. The search for such policies is complicated when political leaders are in some cases not as knowledgeable of the science and how the available resources are needed to create and implement such policies. The lack of comprehensive governance structures often results in disjointed initiatives that do not adequately tackle the systemic aspects of climate issues. In addition, a lack of strong data and evidence to inform policy choices can breed uncertainty (Polisena et al. 2019). While the authors discuss barriers to disinvestment decisions specifically in health technology assessment, the analogous two themes of political challenge and relevant data apply to climate policy, wherein a political economic perspective suggests that the uncertainties under which policymakers operate may lead them to shy away from ambitious targets for addressing climate change.

Socio-Economic Factors

Political will is also tied to socioeconomic issues. Especially in developing countries, areas cannot ignore jobs, unemployment and social inequality in the face of environmental issues. Sengupta et al. (2019) and Oyenuga and Omale (2024) point out that a lack of political will is often associated with competing priorities such as health, education and economic development being seen as more important than climate action. Consequently, sustainable technologies and practices are under-invested in, and the hurdles to net zero get further entrenched.

Community Acceptance of Green Initiatives

Acceptance of green initiatives by local communities is an important determinant of the successful implementation and sustainability of environmental projects. How much the local populations buy into these initiatives can heavily affect both their effectiveness, but also how long those initiatives can hang around and how well they can effect change for global environmental sustainability. Here, we discuss the practical implications of these varying concepts of community acceptance, including the levels, socio-economic factors, and the role local governance can play in the creation of a supportive environment for greening initiatives.

Awareness and Education

The awareness of the benefits of green initiatives is one of the key factors for the success of community acceptance. According to Thang and Thanh (2023), a more sustainable approach would be a stronger commitment from local communities if they would become more aware of the benefits associated with green tourism. Community members become more supportive of and engage more in environmental efforts as they become more aware of the potential environmental, economic, and social benefits of green initiatives. Elaboration of educative programs describing awareness regarding how the people are able to enjoy the leverage brought by green initiatives and information on economy upliftment would convince and motivate local population to continue the guadance and practice sustainable living (Oyenuga and Akwen 2025).

Participation and Engagement

Community acceptance is achieved primarily through active involvement in decision-making processes. According to Oikonomaki (2024), initiatives lead by citizens for urban greening can supplement institutional action and build stronger community support. Using the knowledge, skills and resources of residents to plan and implement green initiatives in collaboration with local government can both empower residents and better align the initiatives with the needs and values of the community. That participatory attitude can reduce the sense of threat and opposition and, in turn, facilitate the acceptance and excitement about green projects.

Socio-Economic Considerations

None of these factors are wholly socioculturally determined but represent an underlying socioeconomically determined acceptance of green community reshaping. Maantay and Maroko (2018) argue that greening initiatives need to integrate social equity objectives to avoid environmental gentrification that can displace low-income residents. Community acceptance can erode where green initiatives are seen as beneficial only to parts of the population. Thus, designing initiatives that meet the needs of all community members is crucial to ensure that greening initiatives provide equitable benefits. Economic effects of green initiatives affect community acceptance projects that tout job creation, better local economies and increased property values tend to attract more support. Kondo et al. (2016) discovered that community reuse of underutilized vacant lots for gardens or orchards complemented environmental beautification efforts that reinforce community neighbourhood identity and self-attachment, which can reduce crime rates. Positive results like this can help build community buy-in and support for future green initiatives.

Governance and Policy Frameworks

Good governance, policies and frameworks play a vital role in encouraging people to accept greener approaches. Local governments serve as a crucial intermediary, enabling community participation and ongoing improvement aligned with residents' interests. As pointed out by Garrett (2023), processes of urban greening often depend on the work of volunteers or low-wage workers and may contribute to economic marginalization. These dynamics must be addressed by policymakers who are committed to creating inclusive policies that empower local communities and recognize their contributions toward sustainability efforts. Moreover, agencies on different levels should work closely with the community and potential interest groups to create hours to suit everyone and to gather opinions on the implementation of the services. Upham et al. (2015) highlight the need for understanding public perceptions and attitudes towards energy systems, which can inform more effective communication strategies and promote positive sentiment towards green energy investments and initiatives in the community.

Social Resistance to Changing Consumption Habits

One of the most important challenges towards sustainable development and environmental goals is social resistance against changing consumption behaviours (Oyedele et al. 2023). Even while there is increasing recognition that people need to live more sustainably, many people and communities are loth to change the way they consume. Such resistance may arise from a variety of reasons, such as rigid behaviours, so-called "inconvenience" and economically beneficial misperception of sustainable consumption as a priority in the face of tension to lifestyle sustainability communication.

Cultural Norms and Social Identity

Cultural norms guide consumption behaviours. Consumer behaviour is also anchored in sociocultural values and traditions, thus making it complicated to change established habits (Peattie 2010; Marcus et al. 2020). Some consumption patterns in various cultures are tied to social status or identity, causing humans to resist changes that may affect their perceived social status. This is especially true in the luxury market, where consumers would go for high-status items rather than more sustainable options. Furthermore, social identity can reduce individual resistance to novel consumption practices. Research by Thomas et al. (2017) has found that people are more apt to adjust their diet when they feel their peers are doing the same. This shows that social norms can enable or hold back the adoption of sustainable consumption practices. Knowing this was about their social group when someone feels that their consumption choices are at odds with those in their social group, they are resistant to change to maintain a feeling of conformity and acceptance in their social community.

Economic Considerations

Economic factors play a big role in creating resistance to changes in consumer habits. Many consumers believe sustainable products to be more expensive or less accessible compared to their traditional variants. Bruijn et al. (2011) highlight that implicit associations with higher costs can discourage the adoption of healthier or more sustainable options among consumers who are aware of long-term benefits. Such a perspective is intensified among low-income communities, who find themselves unable to expend effort toward making sustainable consumption choices due to stricter financial limitations. Resistance may also be exacerbated by the perceived inconvenience of changing consumption habits. People are less likely to support measures to encourage plant-based diets or limit single-use plastics if they think implementing those measures will involve lifestyle changes or personal sacrifice. More recently, Trujillo (2023) and Omale et al. (2023) point out that the COVID-19 pandemic contributed to changing consumption patterns, but also further revealed the barriers to shifting towards more sustainable ways of consuming, given that many shoppers fell back into old patterns for the sake of convenience.

Communication and Awareness

Overcoming social resistance to changing consumption habits requires effective communication. People often do not have enough information about the environmental impacts of their consumption choices and the benefits of more sustainable practices. Yang et al. (2020), the gap between knowledge and action may be narrowed through messaging, as clear actionable suggestions encourage people to engage with the material on a deeper level, leading to change. Emphasising the benefits of sustainable consumption, for instance, on health and community well-being, can increase acceptance and stimulate behavioural change (Joshi et al. 2024). Additionally, utilizing social pressure through community promotion may facilitate the acceptance of sustainability practices. As shown by Garcia et al. (2021), social norms are subject both to political framing and social practice behaviours and can therefore greatly influence dietary change, a strong indication that community actions can encourage more sustainable consumption. This also saves time individuals would spend resisting community norms by creating environments that are aligned with sustainable choices.

Governance Issues: Corruption and Inefficiency

Governance structures are often characterised by corruption and inefficiency, which undermines the public administration and the successful implementation of policies designed to promote sustainable development. The problems not only sap public confidence in state institutions but also hinder economic performance and multiply social inequalities.

Corruption is typically described as the use of public authority for private interest, in forms ranging from bribery to embezzlement, to nepotism. Corruption, according to Lecuna (2012), is more likely to occur in decentralized forms of governance as smaller jurisdictional units tend to have fewer oversight mechanisms that prevent such practices. In such contexts, corrupt private elites are better positioned to corrupt public servants, thus creating a vicious cycle of corruption that captures the political system. Moreover, Reinsberg et al. (2019) illustrate how privatization serves to deepen corruption in developing states. Overall, the privatization of public services, when not backed by proper regulatory frameworks, leads to chances of corrupt practices, motivating private institutions to use them for their benefit, by ignoring the public interest. This highlights the need to have strong governance structures that can oversee and regulate both public and private sectors (Nwokolo et al. 2023a).

Inefficiency in Governance

Bureaucratic red tape, lack of accountability, and inefficient resource allocation are all reasons why governance becomes ineffective. Wu et al. advocate the use of public sector innovation and e-government initiatives as a means to address inefficiency and corruption. But technology alone does not deliver (Nwokolo et al. 2023b; Apata et al. 2024) transparent and accountable commitments must accompany technology and are essential in producing meaningful results. Governance is so inefficient that it presents opportunities for corruption, as the lack of oversight and poor management practices create fertile ground for corruption to rot. Corruption and good governance are related, yet they have an intricate relationship. As Park and Kim (2019) observe, though, e-government is an imperfectly effective anti-corruption tool, but not a cure. Without concurrent elements of accountability and transparency, open government activities may not be enough to reduce corruption. This indicates that instead of a dual approach to tackling corruption and malfeasance, we should opt for a holistic approach dealing with both nepotism and inefficiency.

Opportunities for Overcoming Political and Social Challenges

Tackling the political and social hurdles will be critical to the successful deployment of sustainable initiatives and policies to meet environmental objectives. Although it may feel intractable, there is plenty of room for collaboration, community-building, and governance. These various challenges can be overcome with a variety of techniques and opportunities, some of which are also used in different contexts as discussed below:

1. **Community Engagement and Empowerment**: Community engagement is one of the best strategies for tackling social and political issues. Having local voices in policy discussions helps people feel like they have a stake in the ground; when people feel responsible for making cost-effective sustainable practices, they are more likely to commit to them. Sadler and Pruett (2015) point out that community-driven programs, like land banks, are successful in doing just that and not only reduce urban blight but also develop community pride. Engaging residents in greening and urban renewal activities helps communities develop a cultivation of commitment to sustainability (Nwokolo et al. 2024). In addition, local communities should be involved as it can contribute to sustainable development in green tourism (Anh 2023). Local support for environmental initiatives can also be strengthened by making the payoffs of sustainable tourism concrete and directly connected to local well-being. Not only does this participatory approach make programs more effective, it also builds social capital and resilience.
2. **Leveraging Technology for Transparency and Accountability**: This is where technology can help provide an innovative solution in bringing governance and eradicating corruption, which are usually the biggest hurdles to implementing policies efficiently (Nwokolo et al. 2023c). For example, e-government initiatives can enhance transparency and accountability of public administration. Wu et al. state that e-government can significantly reduce opportunities for engaged corrupt practices as transparent processes and systems will encourage less corruption in the citizenry; thus, transparency is essential for it to work. When governments deploy technology to make processes more seamless and access to information available, they can bolster public trust and engagement in the political system. Additionally, participatory platforms that enable citizens to report issues, share feedback, and interact with local authorities can bolster accountability. For example, such initiatives help to engage citizens in environmental monitoring to retain local governance and directly hold decision-makers accountable for their actions.
3. **Building Partnerships and Collaborative**: There is a need to move beyond politics; a cooperative administration is required. Collaboration among multiple stakeholders, including government agencies, civil society organizations, and the private sector, is crucial to solving political and social problems. They can help in sharing resources, data, skills and collaboration that is needed to tackle difficult problems. In the discussion of their research, Oikonomaki (2024) summarises how citizen-led urban greening initiatives can enhance institutional initiatives; creating public support and interest in sustainable practices. Bickerdike et al. (2017), demonstrates the potential for partnerships between healthcare providers and community-based organizations to tackle some of the social determinants of health. This enables stakeholders to combine social services concerning health and environmental challenge challenges and make comprehensive approaches.
4. **Education and Awareness**: Campaigns are key to awareness and, crucially, that even small sustainable efforts can be effective against inertia and fear of change. Providing educational programs that raise community awareness of the environmental, economic, and social benefits of sustainable lifestyle changes can lead

to increased acceptance and engagement. In addition, Thang and Thanh (2023) stressed the importance of educating target markets on the benefits of sustainable tourism, making them more proactive in nature-based tourism sustainability measures. Educating people about the environment allows them to become more empowered in the choices they make about the things they consume and the lifestyles they lead. An education that includes a culture of sustainability thinking could present the backdrop of society which will be necessary to foster a future of environmental guardianship.

5. **Policy Innovation and Reform**: Policymakers should be willing to think outside the box and develop new ways to promote environmental concerns through economic and social policies. Ibrahim (2023), for example, notes that sustainable practices are tested over time, and must be closely observed, evaluated, and modified so that they remain effective. It allows further context into the findings by involving engagement with multi-actors which can aid in the barrier to be identified for implementation and its strategies to overcome them. Governments can also encourage collaborative spaces for policy innovation, which will lead to inclusive frameworks for sustainable development (Singh et al. 2024a).

Implications

These challenges signal a growing disconnect between climate science and political and social frameworks, with implications for the current and future effectiveness and sustainability of climate action efforts. These implications are critical for policymakers, stakeholders and communities seeking insight into the complexities of net zero implementation.

The policy itself is undermined by corruption and ineffectiveness at the political level, which are mortal enemies of the operating framework of net zero itself. Corruption can divert resources from sustainable development by embezzling funds meant for these projects for personal enrichment. This not only stalls climate action but also undermines public trust in government institutions. As a result, the loss of trust may lead to lower public engagement and backing for climate action, which in turn makes it all the more challenging to reach net zero objectives (Singh et al. 2024b). In addition, the quality of governance should directly impact how those environmental policies are put into practice. Kulin and Sevä (2019) note that they see a clear pathway from normative beliefs about government responsibility to a set of evidence-based policy recommendations. Effective governance is important to make that happen. This downward spiral of inefficiency and corruption creates a vicious cycle that makes successful policy implementation less likely, which in turn further complicates climate action efforts, particularly in poorly governed countries.

Also, socio-political aspects such as the corruption index and trust in the political system have a high influence on the community's engagement in climate-associated initiatives. Indeed, as highlighted by Rafaty (2018), perceptions of corruption are linked to weaker climate policies, particularly in democracies. Supporting or engaging with climate initiatives becomes less likely when citizens see their government as corrupt, and initiatives less effective or mismanaged. The engagement of citizens in these plans is critical; if people do not understand or are not interested in the plans, it may lead to issues in the policies being embraced and practised. Further, climate policies tend to rely heavily on public support and engagement to be successful.

Furthermore, the political and social challenges that accompany net zero policies may also create problems for economic development. Networks of corruption and inefficiency may stall or deter investing in things like sustainable technologies and infrastructure, as investors often see the potential of investing under unstable governance as high risk. According to Lindvall (2021), too little governance leads to political risk, which creates disincentives to invest in green initiatives. This leads to missed opportunities for economic growth and job creation in the emerging green economy (Oyetunde et al. 2023). Additionally, neglecting social equity issues can worsen pre-existing disparities and culminate in social upheaval and pushback against climate measures. Barua et al. (2021) argue that good governance is critical to meet the needs of marginalized communities and to ensure that the benefits of sustainable development are shared equitably.

Lastly, political and social challenges have consequences for social equity and inclusion. As Ziya (2024) has observed, the quality of governance and control of corruption are key determinants of access to a just transition to sustainable economies. A net zero path-way that omits the perspectives and needs of vulnerable populations will only perpetuate, reproduce or even amplify pre-existing social injustices and inequalities. This could engender social resistance and weaken the legitimacy of climate actions. By designing and implementing policies around a net zero transition with a focus on social justice, acceptance and support for them will be more forthcoming. Incorporating marginalized communities in decision-making can increase the legitimacy of climate projects, as well as build a sense of ownership by stakeholders.

Conclusion

This emphasis on carbon action must be coupled with a long-term societal change to combat the political impetus behind all aspects of net zero policy action and presents an obstacle to net zero objectives that must be navigated to see any meaningful change in climate policy action. A lack of political will, strong social norms and resistance to changing consumption habits are just a few of the challenges that can undercut the success of climate initiatives and slow the transition to a sustainable future. By identifying these challenges, we also open ourselves up to innovative solution sets and collaborative approaches that enable acceptance and support for net zero policies. The most important step for overcoming resistance to change is focused on improving engagement and participation in the community. Local initiatives that engage community members help promote ownership over initiatives and provide the localized capacity and logistical support needed. Raising awareness and understanding about the positive outcomes that net zero policies can produce is a way to ensure that communities feel another layer of incentivization or motivation to implement such changes and engage in climate action (Tangney and Howes 2016; Hornung 2022).

Second, using technology to enhance transparency and accountability in governance is one way of addressing corruption and inefficiency which are major impediments to effective implementation of policies. Reduced public trust in government institutions is also a common obstacle to citizen engagement in climate initiatives, which could be addressed through participatory platforms, e-government initiatives, and financial incentives. Nurturing a culture of accountability is a way for governments to create an enabling environment for successful net-zero policies (Latkin et al. 2021; Colvin and Jotzo 2021). Working towards shared climate solutions, especially with broad coalitions of government agencies, civil society organizations and the private sector will provide support for complex climate actions.

These critical tools can be used by building collaborative networks that help share resources, and exchange ideas, and initiatives to address political and social challenges. Through a collaborative effort, stakeholders will be able to create integrative approaches that not only support sustainable development but also improve the ability to mitigate climate change impacts (Mukheibir et al. 2013; Basseches et al. 2022).

Finally, socioeconomic considerations and measures to achieve net zero need to be equity and socially inclusive if they are to have public buy-in. Social equity should be taken into consideration by policymakers as they formulate and implement climate actions, to ensure that accomplishments about sustainability become accessible to all people living in a society. By recognizing and responding to the needs of marginalized populations, stakeholders can move towards a more equitable and sustainable future (Tangney and Howes 2016; Crowley 2021).

References

Anh T (2023) The impact of green tourism on the Hmong community in Vietnam. Multidiscip Sci J 6(4):2024034. https://doi.org/10.31893/multiscience.2024034

Apata AO, Oyenuga MO, Abeoba M, Ugom A (2024) Internet of Things (IoT) solutions for smart transportation infrastructure and fleet management. J Propuls Technol 45(4):1492–1505. https://propulsiontechjournal.com/index.php/journal/article/view/8325

Barua P, Mitra A, Eslamian S (2021) Disaster management strategies and relation of good governance for the coastal Bangladesh. Resour Environ Econ 3(2):269–279. https://doi.org/10.25082/ree.2021.02.002

Basseches JA, Bromley-Trujillo R, Boykoff MT, Culhane T, Hall G, Healy N, Hess DJ, Hsu D, Krause RM, Prechel H, Roberts JT, Stephens J (2022) Climate policy conflict in the U.S. states: a critical review and way forward. Clim Change 170(3–4). https://doi.org/10.1007/s10584-022-03319-w

Bickerdike L, Booth A, Wilson P, Farley K, Wright K (2017) Social prescribing: less rhetoric and more reality. A systematic review of the evidence. BMJ Open 7(4):e013384. https://doi.org/10.1136/bmjopen-2016-013384

Bruijn G, Keer M, Conner M, Rhodes R (2011) Using implicit associations towards fruit consumption to understand fruit consumption behaviour and habit strength relationships. J Health Psychol 17(4):479–489. https://doi.org/10.1177/1359105311421049

Colvin R, Jotzo F (2021) Australian voters' attitudes to climate action and their social-political determinants. PLoS One 16(3). https://doi.org/10.1371/journal.pone.0248268

Crowley C (2021) Fighting the future: the politics of climate policy failure in Australia (2015–2020). Wiley Interdiscip Rev Clim Change 12(5). https://doi.org/10.1002/wcc.725

Garcia A, Higgs S, Lluch A, Darcel N, Davidenko O (2021) Associations between perceived social eating norms and initiation and maintenance of changes in dietary habits during the first COVID-19 lockdown in France. Foods 10(11):2745. https://doi.org/10.3390/foods10112745

Garrett C (2023) Greenwork: the devaluation of labor when caring for nature. City Community 22(4):330–351. https://doi.org/10.1177/15356841231182071

Hornung J (2022) Social identities in climate action. Clim Action 1(1). https://doi.org/10.1007/s44168-022-00005-6

Ibrahim I (2023) A conceptual model for sustainable green port practices: a case study of Northport (Malaysia) Berhad. Inf Manag Bus Rev 15(3(SI)):267–279. https://doi.org/10.22610/imbr.v15i3(si).3483

Joshi A, Singh R, Kumar V, Rani S (2024) Net zero strategy: exploring opportunities, confronting challenges for developing countries. In: Approaches to global sustainability, markets, and governance, pp 233–252. https://doi.org/10.1007/978-981-97-5756-5_11

Kondo M, Hohl B, Han S, Branas C (2016) Effects of greening and community reuse of vacant lots on crime. Urban Stud 53(15):3279–3295. https://doi.org/10.1177/0042098015608058

Kulin J, Sevä I (2019) The role of government in protecting the environment: quality of government and the translation of normative views about government responsibility into spending preferences. Int J Sociol 49(2):110–129. https://doi.org/10.1080/00207659.2019.1582964

Latkin C, Dayton L, Lee D, Yi G, Uzzi M (2021) Correlates of levels of willingness to engage in climate change actions in the United States. Int J Environ Res Public Health 18(17):9204. https://doi.org/10.3390/ijerph18179204

Lecuna A (2012) Corruption and size decentralization. J Appl Econ 15(1):139–168. https://doi.org/10.1016/s1514-0326(12)60007-5

Lindvall D (2021) Democracy and the challenge of climate change. https://doi.org/10.31752/idea.2021.88

Maantay J, Maroko A (2018) Brownfields to greenfields: environmental justice versus environmental gentrification. Int J Environ Res Public Health 15(10):2233. https://doi.org/10.3390/ijerph15102233

Marcus GO, Oyenuga MO, Ahungwa AI (2020) Effects of sales promotion on consumer buying behaviour of food seasoning among Nigerian households: a case study of Nestle Maggi Naijapot. Bp Int Res Crit Inst J 3(1):134–139. https://doi.org/10.33258/birci.v3i1.724

Mazhambe Z (2023) Challenges affecting strategy implementation at the food reserve agency in Zambia. J Public Adm Gov 13(2):10. https://doi.org/10.5296/jpag.v13i2.21097

Mukheibir P, Kuruppu N, Gero A, Herriman J (2013) Overcoming cross-scale challenges to climate change adaptation for local government: a focus on Australia. Clim Change 121(2):271–283. https://doi.org/10.1007/s10584-013-0880-7

Nwokolo SC, Singh R, Khan S, Kumar A, Luthra S (2023a) Global Investment and Development in Africa. In: CSR, sustainability, ethics & governance, pp 15–58. https://doi.org/10.1007/978-3-031-44514-9_2

Nwokolo SC, Singh R, Khan S, Kumar A, Luthra S (2023b) Remedies to the Challenges of Renewable Energy Deployment in Africa. In: CSR, sustainability, ethics & governance, pp 59–74. https://doi.org/10.1007/978-3-031-44514-9_3

Nwokolo SC, Singh R, Khan S, Kumar A, Luthra S (2023c) Technological pathways to net-zero goals in Africa. In: CSR, sustainability, ethics & governance, pp 93–210. https://doi.org/10.1007/978-3-031-44514-9_5

Nwokolo SC, Singh R, Khan S, Kumar A (2024) Key players in net zero technology. In: Elsevier eBooks, pp 193–215. https://doi.org/10.1016/b978-0-443-31486-5.00007-0
Oikonomaki E (2024) Promoting green transformations via smart engagement: an assessment of 100 citizen-led urban greening projects. https://doi.org/10.20944/preprints202403.0451.v1
Omale SA, Oyenuga M, Gurin IM (2023) Effects of COVID-19 pandemic on organizational transformation and employees' performance. J Manag Econ Ind Organ 7(1):16–33. https://doi.org/10.31039/jomeino.2023.7.1.2
Oyedele MO, Marcus GO, Ahungwa AI (2023) Do consumers care about green marketing practices? Insight from a developing nation. Bp Int Res CritS Inst J (BIRCI J) 6(3):(1424–1436). https://bircu-journal.com/index.php/birci/article/view/7665
Oyenuga MO, Omale SA (2024) Is Africa jinxed? Exploring the challenges of technology access and adoption in Africa. Afr J Econ Sustain Dev 7(4):142–161. https://doi.org/10.52589/AJESDULN1LRNF
Oyenuga MO, Akwen PI (2025) Tourism and intercultural understanding in developing nations. Afr J Cult Hist Relig Tradit 8(1):1–14. https://doi.org/10.52589/ajchrt-okuxnmsq
Oyetunde TO, Oyenuga MO, Adoga GJ (2023) Green management practices and organisational efficiency in higher institution. Bp Int Res CritS Linguist Educ (BirLE) J 6(3):276–290. https://bircu-journal.com/index.php/birle/article/view/7716
Park C, Kim K (2019) E-government as an anti-corruption tool: panel data analysis across countries. Int Rev Adm Sci 86(4):691–707. https://doi.org/10.1177/0020852318822055
Peattie K (2010) Green consumption: behaviour and norms. Annu Rev Environ Resour 35(1):195–228. https://doi.org/10.1146/annurev-environ-032609-094328
Polisena J, Trunk G, Gutiérrez-Ibarluzea I, Joppi R (2019) Disinvestment activities and candidates in the health technology assessment community: an online survey. Int J Technol Assess Health Care 35(03):189–194. https://doi.org/10.1017/s0266462319000229
Qasim A (2023) Firms and technology adoption: the role of political institutions and market size. Pak Dev Rev 61–85. https://doi.org/10.30541/v62i1pp.61-85
Rafaty R (2018) Perceptions of corruption, political distrust, and the weakening of climate policy. SSRN Electron J. https://doi.org/10.2139/ssrn.3175064
Reinsberg B, Stubbs T, Kentikelenis A, King L (2019) Bad governance: how privatization increases corruption in the developing world. Regul & Gov 14(4):698–717. https://doi.org/10.1111/rego.12265
Sadler R, Pruett N (2015) Mitigating blight and building community pride in a legacy city: lessons learned from a land bank's clean and green programme. Community Dev J 52(4):591–610. https://doi.org/10.1093/cdj/bsv052
Sengupta S, Barman P, Lo J (2019) Opportunities to overcome implementation challenges of infection prevention and control in low-middle income countries. Curr Treat Options Infect Dis 11(3):267–280. https://doi.org/10.1007/s40506-019-00200-w
Singh R, Crowther D, Ravesangar K (2024a) Harnessing human capital for a sustainable future: HR's pivotal role toward net zero policies. In: Approaches to global sustainability, markets, and governance, pp 27–47. https://doi.org/10.1007/978-981-97-5756-5_2
Singh R, Gupta A, Kandpal V (2024b) ESG excellence in the digital world. E-Bus Solut 193–204. https://doi.org/10.1002/9781394234035.ch10
Tangney P, Howes M (2016) The politics of evidence-based policy: a comparative analysis of climate adaptation in Australia and the UK. Eviron Plann C Gov Policy 34(6):1115–1134. https://doi.org/10.1177/0263774x15602023
Thang ND, Thanh NN (2023) The relationship between awareness and commitment to participate in the green tourism model of local communities: evidence from the Hmong people community in Lao Cai Province-Vietnam. Int J Prof Bus Rev 8(6):e02540. https://doi.org/10.26668/businessreview/2023.v8i6.2540
Thomas JM, Ursell A, Robinson EL, Aveyard P, Jebb SA, Herman CP, Higgs S (2017) Using a descriptive social norm to increase vegetable selection in workplace restaurant settings. Health Psychol 36(11):1026–1033. https://doi.org/10.1037/hea0000478

Trujillo C (2023) The future of sustainable consumption after the pandemic, optimism or pessimism? https://doi.org/10.5772/intechopen.107140

Upham P, Oltra C, Boso À (2015) Towards a cross-paradigmatic framework of the social acceptance of energy systems. Energy Res Soc Sci 8:100–112. https://doi.org/10.1016/j.erss.2015.05.003

Yang M, Chen H, Long R, Hou C (2020) Overview, evolution and thematic analysis of China's green consumption policies: a quantitative analysis based on policy texts. Sustainability 12(20):8411. https://doi.org/10.3390/su12208411

Ziya A (2024) Role of corruption and governance quality on green electricity transition: learning through the lens of economic complexity and sustainability. Politická Ekonomie 72(2):375–402. https://doi.org/10.18267/j.polek.1439

Chapter 19
Financing the Transition to Net Zero

Abstract A financial revolution is critical to accelerate the transition to a net zero economy to build a concrete instrument to together address climate finance and investment, and to reshape one of the biggest challenges of our time. This chapter explores the complex challenges and opportunities of financing the transition to net zero. It underlines the importance of mobilizing public and private capital while involving relevant stakeholders and technologies to build sustainable financing models. It also mentions the role of policy frameworks for investment in low-carbon technologies and infrastructure. The Chapter attempts to draw recommendations on financing mechanisms that can mobilise private and public capital flows for climate financing that can not only accelerate the transition to a net zero economy but can also ensure social equity and environmental sustainability through research on climate financing.

Introduction

Today, with the pressure to mitigate the effects of climate change increasing by the day, the transition to a net zero economy is one of the biggest imperatives facing governments, businesses and communities all around the (Joshi et al. 2024) world. To decarbonize the economy by mid-century will require a radical rethinking of how energy is produced, how it is consumed and the economic structures that underpin these activities. But concerted action to guide this transition is fraught with challenges, not the least of which is securing sufficient financing to facilitate the required investments in low-carbon technologies and sustainable infrastructure. These initiatives all share a complex funding landscape, with the potential for either public or private funding sources, investment risks and varying levels of commitment across various regions. As Vijay and Sathish (2024) note, advancing and delivering sustainable projects brings the best and worst of the translation of the economy to net zero. To address the issue of profitability and the long-term viability of sustainable businesses, lawmakers must provide an ecosystem where businesses are incentivized to invest in sustainable solutions. Also, net zero is not just a technical challenge; it is also a social and political challenge. Bray et al. (2022) highlight that transitions in

M. 0. Oyenuga et al., *Smart Technologies for Climate Change and Net Zero Policies: Practical Approaches Towards Sustainability*, Climate Change Management,
https://doi.org/10.1007/978-3-031-92221-3_19

energy systems are often time-consuming and therefore require immediate work on the development of skills and workforce training to guarantee a just transition.

Bringing the public on board and leaving no one behind will require inclusive policies that can help people deal with the socio-economic ramifications of the transition. This chapter further explores the multiple dimensions of net zero transition funding, including the need for new financing models, sustainable universal coalitions, and policy architecture.

This chapter aims to contribute to the discourse on the need for sustainable financing, with a specific focus on developing effective models and practices that would ensure a sustainable transition towards a net zero economy, while addressing both equity and resilience in that transition through effective financing by analyzing best practices that have emerged from the literature across different contexts.

International Climate Finance: Opportunities and Limitations

International climate finance is the financial support that helps countries, especially developing countries, to implement climate change mitigation and adaptation projects. Greening the financial system finance includes the finance flows in both directions, from developed and developing countries to meet funding needs to cooperation and collaboration in needed areas, and to clean energy technologies needed for transitioning to low-carbon and climate-resilient economies, including a range of financial instruments such as grants, loans, and investments. But, alongside the important opportunities presented by international climate finance, some key constraints can undermine its efficacy and accessibility.

International Climate Finance Opportunities

A significant source of funding for mitigating greenhouse gas emissions and adapting to climate impacts is international climate finance. Mobilization of climate finance is essential for Caribbean Small Island Developing States (SIDS) to deliver on their Nationally Determined Contributions (NDCs) under the Paris Agreement (Mohan 2023). Such funding could help those countries with renewable energy projects, infrastructure development and disaster preparedness. Furthermore, the rise of alternative financing options, such as green bonds and climate adaptation funds, provides new channels for capital influx. This can help pull some private sector funds and create a "template" to fund green projects to achieve climate results while creating sustainable land practices (Zhou 2024).

Moreover, climate finance also provides an opportunity for adaptation in the most vulnerable communities by funding initiatives that improve infrastructure, health care and agricultural practices, where health considerations can be integrated into climate finance strategies facilitating comprehensive approaches to adapt to climate (Borghi 2024). Climate finance encourages countries to cooperate and share knowledge and technology. Providing technical assistance and capacity-building support to developing nations is one way through which developed nations can enable them to enhance their governance frameworks in a manner that better positions them to access and deploy climate financing to further their climate goals. Such an integrated model can strengthen the capacity of climate actions.

Limitations of International Climate Finance

Arguably the greatest setback of global climate finance is the fluctuation of funding streams. The Caribbean region has witnessed a decline in international climate finance since 2019 (Mohan 2023), and the economic ramifications of the COVID-19 pandemic have only worsened (Oyedele and Iember 2021; Omale et al. 2021).

Such fluctuations can be of such magnitude that they impact the planning and execution of climate projects, wasting time and resources. Thus, the complexity of climate finance mechanisms can also impede access for developing countries where these flows could be used. The reality is that most countries lack the institutional capacity and technical expertise to navigate the complexities of international funding processes. As noted by Bhandary (2024), the very structure of climate finance mechanisms often erects obstacles that can impede adequate resource mobilization, especially for smaller and less developed nations.

Moreover, how climate finance would be allocated is also an issue that pertains to equity and fairness, especially in consideration of the principles of "common but differentiated responsibilities". Pickering et al. (2015), it is important to establish mutually held norms about equity to better distribute resources more equitably among recipient countries. If these issues are not addressed, there is a risk of further entrenching existing inequities in access to climate finance. Corruption, inefficiency or the lack of political will could also mean climate finance was not effective enough. The contentious definition of what constitutes "climate-related" finance will, as Bracking and Leffel (2021) explain, entail a misallocation of resources as well as institutions that act on climate policies that cannot be deemed effective. Strengthening governance and accountability systems will be needed if climate finance is to be used effectively and transparently.

Green Bonds and Sustainable Investment Strategies

Green bonds are a powerful form of finance for a sustainable future and are being used widely to contribute to climate change action and sustainable development. These bonds are earmarked exclusively for projects that create positive environmental benefits, like renewable energy, energy efficiency, sustainable agriculture, and clean transportation.

The Rise of Green Bonds

In the last decade, the green bond market has rapidly expanded on the premise of climate risk awareness and the sustainable financing solutions required. According to Liaw (2020), green bonds have emerged as one solution to the growing need to finance climate and environmental solutions, providing a source of capital that is appealing to investors interested in ensuring their portfolios reflect their sustainability goals. Green Bonds have great relevance to the financial market's growth also because they provide a structured framework for market development as the green bond principles (GBP), based on guidelines for issuers (market facilitators that consolidate and generate data to measure results).

Mostly as government-issued bonds, they trade at a premium with buyers that have ESG mandates. An issuing organisation of substance can bond itself under a greenish for which either its core business does not hinder sustainability or it maintains a focus towards becoming more sustainable. This will ensure a positive economic feedback loop on green-issue investment forms, echoing the earlier signalling effect of a sustainable economy in turn drives investment growth in the green bonds.

Sustainable Investment Strategies

Sustainable investment strategies include an array of approaches to incorporate environmental, social and governance factors into investing, and the data feeds a burgeoning demand for more sustainable investment strategies. They want to adopt a market-based, financial return perspective toward the generation of social and environmental impact. Growing evidence of firm time preferences for sustainability (Auer 2021) also suggests a willingness to mix economic and non-economic, or traditional and newer investment forms. One of the main components of sustainable investment strategies is the incorporation of environmental, social, and governance (ESG (Singh et al. 2024a)) criteria within the investment process. It will allow the investors to analyze the potential risk and reward their investment decisions, leading to a good investment strategy. As pointed out by Escrig-Olmedo et al. (2017) sustainable investment is a broad concept which encompasses responsible investment and socially responsible investment (SRI). By integrating these factors into their investment processes, investors can align their portfolios with their values and respond to some of the greatest challenges facing our planet, including climate change, inequality and resource scarcity.

The Role of Private Capital in Climate Financing

The transition to a sustainable, low-carbon economy necessitates substantial financial resources, and private capital has emerged as a critical component in climate financing. As public funding alone is insufficient to meet the growing demands for climate action, leveraging private investments is essential for achieving ambitious climate goals, such as those outlined in the Paris Agreement.

Opportunities for Private Capital in Climate Financing

Republican politicians put global warming and climate crisis on the back burner, while private capital can mobilize trillions into well-targeted climate solutions with proper policy support. By mobilizing private sector investments in climate action, not only the developing world, which faces critical adaptation and mitigation needs, can be assisted, but also climate action can be integrated into private sector development strategies (Adhikari and Chalkasra 2021). The private sector (Nwokolo et al. 2024) has much of the capital needed to finance efforts to meet net zero targets, including renewable energy projects, sustainable infrastructure, and other sustainability efforts. And blended finance, which marries concessional public funding (of various sorts) with private investments, is another very promising approach to mobilizing capital for climate. Tan (2021) outlines governance for private financing for development but does not specifically define governance related to blended finance for climate projects through time. However, blended finance does pool risk for private investors to bring in more funding for sustainable development goals, this should never be belittled.

Simultaneously, exciting new pathways have emerged for private investment in climate action with the introduction of new financial instruments, including green bonds and climate-linked loans. Green bonds, in particular, have created quite a stir, as a way for companies and even governments to directly raise capital from investors to co-finance green projects. As pointed out by Price (2022), the issuance of green bonds by the Pacific Islands shows the potential of private capital in renewable energy initiatives, whilst contributing to economic activity regeneration. Sustainability has become a key component of corporate business strategy (Oyetunde et al. 2023) as lots of companies are committing to net zero emissions and embedding climate into the way they run their businesses (Alam et al. 2024). This liberalization represents an opportunity for private capital to provide climate financing, as firms seek investment in sustainable technology and practices. Climate financing complements corporate sustainability objectives, which can unleash significant capital into green projects (Chen 2023).

Challenges in Mobilizing Private Capital

Investors could be hesitant to deploy capital because of some degree of uncertainty about regulatory structures, market conditions, and project sustainability over the long term. Nevertheless, effective de-risking strategies will be a key enabler of private investment in climate solutions and multilateral development banks can be a big part of the solution. Moreover, even with the availability of capital in terms of equity capital and/or debt capital, several project developers might not be aware of opportunities in climate financing or lack the expertise to assess potential projects.

Pauw et al. (2015) highlight that low awareness of political aspects and conceptualizations of climate adaptation in the private sector can deter investments. A gap that can be closed with capacity building and climate finance awareness and that allows the private sector to get so much more involved. It's also important that the mobilization of private capital addresses equity and inclusion concerns and climate action by the private sector must be designed to address social inequality and benefit the marginalised (Abdoul-Azize et al. 2021). Equal access to climate financing for all stakeholders, especially marginalized groups, is critical to building a broad base of support for climate initiatives. Good governance and policy frameworks are key to providing an enabling environment for private capital for climate financing. Public policies need to establish clear regulations and incentives that would encourage private investments while maintaining accountability and transparency over fund utilization.

Public Funding: National Budgets and Climate Priorities

Public finance is critical to addressing climate change and to our transition to a sustainable, low-carbon economy. In our view, national budgets are indispensable instruments for governments to prioritise long-term climate-oriented investments by reforming and realigning spending away from activities that continue to contribute to greenhouse gas emissions and to activities that would foster resilience against climate change hereafter.

Importance of Public Funding for Climate Action

Public funding is important to implement climate policies and to finance projects; they are particularly needed in developing countries where financial means are usually scarce (Oyenuga and Omale 2024). As noted by Dasandara et al. (2022), climate finance is critical to realizing emissions reductions, fortifying communities, and enabling adaptation. By investing in climate-friendly policies (Kumar et al. 2024), national budgets can also serve as mechanisms to mobilize resources for renewable energy projects, sustainable infrastructure, and disaster risk reduction initiatives. Making the above investments will enable governments to lay the groundwork for long-term, sustainable development immediately needed to fight climate change. Moreover, public funding can unlock private-sector investment in climate solutions. Even if government coffers are empty, investing in weather adaptation, public investments such as levees, seawalls and flood-proofing compounds boost private capital by signalling to investors that the economy is serious about sustainability. Leveraging public funds can amplify the action of climate financing (Abdoul-Azize 2021) due to the role of non-state actors in driving government climate action.

Integrating Climate Priorities into National Budgets

This timely embedding of climate priorities into national budgets will ensure that climate action is both funded and prioritized (Nwokolo et al. 2023b). This refers to the alignment of budgetary allocations with national climate priorities, including those outlined in Nationally Determined Contributions (NDCs) as per the Paris Agreement. A well-designed climate finance strategy should integrate all relevant public and private sources of funding (Adhikari and Chalkasra 2021). Governments need to present clear frameworks on how their resources will be allocated in climate projects; climate budgeting is thus an essential first step towards an all-integrated climate framework.

This involves defining priority areas for investment, establishing measurable targets, and tracking progress. So, by following a climate budgeting approach, the government can hold their public money accountable ensure that there is no embezzlement and ensure transparency in the use of public funds as well, which will eventually instil public confidence in the climate-related initiatives taken.

Letting into the room this public money allocation balance toward climate priorities has important consequences for effective climate governance. First, it affirms the commitment of governments to act on climate change and to play a leadership role in the global agenda on climate (Nwokolo et al. 2023a, b, c). National budgets aligned with climate action reinforce the credibility of governments in global climate negotiations. Secondly, public funding can catalyse the building of local capacities and expertise concerning climate action; As emphasized by Berry et al. (2018), direct investments into vulnerable communities, based on realistic projections of climate-deflation impacts, will be effective in fostering resilience and adaptive capacity, while climate finance will also need to be directed toward the most impacted groups by climate change. The emphasis on equity and inclusion is critical to garnering social acceptance of climate actions and preventing any community from being left behind in the shift to a sustainable future. Lastly, when utilized effectively, public funding can spur innovation and technology development for climate action.

Governments can also invest in advanced research and development to support the development of new technologies and solutions (Apata et al. 2024) for climate change. Such investment in innovation results in job creation, economic growth, and increased competitiveness of the South African green economy in the global marketplace.

Barriers to Accessing and Utilizing Climate Finance

Unblocking climate finance is essential for countries, particularly developing countries, to take climate action and meet their Nationally Determined Contributions (NDCs) under the Paris Agreement. Nonetheless, there are several room hurdles in mobilizing and effectively using the climate finance, though it has emerged as the key to achieving sustainable development and climate resilience.

A key obstacle to accessing climate financing is the lack of an institutional structure to mobilise and disburse funds. As noted by Román et al. (2017), the structure and impact of climate finance are fundamentally dependent on the ability of developing countries to mobilize available resources in line with national development strategies. Such inefficiencies may stem from weak governance structures, absence of regulatory frameworks and coordination failures of government agencies. The fractured character of those programs can result in lost chances to invest in climate initiatives if would-be financial backers are put off by perceived hazards associated with rickety rule-making. A further hurdle to accessing climate finance is the ability to design, implement, and monitor climate projects. For instance, in developing countries, the pipeline for developing high-quality, bankable projects that provide the necessary detail for international climate finance commodity mechanisms is often limited by a lack of in-country technical capacity and human capital resources (Singh et al. 2024a, b). Bhandary et al. (2021) highlight that the lack of quantifiable incentives and firms' reluctance to internalise environmental externalities are additional barriers to the mobilization of private finance. These necessary climate solutions will not be effectively implemented without substantial capacity-building efforts to unlock the funding that will otherwise not be available.

Besides, Climate finance needs to be transparent and subject to accountability but so many nations find it hard to ensure that money is being spent properly and effectively. The clear absence of Monitoring, Reporting and Verification (MRV) systems, as outlined by Chelminski (2022), makes it impossible to compare funded greenhouse gas emissions and ascertain the effectiveness of mitigation strategies financed.

Implications

The transition to a net zero economy is a multi-faceted challenge requiring large-scale investment and innovative financing solutions. The consequences of funding this transformation are monumental, influencing all facets of economic development, social equity and environmental sustainability. In this sense, those implications are important for policymakers, investors and stakeholders as they chart a course towards net zero emissions. The transition to net zero can drive economic growth and new job creation (Nwokolo et al. 2023b). The transition to sustainable practices can create industries segmenting from the traditional measures and reprogramming them (Verfuerth 2023).

Scaling up climate finance can help governments face their environmental challenges without forgoing economic progress. The financing of net zero cannot be all-gains, no-pain: there also needs to be a strong emphasis on the social equity of transition. Based on the work of Verfuerth (2023), such an approach is essential to ensure that vulnerable groups, who may disproportionately pay the cost of climate policies, have their needs and concerns adequately considered. Structuring climate finance to maximize benefits for low-income and marginalized communities can also increase the social acceptability of climate policies and encourage public support for, and participation in, climate action. Such an equity focus can alleviate any potential social friction and amplify the sense of ownership felt by the communities.

One important avenue toward that goal is climate finance, which will be less than a tenth of the magnitude of private sector investment by financial flows but a critical way to facilitate vulnerable developing countries to adapt better and address different aspects of climate change. This could mean allocating resources for adaptation efforts that build upon existing infrastructure, craft disaster response plans, or encourage land management practices, to name a few that help stakeholders build resilience in communities. As emphasized by Bray et al. (2022), tackling climate change requires the need for labour and asset mobilization for a just transition.

The implications for governance and institutional capacity are significant for financing this transition to net zero. Similarly, requisite governance frameworks are essential for the transparent and effective allocation of resources. Murau (2023) has excellently underlined that financing is needed for the green transition and that money needs to be re-architected so that it goes to where it will do good. There must be an enhancement of institutional capacity for planning, managing, monitoring and evaluating the action taken on climate so that the climate finance being deployed serves the best purpose.

On the other hand, moving to net zero is a national issue but it is also a global issue that we need international support and partnership. Understanding local climate adaptation needs would streamline global knowledge sharing and partnerships, free of charge. As noted by Seto et al. (2021) this will be an integral step toward reducing the transboundary nature of climate change that human activities have created and ensuring net zero emission through coordinated actions across borders. The article argues that it was the role of governments to build a more sustainable and resilient global economy through collaboration with private sector actors and civil society.

Conclusion

The implications of financing the net zero transition are immense, spanning economic growth, social equity, resilience, governance and international cooperation. Governments, the private sector, and donors can lead progress by putting climate finance at the top of their agenda and making effective policy choices to mobilize resources needed to achieve net zero emissions. However, the challenges of financing this transition are also important for ensuring that the benefits are equitably shared and vulnerable communities are not excluded. The years ahead must focus on improving both to be able to formulate a climate finance strategy to is holistic and inclusive, which will be crucial towards a sustainable and resilient future for all of us.

References

Abdoul-Azize H, Salé A, Soulé M (2021) Private telecommunication companies and climate finance: evidence from the Niger Republic. Int J Environ Clim Change 374–385. https://doi.org/10.9734/ijecc/2021/v11i1230588

Adhikari B, Chalkasra L (2021) Mobilizing private sector investment for climate action: enhancing ambition and scaling up implementation. J Sustain Financ & Invest 13(2):1110–1127. https://doi.org/10.1080/20430795.2021.1917929

Alam MN, Turi JA, Bhuiyan AB, Kharusi SA, Oyenuga M, Zulkifli N, Iqbal J (2024) Factors influencing intention for reusing virtual reality (VR) at theme parks: the mediating role of visitors satisfaction. Cogent Soc Sci 10(1). https://doi.org/10.1080/23311886.2023.2298898

Apata AO, Oyenuga MO, Abeoba M, Ugom A (2024) Internet of Things (IoT) solutions for smart transportation infrastructure and fleet management. J Propuls Technol 45(4):1492–1505. https://propulsiontechjournal.com/index.php/journal/article/view/8325

Auer B (2021) Implementation and profitability of sustainable investment strategies: an errors-in-variables perspective. Bus Ethics Environ & Responsib 30(4):619–638. https://doi.org/10.1111/beer.12334

Berry P, Enright P, Shumake-Guillemot J, Prats E, Campbell-Lendrum D (2018) Assessing health vulnerabilities and adaptation to climate change: a review of international progress. Int J Environ Res Public Health 15(12):2626. https://doi.org/10.3390/ijerph15122626

Bhandary R (2024) The role of institutional design in mobilizing climate finance: empirical evidence from Bangladesh, Brazil, Ethiopia, and Indonesia. PLoS Clim 3(3):e0000246. https://doi.org/10.1371/journal.pclm.0000246

Bhandary R, Gallagher K, Fang Z (2021) Climate finance policy in practice: a review of the evidence. Clim Policy 21(4):529–545. https://doi.org/10.1080/14693062.2020.1871313

Borghi J (2024) Climate finance opportunities for health and health systems. Bull World Health Organ 102(5):330–335. https://doi.org/10.2471/blt.23.290785

Bracking S, Leffel B (2021) Climate finance governance: fit for purpose? Wiley Interdiscip Rev Clim Change 12(4). https://doi.org/10.1002/wcc.709

Bray R, Montero AM, Ford R (2022) Skills deployment for a 'just' net zero energy transition. Environ Innov Soc Trans 42:395–410. https://doi.org/10.1016/j.eist.2022.02.002

Chelminski K (2022) Climate finance effectiveness: a comparative analysis of geothermal development in Indonesia and the Philippines. J Environ & Dev 31(2):139–167. https://doi.org/10.1177/10704965211070034

Chen J (2023) Assessing privatization efficiency in cis countries: financial, non-financial, and GDP-linked indicators. Eur J GovMent Econ 12(2):157–174. https://doi.org/10.17979/ejge.2023.12.2.9850

Dasandara M, Ingirige B, Kulatunga U, Fernando T (2022) Climate financing barriers and strategies: the case of Sri Lanka. J Financ Manag Prop Constr 28(2):242–259. https://doi.org/10.1108/jfmpc-12-2021-0069

Escrig-Olmedo E, Rivera-Lirio J, Torres M, Izquierdo M (2017) Integrating multiple ESG investors' preferences into sustainable investment: a fuzzy multicriteria methodological approach. J Clean Prod 162:1334–1345. https://doi.org/10.1016/j.jclepro.2017.06.143

Joshi A, Singh R, Kumar V, Rani S (2024) Net zero strategy: exploring opportunities, and confronting challenges for developing countries. In: Approaches to global sustainability, markets, and governance, pp 233–252. https://doi.org/10.1007/978-981-97-5756-5_11

Kumar K, Singh R, Sharma A, Lehri BPS (2024). Climate change adaptation strategies for achieving net-zero economy. In: Approaches to global sustainability, markets, and governance, pp 163–183. https://doi.org/10.1007/978-981-97-5756-5_8

Liaw K (2020) Survey of green bond pricing and investment performance. J Risk Financ Manag 13(9):193. https://doi.org/10.3390/jrfm13090193

Mohan P (2023) Financing needs to achieve nationally determined contributions under the Paris agreement in Caribbean small island developing states. Mitig Adapt Strat Glob Change 28(5). https://doi.org/10.1007/s11027-023-10062-9

Murau S (2023) Monetary architecture and the green transition. Environ Plan Econ Space 56(2):382–401. https://doi.org/10.1177/0308518x231197296

Nwokolo SC, Singh R, Khan S, Kumar A, Luthra S (2023a) Remedies to the challenges of renewable energy deployment in Africa. In: CSR, sustainability, ethics & governance, pp 59–74. https://doi.org/10.1007/978-3-031-44514-9_3

Nwokolo SC, Singh R, Khan S, Kumar A, Luthra S (2023b) Global investment and development in Africa. In: CSR, sustainability, ethics & governance, pp 15–58. https://doi.org/10.1007/978-3-031-44514-9_2

Nwokolo SC, Singh R, Khan S, Kumar A, Luthra S (2023c) Africa's awakening to climate action. In: CSR, sustainability, ethics & governance, pp 299–310. https://doi.org/10.1007/978-3-031-44514-9_9

Nwokolo SC, Singh R, Khan S, Kumar A (2024) Key players in net zero technology. In: Elsevier eBooks, pp 193–215. https://doi.org/10.1016/b978-0-443-31486-5.00007-0

Omale SA, Oyenuga MO, Oriaku C (2021) Effect of organizational culture on the performance of hospitality industry in Covid-19 era. Abhigyan Manag J 39(3):24–32. https://doi.org/10.56401/abhigyan/39.3.2021.24-32

Oyedele OM, Iember AA (2021) Covid-19 and the Future of Higher Education. IEEE Technol Policy Ethics 6(4):1–3. https://doi.org/10.1109/ntpe.2021.9778140

Oyenuga MO, Omale SA (2024) Is Africa jinxed? Exploring the challenges of technology access and adoption in Africa. Afr J Econ Sustain Dev 7(4):142–161. https://doi.org/10.52589/AJESDULN1LRNF

Oyetunde TO, Oyenuga MO, Adoga GJ (2023) Green management practices and organisational efficiency in higher institution. Bp Int Res CritS Linguist Educ (BirLE) J 6(3):276–290. https://bircu-journal.com/index.php/birle/article/view/7716

Pauw P, Klein R, Vellinga P, Biermann F (2015) Private finance for adaptation: do private realities meet public ambitions? Clim Change 134(4):489–503. https://doi.org/10.1007/s10584-015-1539-3

Pickering J, Jotzo F, Wood P (2015) Sharing the global climate finance effort fairly with limited coordination. Glob Environ Polit 15(4):39–62. https://doi.org/10.1162/glep_a_00325

Price R (2022) Private sector investment in the clean energy sector in the Pacific islands. https://doi.org/10.19088/k4d.2022.132

Román M, Arto I, Ansuategi A (2017) Why do some economies benefit more from climate finance than others? A case study on north-to-south financial flows. Econ Syst Res 30(1):37–60. https://doi.org/10.1080/09535314.2017.1334629

Seto KC, Churkina G, Hsu A, Keller M, Newman PW, Qin B, Ramaswami A (2021) From low- to net-zero carbon cities: the next global agenda. Annu Rev Environ Resour 46(1):377–415. https://doi.org/10.1146/annurev-environ-050120-113117

Singh R, Gupta A, Kandpal V (2024a) ESG excellence in the digital world. E-Bus Solut 193–204. https://doi.org/10.1002/9781394234035.ch10

Singh R, Crowther D, Ravesangar K (2024b) Harnessing human capital for a sustainable future: HR's pivotal role toward net zero policies. In: Approaches to global sustainability, markets, and governance, pp 27–47. https://doi.org/10.1007/978-981-97-5756-5_2

Tan C (2021) Audit as accountability: technical authority and expertise in the governance of private financing for development. Soc Leg Stud 31(1):3–26. https://doi.org/10.1177/0964663921992100

Verfuerth C (2023) A people-centred approach is needed to meet net-zero goals. J Br Acad 11:97–124. https://doi.org/10.5871/jba/011s4.097

Vijay V, Sathish AS (2024) Advancing towards a sustainable future: a systematic review of sustainable net zero economy implementation studies. Qubahan Acad J 4(1):137–149. https://doi.org/10.58429/qaj.v4n1a279

Zhou K (2024) Carbon finance and funding for forest sector climate solutions: a review and synthesis of the principles, policies, and practices. Front Environ Sci 12. https://doi.org/10.3389/fenvs.2024.1309885

Chapter 20
Energy Transition: Fossil Fuels Versus Renewables

Abstract The shift from fossil fuels to renewables is a top priority in the world to fight against climate change and promote sustainable development. This chapter explores the politics of energy transition, detailing fossil fuel and renewable comparative advantage and risk. It highlights the importance of technological innovation, policy frameworks, and market mechanisms in facilitating the transition. Additionally, there is a conversation on the geopolitical implications of the energy transition, particularly in terms of energy security and economic development. This transition will be examined in the chapter, particularly looking at the cases of countries that have transitioned, to explore whether additional innovations and emerging trends could pave the way for an environmentally sustainable and economically viable energy transition.

Introduction

As the alarm of the climate change crisis grows urgent, the world energy system is in a massive transformation. This transition, from fossil fuel to renewable energy is a prerequisite not only for greenhouse gas emission reduction but also for sustainable economic growth and energy security.

The transition to renewable energy has been a rallying-pissing point on international and national policies to vasectomy low-carbon energy to achieve a key component of the commitments made as part of international climate agreements, such as the Paris Agreement. Based on fossil fuel which continues having lead industrial development it is still present in the world energy mix. But their environmental impact, along with the reality that these resources are finite, have caused an energy strategy rethink. As highlighted by Wang et al. Specifically, when it comes to the absence of fossil fuels (Wang et al. 2023) to lean on, this endeavour toward a sustainable energy future takes a second look at strategies that energy supply that must be shifted to renewables. This transition is not only a technical issue it is also a socio-political project that has to involve economic, environmental and social goals.

M. 0. Oyenuga et al., *Smart Technologies for Climate Change and Net Zero Policies: Practical Approaches Towards Sustainability*, Climate Change Management,
https://doi.org/10.1007/978-3-031-92221-3_20

Technologies for renewable energy resources like solar, wind, hydro, and biomass thus hold the promise of providing alternative fossil fuels (Apata et al. 2024).

Though they are both critical for reducing carbon emissions and for job creation, energy independence, and resilience to climate impacts, the transition wouldn't be easy, with obstacles that include technological shortcomings, economic considerations and a societal aversion to altering consumption patterns. Policy frameworks, technological innovation and market dynamics all provide the transformative vehicle which takes us from fossil fuels (primary energy) to the energy of the future.

The Case for Renewables

Solar energy, wind energy, hydroelectric power, and geothermal energy are all renewable energy sources that provide a cleaner alternative to fossil fuels. They are plentiful, becoming ever more cost-effective, and can dramatically reduce greenhouse gas emissions (Nwokolo et al. 2024a). Transitioning to renewable energy facilitates new job growth in energy efficiency and clean transport technologies, promoting innovation and potential long-run economic growth. In addition to their ability to accelerate climate change mitigation efforts, renewable energy resources have the added environmental benefits of reduced air and water pollution, making them even more attractive as the issue of global warming faces mankind shortly. Technological improvements and falling costs also drive the move toward renewables. As highlighted by Daneeva et al. (2020) these technological developments provide new opportunities for integrating renewable energy within existing energy systems through new developments involving energy storage, grid management, and energy efficiency. It is in this technological space, supplemented by warm policy ecosystems, that we can progress to a managed disengagement from fossil fuels.

The Economics of Fossil Fuel Dependence

Fossil fuel reliance continues to be a defining feature of global energy systems, especially in many of the developing and emerging economies. This dependence on fossil fuel raw materials including coal, oil, and natural gas has enormous economic consequences, impacting everything from nation-state budgets to international trade flows. Though fossil fuels have long served as a basis for economic growth and development, their ongoing employment presents considerable dangers.

Economic Benefits of Fossil Fuel Dependence

The story of fossil fuels has been the story of fuelling industrialisation and growth around the world. Countries blessed with fossil fuel resources such as Saudi Arabia and Russia have utilized their energy wealth for meaningful economic development. Income from the extraction and export of fossil fuel contributes significantly to national budgets and finances public services and infrastructure projects. As highlighted by Xue et al. (2021), many South Asian economies have relied heavily on fossil fuels to fuel their energy needs, and to do so has promoted economic expansion albeit with high environmental costs associated with this dependency. The fossil fuel industry supports millions of jobs globally, from extraction and refining to transportation and distribution. Where fossil fuel jobs are still a part of the local economy and no significant alternative energy source is yet viable, coal jobs are often considered essential for the area's economic well-being. Fossil fuels were also previously regarded as a reliable source of energy, paving the way for energy security and stability. Fossil fuel-rich nations tend to wield a certain amount of geopolitical power because their oil and gas exports are indispensable to global markets.

Economic Drawbacks of Fossil Fuel Dependence

While renewable energy has obvious benefits, the transition is not without challenges. Managed by chronic oppression, fossil fuels remain at the core of the world's energy system. The most-sacrificed decisions will naturally come under the greatest political and economic pressure to be reversed, especially in fossil fuel-rich countries. According to Phan, the transition toward fossil fuels is being slowed down as the oil demand is still so high worldwide. Moreover, not transitioning from fossil fuels will have geopolitical ramifications you can't ignore. Countries that rely on fossil fuel exports for revenue could face economic uncertainty as global demand shifts.

This can create tensions between nations and make international cooperation on climate action more complex. The shift to renewables needs to be managed carefully so that energy security is enhanced while carbon emissions are reduced. The reliance on fossil fuels makes economies vulnerable to substantial volatility in global oil and gas prices. When prices fall, it can trigger economic downturns, budget deficits, and cuts to public spending. As noted by Cuma et al. (2016), economic drawbacks may arise in dependence on fossil fuels in countries that have no or little diversity in the energy mix. This adds to the difficulty of long-term planning and investment in sustainable alternatives.

Environmental Effects of Fossil Fuel Dependence

This describes the widespread air and water pollution, habitat destruction, and climate change caused by fossil fuel dependence the environmental damages can have substantial economic costs, including, for example, the effects of pollution on public health and health care costs. The economic benefits of fossil fuels can be compromised by the ecosystem degradation and health risks associated with fossil fuel extraction and consumption (Adeojo and Osinibi 2016). Dependence on fossil fuels has a direct impact on geopolitical dynamics amongst countries, with interaction at the axis of competition and resource exploitation with pressure to move to alternative and renewables-based energy sources. Fossil fuel wealth benefits accrue to a small elite while local people face displacement, health warnings, and loss of livelihoods. The shrinking of fossil fuel industries can further exacerbate the areas in which employment, livelihoods and economic activity are concentrated, which rightly raises concerns about social cohesion and equity (Banerjee and Schuitema 2023).

The Energy Security Debate in Transitioning to Renewables

A transition from fossil fuels to renewable energy sources is essential to all worldwide efforts to combat climate change and reach sustainable development. But this transition comes with some vital questions around energy security, which is defined as the availability of energy at all times in various forms, in sufficient quantities to match the demands of people. The energy security discussion in terms of moving towards renewables needs to reflect multiple layers, ranging from renewable sources' reliability to geopolitical implications associated with changing energy dependencies, to economic effects associated with such shifts.

The Need for Transition

There are economic disadvantages when countries depend solely on fossil fuels. The transition is vital as it contributes to combatting climate change but also builds urban climate resilience in the long run. According to Johnson (2017), investment in renewable energy technologies stimulates economic growth, creates new job opportunities, and reduces dependence on volatile fossil fuel markets. Furthermore, the low carbon economy is a more diverse energy economy and less exposed to geopolitical risks. With more and more countries realizing the economic and environmental need to stop fossil fuels, the development of overall policies and investment is crucial to promote the transition.

Reliability of Renewable Energy Sources

Supply reliability is one of the key points of concern around energy security in the renewables transition. While fossil fuels yield a constant and controllable energy output, renewable energy sources like solar and wind are by nature variable and reliant on environmental factors. As mentioned by Ríos-Ocampo et al. (2021), a transition to high shares of renewables in electricity markets can cause problems with the stability of supply and its management of the grid. Such variation creates a need for large energy storage systems and more transmission lines to avoid supply problems which can be technically and financially challenging. In addition to renewable energy, energy storage technologies are becoming increasingly important in decarbonization pathways and energy security needs to be addressed accordingly. Storage technologies, such as batteries and pumped hydro, can play a role in some cases to help smooth the use of a spectrum of renewable resources, but they depend on a lot of investment and technical development. According to Cergibozan (2022), the successful incorporation of renewables in the energy mix will be determined by the capability to effectively create and produce storage options.

Phasing Out Fossil Fuels: Political and Economic Challenges

The earth is facing transformational change which offers big challenges and big opportunities. However, the political and economic hurdles in this transition can be substantial, creating significant backsliding. By examining the experiences of other regions, policymakers, stakeholders, and communities can better strategize in the face of the complex reality of moving away from fossil fuels and towards cleaner energy resources.

Political Challenges

The shift from fossil fuels to renewables is also full of immense political obstacles. Countries where fossil-fuel exports make up a large customer for the world will be big losers as oil and gas demand decline everywhere. According to Aliyev (2024), states in areas like the South Caucasus must also revise their energy policies according to their essential need for energy security while still transitioning to renewables. This paradigm shift can result in fluctuations in global power relations since nations that adopt renewable technologies can attain geopolitical advantage, while those that do not adapt could face economic downturn. In addition, the shift to renewables may change conventional energy trade patterns. Fossil fuel importer countries could diversify their energy portfolio through local investments in renewable energy production. As the world begins to navigate new energy landscapes, tensions between nations are likely to increase in ways that might change energy independence and create new energy security concerns. As pointed out by Paravantis and Kontoulis (2020), the geopolitical dynamics attached to energy security need to be considered in the transition to renewables.

One of the biggest political challenges to transitioning away from fossil fuels is the absence of political will among the leaders and policymakers (Oyenuga and Omale 2024). In numerous nations, fossil fuel industries produce political clout that can tend to the type of policies that perpetuate the status quo. The well-dreamed interests of fossil fuel producers can make it difficult to bring about policies that challenge dependence on energy sources associated with economic, environmental and human costs. This can take the form of fossil fuel subsidies, regulatory obstructions to renewable energy development inadequate funding for climate initiatives. Poor energy transition at both national and global levels can be negated by inconsistencies and complexity of policies. But this fragmented approach has resulted in many countries pursuing energy transition plans that are thin on coherence at best and, at worst, are inadequate in addressing the systemic nature of the problems at hand through climate policies that are relevant across sectors and levels of government are required to create a predictive pathway to phase out fossil fuels.

Economic Challenges

The economics of moving to renewable energy are complex. The transition has the potential to create new employment opportunities and boost economic growth within the renewable energy sector, but can also result in job losses in fossil fuel industries. According to Eze (2023), this transition needs to be managed to protect workers in fossil fuel jobs, and that support comes through retraining and reskilling programs. This is necessary for advancing a just transition that alleviates social disruption and strengthens economic resilience. First, renewable energy initial capital costs are sometimes large. Governments and private investors (Nwokolo et al. 2024b) must cooperate to establish financing mechanisms that support the transition while ensuring that energy remains affordable for consumers. Although the cleanup is vital to the battle against climate change, the shift to smoother fuel types mustn't impact energy sustainability and accessibility for all (Ugwu 2024). Transitioning away from fossil fuels to renewables requires a gigantic financial spend. Nevertheless, countries experience significant investment disparities in financing renewable energy projects.

The key hurdle developing economies face in the creation of renewable energy infrastructure is the infrastructure itself this initial capital requirement (Joshi et al. 2024). Without adequate funding and financial innovation, this transition could be delayed or stalled. In addition, the shift away from fossil fuels could lead to massive unemployment in traditional energy sectors, representing a financial threat to workers and communities reliant upon these industries. And we need to manage the transition with care to ensure that those workers impacted by this change are supported through retraining and reskilling programs. Neglecting these economic considerations can exacerbate social unrest and invite resistance to the transition. Transition to renewable energy for both the economy and the environment. As we move away from our dependence on fossil fuels, volatility in the prices of fossil fuels can create uncertainty regarding economic trends. The transition also has the potential to disrupt existing energy markets, forcing them to adapt and accommodate new technologies.

Balancing the Transition

It is necessary to mitigate the challenges of energy security and ensure that renewable energy transition does not jeopardise access to reliable and affordable energy. As Yang et al. and (Nwokolo et al. 2023) noted, one must also navigate the geopolitics of energy transition to ensure we do not exacerbate pre-existing conflicts and inequalities. This must also include a consideration of social equity in this transition, ensuring that all communities can access and get the benefit of this renewable energy. According to Capellán-Pérez et al. (2017) addressing the needs of marginalized populations is central to garnering public support and attendance for climate initiatives. Inclusive policies that consider the socio-economic impact of the transition enable stakeholders to co-create a just, equitable, and sustainable energy transition going forward.

The Role of International Organizations in Supporting the Transition

To the power system from 1e international organizations that enable essential services in support of renewable energy transition such as funding, expertise policy advice. The organisations are increasingly being recognised for their contribution to promoting a more sustainable energy sector while helping to reduce fossil fuel consumption—as the world grapples with the desperately urgent need to end climate change and for international solidarity.

International organizations primarily exist to mobilize financial means for renewable energy projects. Institutional assistance for the generation of renewable energy infrastructure is already supported through funding and investment, to developing nations by institutions such as the World Bank, the International Renewable Energy Agency (IRENA) and regional development banks. For instance, the World Bank has provided development and financing for solar and wind projects in developing countries to bridge the financing gap that often makes the transition to clean energy more challenging (Quitzow et al. 2019). Additionally, several international organizations have advocated innovative financing mechanisms—so-called green investments, climate funds, etc.—to attract private investment in renewable energy.

They not only serve as funding sources for projects but also as indicators of the commitment of governments and organizations to sustainable development. Indeed, Quitzow et al. (2019) highlight how the growing emphasis on renewables among international agencies has put upward pressure on ODA flows to clean energy, yet thcre is still a long way to go to ensure equitable access to these resources.

These organizations help countries formulate coherent and effective energy policies by providing guidelines, best practices, and technical assistance. IRENA has helped lead the way in supporting renewable energy law and policy frameworks and their alignment with the Sustainable Development Goals (SDGs) and national climate commitments (Overland and Reischl 2018). Since many forums and conferences are organized by an international organization that collects policy-makers, industry leaders, and civil society actors to discuss challenges and share experiences on renewable energy transitions in the world, some other resources may be available in these organizations. These will promote lessons learnt and cross-pollination between the platforms, and enable countries to learn from one another and replicate how they implemented successful approaches to implement renewable energy projects.

International organizations are also heavily involved in capacity building, which is another crucial effort for the transition to renewable energy. So placing renewable energy projects in developing countries often comes with huge challenges for the technical expertise and institutional capacity that these countries possess. For instance, the United Nations Development Programme (UNDP) offers technical assistance and training programmes to strengthen local capacities (Quitzow et al. 2019; Singh et al. 2024). Also, there are global initiatives such as Mission Innovation that accelerate public and private innovation in clean energy and show examples of how international collaboration can enhance renewable energy technology (Sandria 2023). Bridging this gap will require the support of tailored and accessible financial resources (Oyetunde et al. 2023) that allow countries to maximise access to climate finance and reach their renewable energy targets. Also, many times international organizations serve as a mediator during negotiations about climate agreements, this serves to align the interests of different countries and promote collective action. The motivation for a global shift to renewable energy is indebted to international accords which institutionalise the process (Overland and Reischl 2018).

Case Studies from Countries Transitioning from Fossil Fuels

Each country faces its own political, economic and social challenges and opportunities as it seeks to decarbonize its economy and put in place sustainable energy practices.

Bulgaria

Moving to Green Energy, Bulgaria has been in a significant policy shift, amid investments in green technologies, transitioning to renewable energy. A variety of support policies directed at renewable energy resources, predominantly solar and wind energy, have been established in the country (Byanova 2024). Yet, Bulgaria faces the dual challenge of being shaken by its historical dependence on coal and the need to modernize its energy infrastructure. The government has attempted to develop a national planner of energy to enhance and integrate renewable energy into the national grid, by ensuring energy security as well as affordable prices.

China

China has a heavy focus on reducing carbon emissions and improving energy security which places it as the top country for both domestic and international investments in renewable energy. According to Chen et al. (2013), renewable energy technology covering solar and wind has been regarded by the Chinese government as a top priority in its national development strategy for enhancing energy security and fostering economic development. The government has introduced policies luring the private sector to engage in renewable energy projects, which have seen considerable technology and capacity improvements. However, obstacles continue, for example in managing economic growth alongside environmental sustainability and facing the social challenges of moving away from coal use.

Indonesia

Indonesia's energy transition is an uphill battle; as an emerging economy, it runs the paradox of needing to transition away from fossil fuels while also needing to expand its energy sources. Indonesia has a great opportunity to utilize renewable energy, particularly solar energy and geothermal energy (Adrian et al. 2023). Energy-rich nations have been ambitious with their targets to move towards the renewable mix. Yet pressure from the fossil fuel industry, along with the considerable investment required for a successful transition, complicates the transition. The aim must be to make it inclusive, making sure that every part of society benefits.

The Western Balkans

Regional Cooperation for a Green Agenda Implemented in the Western Balkans as a framework for the adoption of a green transition and resilient climate. According to Lobanova (2024), decarbonization, circular economy practices, and sustainable food systems are core components of the Green Agenda. However, the region's reliance on coal and the investments required to expand any renewable energy infrastructure amount to major hurdles. Implementing the Green Agenda also calls for regional cooperation and commitment of the countries to end coal subsidies and carbon pricing mechanisms.

The European Union

The EU has adopted a broad and ambitious package of measures to support the transition to renewable energy and to reduce GHG emissions, with concrete targets for the share of renewables in the energy mix, for example. The EU's commitment towards climate change is evidenced by the policies and financing mechanisms that support respective renewable energy projects across member states (Rokicki and Perkowska 2020). However, the transition will not be smooth as countries contend with the imperative of energy security, economic competitiveness and social equity. The EU method emphasizes the obligation of member states to collaborate and realise a common climate target.

Implications

Fossil fuel transition to renewable energy sources these transitions are a fundamental but also complex process with far-reaching effects on various energy systems, economies, and the environment on a global scale. As the urgency to address the climate crisis grows, countries around the world are already facing the trials and tribulations of this transition. Fossil fuels for centuries have all but powered the expansion and development of human economies, but their environmental cost and limited supply have also introduced a new question about energy systems.

The economic consequences of this shift from fossil fuels to renewables are significant. The fossil fuel industry, on the one hand, has been an engine for economic growth, job creation and government revenues in many countries. However, amid growing volatility in fossil fuel markets and an increasing awareness of the environmental costs associated with extracting and using fossil fuels, the long-term sustainability of this model is increasingly under scrutiny. Transitioning from fossil fuels to renewable energy sources can also lead to the blossoming of new sectors of economic opportunity, creating jobs in emergent sectors and reducing dependence on foreign fuels, increasing energy security. And shifting to renewables can reduce long-term costs. As technology has advanced and economies of scale have kicked in, renewable energy technologies such as solar and wind have plummeted in price—in some cases, more than 90%.

It is continuing, and renewables are becoming evermore a match for fossil fuels. Furthermore, investing in renewable energy infrastructure can also lead to significant economic benefits, as those investments can result in lower operational costs and increased energy efficiency.

This reliance on fossil fuels carries with it a host of environmental implications aspects that are key to the argument over the energy transition. Fossil Fuels are responsible for the greatest portion of greenhouse gas emissions, which fuels climate change and its rewards. The need to mitigate these impacts, alongside the growing consensus that we need to transition to cleaner energy, has never been timelier. But fossil fuels hurt our planet, and in turn, we have to transition toward energy practices that do not threaten ecological health and resilience. When it comes to sustainable energy sources, renewable energy sources have the potential for drastic reductions in carbon emissions and other pollutants. Countries that take advantage of natural resources like sunlight and wind will have less harm to the environment and be a part of a worldwide climate mission (Nwokolo et al. 2024c). But the shift needs to be managed carefully, to avoid undesirable side effects, such as land use tensions and resource exhaustion tied to some renewable technologies.

The social side of stimulating the energy transition is no less important. Transitioning from fossil fuels means job losses in legacy energy sectors, which can pose economic and social problems for impacted communities. Policies to support retraining and reskilling for fossil fuel workers must be part of a just transition and facilitate the transition of workers in fossil fuel industries into new roles in the renewable energy sector. Neglecting these social considerations will cause opposition to renewable energy projects, and will produce new, pernicious inequities. The supporters of renewable energy projects are the ones responsible for their successful establishment. The normal factors of community engagement, stakeholder approval, perceived progress, and project values all greatly influence social acceptance. Such a collaborative approach can ensure acceptance of the initiative to argue for acceptance of the initiative of renewable energy.

Conclusion

The transition from fossil fuels to renewables is the biggest challenge and highest priority that we have in avoiding climate breakdown, and responding to and recovering from the other crises of our time at the political, economic and social levels. Despite these benefits, the road to a sustainable energy future is not without its challenges. It has also underscored the importance of collaboration, the need for investment in technology and infrastructure, and the necessity of fair transition through social equity in navigating the energy transition, which enables these countries to advance towards a more sustainable future with a significant amount of resilience. The shift to renewables will only happen, however, when governments, businesses and communities choose to enact that change and prioritise sustainability in energy decisions.

References

Adeojo O, Osinibi O (2016) Assessing the intersections between renewable energy, sustainable development and the challenges of environmental justice in Nigeria. Interdiscip Environ Rev 17(2):149. https://doi.org/10.1504/ier.2016.076184

Adrian M, Purnomo E, Enrici A, Khairunnisa T (2023) Energy transition towards renewable energy in Indonesia. Herit Sustain Dev 5(1):107–118. ISSN 2712-0554. https://doi.org/10.37868/hsd.v5i1.108

Aliyev S (2024) Comparative assessment of energy security level: the case of the south Caucasus countries. Int J Energy Econ Policy 14(1):651–662. https://doi.org/10.32479/ijeep.14984

Apata AO, Oyenuga MO, Abeoba M, Ugom A (2024) Internet of Things (IoT) solutions for smart transportation infrastructure and fleet management. J Propuls Technol 45(4):1492–1505. https://propulsiontechjournal.com/index.php/journal/article/view/8325

Banerjee A, Schuitema G (2023) Spatial justice as a prerequisite for a just transition in rural areas? The case study from the Irish peatlands. Environ Plan C Polit Space 41(6):1096–1112. https://doi.org/10.1177/23996544231173210

Byanova N (2024) From fossil fuels to renewable energy: tracking Bulgaria's transition. Environ Technol Resour Proc Int Sci Pract Conf 1:90–94. https://doi.org/10.17770/etr2024vol1.7958

Cergibozan R (2022) Renewable energy sources as a solution for energy security risk: empirical evidence from OECD countries. Renew Energy 183:617–626. https://doi.org/10.1016/j.renene.2021.11.056

Chen C, Wang Q, Fang R (2013) An empirical analysis of private participation in infrastructure (PPI) projects in the renewable energy sector of China. Adv Mater Res 869–870:488–494. https://doi.org/10.4028/www.scientific.net/amr.869-870.488

Cuma M, Cengiz A, Tümay M (2016) Çukurova üniversitesinde kentsel elektrikli araç ulaşımı ve şarj İstasyonu uygulama simülasyonu. Çukurova Üniversitesi Mühendislik-Mimarlık Fakültesi Dergisi 31(2):343–354. https://doi.org/10.21605/cukurovaummfd.310314

Eze V (2023) A comparative analysis of renewable energy policies and its impact on economic growth: a review. Int J Educ Sci Technol Eng 6(2):41–46. https://doi.org/10.36079/lamintang.ijeste-0602.555

Johnson A (2017) Renewable energy: non-environmental reasons to make the transition. Pitt Polit Rev 43–48. https://doi.org/10.5195/ppr.2017.95

Joshi A, Singh R, Kumar V, Rani S (2024) Net zero strategy: exploring opportunities, and confronting challenges for developing countries. In: Approaches to global sustainability, markets, and governance, pp 233–252. https://doi.org/10.1007/978-981-97-5756-5_11

Lobanova J (2024) The green transition and energy security in the western Balkans countries, pp 544–569. https://doi.org/10.18485/iipe_gsirescu.2024.ch24

Nwokolo SC, Singh R, Khan S, Kumar A, Luthra S (2023) Remedies to the challenges of renewable energy deployment in Africa. In: CSR, sustainability, ethics & governance, pp 59–74. https://doi.org/10.1007/978-3-031-44514-9_3

Nwokolo SC, Singh R, Khan S, Kumar A (2024a) Key players in net zero technology. In: Elsevier eBooks, pp 193–215. https://doi.org/10.1016/b978-0-443-31486-5.00007-0

Nwokolo SC, Singh R, Khan S, Kumar A (2024b) Technological pathways for Africa' s net-zero economy. In: Elsevier eBooks. https://doi.org/10.1016/c2023-0-52499-1

Nwokolo SC, Singh R, Khan S, Kumar A (2024c) Net zero technology and circular economy. In: Elsevier eBooks, pp 237–247. https://doi.org/10.1016/b978-0-443-31486-5.00009-4

Overland I, Reischl G (2018) A place in the sun? Irena's position in the global energy governance landscape. Int Environ AgreemS Polit Law Econ 18(3):335–350. https://doi.org/10.1007/s10784-018-9388-y

Oyenuga MO, Omale SA (2024) Is Africa jinxed? Exploring the challenges of technology access and adoption in Africa. Afr J Econ Sustain Dev 7(4):142–161. https://doi.org/10.52589/AJESDULN1LRNF

Oyetunde TO, Oyenuga MO, Adoga GJ (2023) Green management practices and organisational efficiency in higher institution. Bp Int Res CritS Linguist Educ (BirLE) J 6(3):276–290 https://bircu-journal.com/index.php/birle/article/view/7716

Paravantis J, Kontoulis N (2020) Energy security and renewable energy: a geopolitical perspective. https://doi.org/10.5772/intechopen.91848

Quitzow R, Thielges S, Goldthau A, Helgenberger S, Mbungu G (2019) Advancing a global transition to clean energy–the role of international cooperation. Econ Open-Access Open-Assess E-J 13(1). https://doi.org/10.5018/economics-ejournal.ja.2019-48

Ríos-Ocampo J, Arango-Aramburo S, Larsen E (2021) Renewable energy penetration and energy security in electricity markets. Int J Energy Res 45(12):17767–17783. https://doi.org/10.1002/er.6897

Rokicki T, Perkowska A (2020) Changes in energy supplies in the countries of the Visegrad group. Sustainability 12(19):7916. https://doi.org/10.3390/su12197916

Sandria A (2023) Exceptions for stopping building coal-fired power plants in Indonesia. E3s Web Conf 467:05007. https://doi.org/10.1051/e3sconf/202346705007

Singh R, Crowther D, Ravesangar K (2024) Harnessing human capital for a sustainable future: HR's pivotal role toward net zero policies. In: Approaches to global sustainability, markets, and governance, pp 27–47. https://doi.org/10.1007/978-981-97-5756-5_2

Ugwu M (2024) Navigating legal and policy challenges in the energy transition: case studies from the United States and Nigeria. Int J Appl Res Soc Sci 6(4):506–517. https://doi.org/10.51594/ijarss.v6i4.988

Wang K, Ouyang L, Wang Y (2023) Study of energy transition paths and the impact of carbon emissions under the dual carbon target. Sustainability 15(3):1967. https://doi.org/10.3390/su15031967

Xue L, Haseeb M, Mahmood H, Alkhateeb T, Murshed M (2021) Renewable energy use and ecological footprints mitigation: evidence from selected south Asian economies. Sustainability 13(4):1613. https://doi.org/10.3390/su13041613

Chapter 21
Capacity Building for Green Technology

Abstract One of the crucial aspects of fostering sustainable development and environmental stewardship is the capacity building for green technology. Capacity building encompasses a range of activities aimed at improving the skills, knowledge, and institutional frameworks required for adopting and implementing green technologies, and is one of the key areas of this chapter. The chapter endeavours to present a holistic overview of capacity building as an enabling catalyst to reinforce the effective deployment of green technologies so that these green technologies can lend their impetus towards moving in tandem with worldwide momentum to combat climate change while upholding the human economy.

Introduction

Concerns surrounding climate change and environmental degradation have required a transition away from pollution towards sustainable development, achievable through the implementation of green technology. Green technology is an umbrella term that encompasses a wide range of innovations that either lessen the adverse effects on the environment or use its resources most efficiently (Nwokolo et al. 2024b). It is also important to ensure that nations are equipped with green technology expertise; many will strive to uphold their climate pledges, for instance with frameworks under the Paris Accord. Capacity building is the process of improving the skills and knowledge (Oyedele et al. 2024) as well as institutional structures, required to facilitate and manage green technologies. This has meant that, especially in developing countries, the transition to green technologies has not been accelerated by available technical knowledge, financial means or the appropriate enabling environment through policy (Oyenuga and Omale 2024; Joshi et al. 2024). Studies such as Yuan (2024) that focus on the National Sustainable Development Agenda Innovation Demonstration Zones in China, describe how technological innovation can drive green development but highlight the important role of some degree of coordination, the role of networks and the need for local capacity building, something that is strongly supported in Kissinger

M. O. Oyenuga et al., *Smart Technologies for Climate Change and Net Zero Policies: Practical Approaches Towards Sustainability*, Climate Change Management,
https://doi.org/10.1007/978-3-031-92221-3_21

but not often addressed in much of the policy documents and papers reviewed. Similarly, Guo et al. (2021) stress the vital contributions of behavioural incentives and government support to the development of green technologies innovation and contend the tripartite cooperation of government, industries, and academia is a precondition.

In addition, the fusion of digital finance and research and development (R&D) investment is essential to facilitating green technology innovation. As is shown in Sun (2024), which highlights the relationships between digital inclusive finance and R&D investment, together with synergies could emerge, which can stimulate the production and commercialization of green technologies, increasing economic resilience. A case in point is the ongoing global industrial revolution in which green technology is recognized as a new competitiveness and sustainability driver.

Developing Technical Skills in Renewable Energy and Digital Technologies

Renewable energy is one of those solutions to mitigate climate change and meet sustainable development objectives. Yet this transition requires a well-trained workforce skilled in not only renewable energy technologies but also digital tools that enhance their productivity and effectiveness and for individuals (Nwokolo et al. 2024a) and organisations to be keen on building the necessary technical education and supporting systems to successfully integrate renewable energy solutions into their infrastructure.

Instructional design plays a pivotal role in effectively imparting knowledge and skills needed for renewable energy technologies including solar, wind, and bioenergy as they advance and the need for skilled individuals who can design, install and maintain these systems grows (Singh et al. 2024) The integration of renewable energy systems is critical for climate resilience and requires innovative solutions and, most importantly, technology development (Ekechukwu 2024). This highlights the importance of adapting training curricula to address the skills gap in hands-on experience with renewable technologies and digital tools in academia and educational institutions. The application of digital twin technology in renewable energy investments, for example, has been found to improve system performance and reliability (Bassey 2024). Hence, this technology enables real-time monitoring and management of systems for renewable energy, and therefore professionals need to hold both technical and digital skills. Additionally, there are various opportunities to upskill in the emerging fields of digital finance, and the innovation of renewable energy technologies, along with many others. According to Zhao and Zhao (2023), the progression of financial markets can assist in assessing innovation projects and handling investment risks, essential for the advancement of technological innovation in renewable energy.

By integrating financial literacy education into existing technical training programs, individuals can acquire the knowledge needed to make informed decisions about financing options, project economics, and risk assessments (Fasanmi et al. 2024). Formal education and training are not enough; building a skilled workforce requires collaborative efforts from the government, the private sector, and educational bodies. For instance, knowledge-sharing and technology transfer initiatives from developed to developing nations can help bolster local capabilities in renewable energy technologies (Eder et al. 2015). Partners with a strong presence in training could also have the capacity to preferentially access sources of funding and mobilize resources to implement country-specific training programs in line with the needs and aspirations of the country. In addition to that, the fast-paced development of digital technologies requires constant learning and adjustment. Digital governance of renewable energy is essential in shining light on improving system efficiency and sustainability (Lin 2023). As a result, there needs to be a continued focus on professional development and upskilling so they are adept at using new digital tools and technologies.

Vocational Training and Education for Climate Jobs

Climate change is one of the most pressing issues of our time, and it translates into a demand for people who understand how and why things work the way they do and can contribute to building sustainable systems. A wide range of professions employed to mitigate climate impacts and promote sustainability can be encapsulated in the term climate employment, and Technical Education and Vocational Training (TVET) is vital for preparing the workforce for this employment. Vocational education provides both theoretical and practical knowledge to prepare the technical know-how required for renewable energy, sustainable farming and environmental management.

Technical and vocational education is one important means of minimizing the effects that promote climate change because it provides people with the skills that they need to combat challenges regarding the environment (Umoru and Okeke 2012). Such a requirement is even more crucial in the developing world, where human capital in green jobs is essential to meet sustainable development goals. Embedding climate education in vocational training can make the curriculum more relevant and prepare graduates to work in the evolving job market.

Furthermore, if innovative approaches, such as STEAM (Science, Technology, Engineering, Arts and Mathematics) approach are applied, the productivity of vocational education and training in the field of green energy will be increased significantly. In addition, independent learning should be combined with creative practices. Indahwati (2023) stated that thinking skills are the main skills, especially in climate practice to develop creative and other skill sets in solving climate problems. The process in which various fields of study and related practices inherently offer tools for solutions at the same time provide research. While vocational training for climate jobs has much promise, there are still significant challenges.

One prominent challenge faced by vocational education is the social status associated with it; vocational education is generally regarded as a secondary option to more traditional academic routes. Improving the social recognition of vocational education is significant for drawing students to these programs (Widianingsih et al. 2018). Reducing this stigma requires movement towards the fruitfulness of vocation education towards both economic improvement and environmental workflow. Additionally, the need for vocational training programs to be aligned with industry requirements is also crucial to guarantee that graduates acquire the skills that are in high demand by employers. According to Zhuang, for effective functioning, good vocational education systems have to be malleable to the labour market's changing requirements, not least because of climate change (Zhuang 2023). Overall, this synergy creates a favourable environment where curricula can evolve in alignment with industry demands, ensuring that graduates are adequately prepared for the job market.

The Role of Universities and Research Institutions

Universities and research institutions are at the heart of creating climate jobs as producers of knowledge, innovation and training. However, despite the various reclamations of higher education, universities around the world continue to have a significant degree of influence; be it through research, education, and community engagement which can lead towards sustainable development as the world tries to address the pressing challenges of climate change. These roles emphasize the wide-ranging contributions of universities and research institutions to education, research, and collaborations with industry and government in advancing climate jobs. The most relevant role of universities in climate jobs is to educate and train students to take on green jobs (Oyetunde et al. 2023). As noted by Karaeva et al. (2022), many universities built green infrastructure and created educational programs on sustainability, renewable energy, and environmental protection. Not only do these programs prepare students for careers in emerging fields, but they also cultivate the environmental stewardship and responsibility that will be crucial in the years to come. For example, universities can place greater emphasis on this because if institutions embed sustainability into their curricula, they can create a workforce that is not only technically efficient but also has attention to the environmental impacts of their work. Research institutions have an important role too, in creating new innovative solutions and in advancing knowledge on the subject of climate change. Cheng et al. (2022) applaud the incorporation of sustainable development principles into scientific research, calling on universities to promote research agendas that address environmental problems.

This type of research fuels innovation in new technologies and methods that can reduce the effects of climate change which lead to jobs in areas such as renewable energy, sustainable agriculture and environmental stewardship, among others. Besides, the transfer of knowledge and technologies from academia and industry cooperation to the application of research results in practice. And universities are anchors of local engagement and development. Purcell et al. (2019) discuss how universities serve as anchor institutions using the resources and technological capacity that they are endowed with to support the sustainable improvement of the communities they are in and public services.

Often seen as considered stakeholder entities in the local economy in which they operate, understanding how universities are perceived in a local context is important. By being anchored there in local needs, particularly about climate action, they not only make the offerings of their local universities more relevant and impactful there but also forge appropriately focused partnerships that can catalyse their locality's generation of jobs in the field of climate. It is true that both universities and research institutions have contributed greatly to this process, and that there are still gaps in systems which are holding those people back from arriving in the climate job market. Barriers that may hinder sustainability in higher education include financial constraints, lack of awareness of sustainability best practices, and resistance to change (Salihoglu and Turhan 2023). Universities can thus take a systemic approach to sustainability strategies across curriculum, research and management of the campus (Tretyakova and Kotomina 2020). Also, this data can help developers track trends and patterns to gain better insight into consumer preferences and attitudes toward sustainable products and initiatives, helping to make stronger recommendations and strategies to support climate change initiatives within universities.

International Knowledge Exchange Programs

The new knowledge exchange projects are vital to helping the qualification to perform climate jobs for students. By sharing research findings, best practices and innovative technologies across countries, these programs help create a global community that is committed to tackling climate change. Engaging in such exchanges helps professionals and students learn from first-hand data on how to implement sustainability and climate resilience, which increases employability in the field. One of the key advantages of international knowledge exchange programs is hearing about climate issues from a broad global context. Because these programs may help transfer knowledge and produce competencies required to work within the complex landscape of climate-related employment, exposure to cultural contexts and practices enables participants to develop a more holistic perspective regarding climate challenges, which helps them to be adaptable and sensitive to culture, important skills in the growing job market. Hossain et al. (2022).

In addition, often these programs also include hands-on training and practical experiences, both of which are critical for developing technical skills in renewable energy, environmental management and sustainable practices. They advocate for on-the-job training programs as a helpful way to assimilate students into the environment that they will be working in, giving them the confidence and readiness to be in the workforce (Catacutan and Tuliao 2020). Some might say that only applies to climate jobs because of the degree of practical skills needed (as well as more theoretical skills like communicating about it), but every job needs those. Moreover, such knowledge exchange initiatives create opportunities for joint research projects, capacity-building programs, and technology transfer between educational institutions, research organizations, and industry stakeholders. Such initiatives can also nurture bilateral research, tech transfer and climate solutions through the encouragement of cross-border partnerships. These exchanges will be collaborative and will also have a fairly important sharing sensc of community that shares responsibility among the participants that are also critically important for ensuring climate action. Nonetheless, obstacles still exist in the way of establishing effective programs for international exchanges of knowledge. While these initiatives show promise, bureaucratic obstacles, limited funding, and varying regulations for education systems can impede their success. Governments, educational institutions and international organisations have to cooperate and develop structuring frameworks to promote knowledge sharing and collaboration. That means (from policy) supporting the funding, clarifying how programming can happen, and raising awareness on how international exchanges build climate jobs.

Building Local Capacity for Sustainable Development

It is imperative to empower local communities to become the solution to environmental issues and foster economic resilience by building local capacity for sustainable development. Local capacity refers to the ability of communities to implement sustainable practices effectively, this includes the skills, knowledge, resources and the institutional framework needed to do so. Capacity building is founded upon education and training at the local level. Education for sustainable development (ESD) can work extremely well in terms of fundamentally changing behaviour if institutional contexts support the transfer of learning into practice (Velasco and Harder, 2014; Oyedele 2025).

Sustainability education programs can provide individuals with the knowledge and skills to engage in green practices in their community. Thus, training programs developed to improve understanding of green human resource management can create awareness and sustainable culture in local organisations using the model of the study by Napathorn regarding GHRM practices in Thailand (Napathorn 2021).

Communities that promote educational curricula and training to incorporate principles of sustainability can develop a workforce that is better equipped to address local environmental issues. Building local capacity requires collaboration across multiple stakeholders. Working with local governments, NGOs, schools and the private sector can synergistically improve the prospects for sustainability practices. All of this looks so impossible yet the main base of academia is our academic institutions where they can partner with local industries and businesses and lead to community-based learning projects and internships where sustainable practice on local economic growth can be used and create a real-life experience for students. Not only does this enable people to learn and build skills, but it also helps to encourage community stewardship and ownership over sustainable practices (Nwokolo et al. 2023). That said, the training and development role in job satisfaction and productivity improvement signifies a demand for capacity building (Omale et al. 2024).

Paposa and Kumar (2019) found that with complementary training programs which can be meticulously organized and offered to make staff economically viable, job satisfaction of academic professionals will drastically rise and eventually reflect positively on their performance and effectiveness in the education system.

In the same vein, Martin (2010) underscored the importance of workplace climate and peer support in facilitating the transfer of training, indicating that environments conducive to support constitute our ability to maximize our learning and development. This could be by investing in training and development to improve and build a skilled workforce, who are motivated and equipped to implement sustainable practices. Nonetheless, the path towards fostering local capacity for sustainable development is still fraught with challenges. Progress can be limited by a lack of awareness, resistance to change, and lack of resources. To Battle with these challenges, there is the need to develop the environment. In particular, this could involve funding for training programs, awareness raising of the necessity of sustainability and creating a culture of collaboration between stakeholders. Re-evaluation and timely reviews of capacity-building programs will also help to alleviate local needs.

Implications

Building capacity concerning green technology has opportunities that impact not just education but the wider socio-economic and environmental environment. Not only do they develop green technology, but they also play an important part in producing work-ready graduates able to address the challenge of climate change. This capacity-building motivates innovation and entrepreneurship (Tong-gong et al. 2023; Fasanmi et al. 2024). Educational institutions have an important role to play in the circular economy, weaving sustainability through their programmes and research agendas. Finally, the transfer and commercialization of clean technologies depend on the interactions among academia, industry, and government. These partnerships if effective, can work towards creating innovative solutions that also safeguard local and global environmental challenges (Chais et al. 2018; Oyenuga and Jeresa 2025).

These initiatives can promote the adoption of green technologies and in turn, can play a vital role in the process of accelerating economic growth and protecting the environment. Additionally, Silva et al. (2023) laid the groundwork for the establishment of research centres that red dot the terrain to inform applied studies of sustainable practices in communities and industries (Silva et al. 2023). This in turn facilitates local capacity building and fosters a culture of sustainability with potentially lasting effects on social behaviour and policy. There are also compelling economic arguments: green technology investments can both promote job creation and strengthen a country's position in the global market. Green practices help companies enhance operational efficiency and lower expenses that, ultimately, are beneficial to their economy. Finally, Rizvi and Garg (2020) state employees should be trained on acquiring green competencies and rewarded for acting sustainably, they help create an organization that provides an avenue of direction towards sustainability. This cycle of mutual benefits of adopting green technology and the continuous requirement to build capacity and innovate leads to development.

Conclusion, Limitations and Future Scope

Overall, green technology capacity-building forms a significant driver to realize the eco-friendly transition in a twenty-first century of human development. Here we have discussed the various dimensions of capacity building and the importance of a holistic approach that takes into consideration every element that plays a role in achieving capacity building—involvement of education, training, policy development, and engagement of stakeholders. Green technologies play an increasingly crucial role in addressing these challenges, making their adoption essential for the global community. Nevertheless, for these technologies to be successful, people, organizations, and communities must continue to know how to use and manage these new tools. Educational institutions contribute to developing a practical talent pool that is knowledgeable enough to promote and apply green solutions. Staff at educational institutions must receive training themselves about the most current green principles, so that the green teachings, sustainability principles, and green technologies of the future can become a part of the curriculum. it will be crucial to expand training in specialized sectors (for example, agriculture, energy, and manufacturing) to ensure that existing workers build the technical skills necessary to reach this new, greener, and sustainable economy. Such programs must inspire a balance of technical training on green technologies in addition to the socio-economic implications of adoption.

Also, enabling policies are needed to create an environment conducive to the adoption of green technologies. It is now the responsibility of policymakers to put research and development front and centre, to incentivize sustainable practices including not just energy use but social responsibility, and to create more public–private partnerships that lead to innovation. It requires joint action from governments, industry stakeholders, and civil society, particularly in breaking down implementation barriers and guaranteeing fair access to green technologies.

Finally, building a culture of sustainability relies heavily on community education and awareness-building initiatives.

References

Bassey K (2024) Economic impact of digital twins on renewable energy investments. Eng Sci Technol J 5(7):2232–2247. https://doi.org/10.51594/estj.v5i7.1318

Catacutan K, Tuliao A (2020) On-the-job training program evaluation of business administration and accountancy departments of the University of Saint Louis. Univ J Educ Res 8(1):143–150. https://doi.org/10.13189/ujer.2020.080117

Chais C, Ganzer P, Olea P (2018) Technology transfer between universities and companies. Innov Manage Rev 15(1):20–40. https://doi.org/10.1108/inmr-02-2018-002

Cheng Z, Xiao T, Chen C, Xiong X (2022) Evaluation of scientific research in universities based on the idea of education for sustainable development. Sustainability 14(4):2474. https://doi.org/10.3390/su14042474

Eder J, Mutsaerts C, Sriwannawit P (2015) Mini-grids and renewable energy in rural Africa: how diffusion theory explains the adoption of electricity in Uganda. Energy Res Soc Sci 5:45–54. https://doi.org/10.1016/j.erss.2014.12.014

Ekechukwu D (2024) A comprehensive review of renewable energy integration for climate resilience. Eng Sci Technol J 5(6):1884–1908. https://doi.org/10.51594/estj.v5i6.1187

Fasanmi OO, Olusanya BB, Oyenuga MO (2024) Entrepreneurial training as a retirement succorance for Nigerian retirees. Twist 19(4):137–147. https://doi.org/10.5281/zenodo.10049652#321

Guo Y, Zou H, Zhu L (2021) Behavioural analysis of subjects for green technology innovation: a tripartite evolutionary game model. Math Probl Eng 2021:1–14. https://doi.org/10.1155/2021/5181557

Hossain M, Hossain M, Hossain M, Hoque A, Khan N, Kamil A (2022) Skill development of undergraduate students through international exchange program. J Educ Soc Res 12(2):40. https://doi.org/10.36941/jesr-2022-0033

Indahwati S (2023) Integration of independent learning and physics innovation in steam-based renewable energy education to improve critical thinking skills in the era of society 5.0 for sustainable development goals (SDGs) 2030. E3s Web of Conferences 450:01010. https://doi.org/10.1051/e3sconf/202345001010

Joshi A, Singh R, Kumar V, Rani S (2024) Net zero strategy: exploring opportunities, confronting challenges for developing countries. In: Singh, R., Crowther, D. (eds) Transition towards a sustainable future. Approaches to global sustainability, markets, and governance. Springer, Singapore. https://doi.org/10.1007/978-981-97-5756-5_11

Karaeva A, Ionescu G, Cioca L, Tolkou A (2022) Role of EU and non-EU universities in achieving environmental sustainability. Matec Web of Conferences 373:00073. https://doi.org/10.1051/matecconf/202237300073

Lin, J.: (2023) Dilemma and countermeasures for digital governance of renewable energy in China under the "dual carbon" goal. https://doi.org/10.20944/preprints202309.0263.v1

Martin H (2010) Workplace climate and peer support as determinants of training transfer. Hum Resour Dev Q 21(1):87–104. https://doi.org/10.1002/hrdq.20038

Napathorn C (2021) The implementation of green human resource management bundles across firms in pursuit of environmental sustainability goals. Sustain Dev 30(5):787–803. https://doi.org/10.1002/sd.2271

Nwokolo SC, Singh R, Khan S, Kumar A, Luthra S (2023) Remedies to the challenges of renewable energy deployment in Africa. In: CSR, sustainability, ethics & governance, pp 59–74. https://doi.org/10.1007/978-3-031-44514-9_3

Nwokolo SC, Singh R, Khan S, Kumar A (2024a) Key players in net zero technology. In: Elsevier eBooks, pp 193–215. https://doi.org/10.1016/b978-0-443-31486-5.00007-0

Nwokolo SC, Singh R, Khan S, Kumar A (2024b) Net zero technology and circular economy. In: Elsevier eBooks, pp 237–247. https://doi.org/10.1016/b978-0-443-31486-5.00009-4

Omale SA, Yusuf SO, Oyenuga MO, Ikemefuna M, Ojo SS, Momodu ID (2024) Organizational learning in the post-COVID-19 era: a prerequisite for stakeholder satisfactions. Twist 19(3):265–272. https://twistjournal.net/twist/article/view/339

Oyedele MO, Sunday AO, Abuh AI (2024) Fostering technological-enhanced training and development for business survival and performance in the new normal. J Propulsion Technol 45(3):1858–1869. https://propulsiontechjournal.com/index.php/journal/article/view/7479

Oyedele OM (2025) The cobra effect: exploring the dark side of innovation and environmental sustainability in developing nations. In: Singh R, Joshi A, Filho WL, Khan S (eds) Zero carbon industry, eco-innovation and environmental sustainability. World sustainability series. Springer, Cham. https://doi.org/10.1007/978-3-031-80220-1_15

Oyenuga MO, Jeresa S (2025) Adaptation and resilience: private, public, and individual change makers. In: Singh R, Filho WL (eds) Climate neutrality through smart eco-innovation and environmental sustainability. Climate change management. Springer, Cham. https://doi.org/10.1007/978-3-031-83250-5_2

Oyenuga MO, Omale SA (2024) Is Africa jinxed? Exploring the challenges of technology access and adoption in Africa. Afr J Econom Sustain Develop 7(4):142–161. https://doi.org/10.52589/AJESDULN1LRNF

Oyetunde TO, Oyenuga MO, Adoga GJ (2023) Green management practices and organisational efficiency in higher institutions. Budapest International Research and Critics in Linguistics and Education (BirLE) Journal 6(3), 276–290. https://bircu-journal.com/index.php/birle/article/view/7716

Paposa K, Kumar Y (2019) Impact of training and development practices on job satisfaction: a study on faculty members of technical education institutes. Manage Labour Studies 44(3):248–262. https://doi.org/10.1177/0258042x19851649

Purcell W, Henriksen H, Spengler J (2019) Universities as the engine of transformational sustainability toward delivering sustainable development goals. Int J Sustain High Educ 20(8):1343–1357. https://doi.org/10.1108/ijshe-02-2019-0103

Rizvi Y, Garg R (2020) The simultaneous effect of green ability-motivation-opportunity and transformational leadership in environment management: the mediating role of green culture. Benchmarking Int J 28(3):830–856. https://doi.org/10.1108/bij-08-2020-0400

Salihoglu G, Turhan Ş (2023) Bursa Uludag university's contribution to society with sustainability projects. IOP Conf Series Earth Environ Sci 1194(1):012002. https://doi.org/10.1088/1755-1315/1194/1/012002

Silva L, Caten C, Gaia S, Souza R (2023) Tool for assessment of the green technology transfer structure in Brazilian public universities. Sustainability 15(8):6873. https://doi.org/10.3390/su15086873

Singh R, Crowther D, Ravesangar K (2024) Harnessing human capital for a sustainable future: HR's pivotal role toward net zero policies. In: Approaches to global sustainability, markets, and governance, pp 27–47. https://doi.org/10.1007/978-981-97-5756-5_2

Sun H (2024) Digital inclusive finance, R&D investment, and green technology innovation nexus. PLoS ONE 19(1):e0297264. https://doi.org/10.1371/journal.pone.0297264

Tong-gong Z, Ma Z, Shang Y (2023) Higher education, technological innovation, and green development—analysis based on China's provincial panel data. Sustainability 15(5):4311. https://doi.org/10.3390/su15054311

Tretyakova E, Kotomina O (2020) Sustainable universities as an essential element of education for sustainable development. E3s Web of Conferences 208:09030. https://doi.org/10.1051/e3sconf/202020809030

Umoru T, Okeke A (2012) The challenges of technical and vocational education in mitigating climate change-induced catastrophes in Nigeria. Afr J Teacher Educ 2(1). https://doi.org/10.21083/ajote.v2i1.1932

Velasco I, Harder M (2014) From attitude change to behaviour change: institutional mediators of education for sustainable development effectiveness. Sustainability 6(10):6553–6575. https://doi.org/10.3390/su6106553

Widianingsih I, Purwaamijaya I, Dwiyanti V (2018) Vocational education: assessment, classification and the realm of research developments. Innov Vocational Technol Educ 14(2):85. https://doi.org/10.17509/invotec.v14i2.14360

Yuan Q (2024) Does technological innovation in national sustainable development agenda innovation demonstration zones promote green development?—the case from Chengde city, china. PLoS ONE 19(5):e0300315. https://doi.org/10.1371/journal.pone.0300315

Zhao X, Zhao J (2023) Digital finance and inequality in renewable energy technology innovation. Energy Environ. 35(7):3673–3702. https://doi.org/10.1177/0958305x231171352

Zhuang J (2023) Promoting social innovation through the development of vocational education. Adv Educ Human Social Sci Res 7(1):60. https://doi.org/10.56028/aehssr.7.1.60.2023

Part V
Sectoral Approaches and Future Outlook

Chapter 22
Sustainable Agriculture and Land Use for Emission Reduction

Abstract Agriculture and land use are key sectors for the challenge of reducing greenhouse gas emissions and fighting climate change globally. It highlights the importance of ensuring that landscape management to mitigate emissions through more sustainable practices includes increased carbon storage, or capturing more carbon in the earth, improved soil quality, and improved resource efficiency. The chapter demonstrates through case studies and discussed strategies, such as crop rotation, agroforestry, and conservation tillage, how these practices can both mitigate emissions and promote food security and resilience in agricultural systems. Additionally, it discusses how to support this transition both through innovative technological solutions and policy frameworks and also includes cases of successful adaptation of these principles in different environments. The goal of this chapter is to provide an overview that allows for the understanding of how sustainable agriculture and land use may facilitate achieving net emission reduction targets, while remaining consistent with our long-term food systems.

Introduction

Nair (2022) states that the agricultural sector is responsible for around 31% of total human-caused emissions given its widespread use of practices that contribute to greenhouse gas emissions through deforestation, the production of livestock and the use of synthetic fertilizers. Global warming is forcing the world to reassess cultural values of food faster than the hasty move towards agricultural uses in overcoming climate change and land use prioritising sustainability. A panoply of sustainable agriculture practices and their suite of systems can yield more food, while reducing the environmental impacts of agriculture, advances which can better achieve food security and aid in emissions reductions. Land use has critical importance from a climate change perspective as one of the main leverage points based on this knowledge, sustainable land use. Land use determines carbon sequestration, biodiversity and ecosystem health. Land and agroecological practices, for example, can increase carbon storage substantially in soils and vegetation (Lamb et al. 2016).

M. O. Oyenuga et al., *Smart Technologies for Climate Change and Net Zero Policies: Practical Approaches Towards Sustainability*, Climate Change Management,
https://doi.org/10.1007/978-3-031-92221-3_22

Furthermore, sustainable land management practices like conservation tillage and crop rotation not only mitigate emissions but also enhance soil health and resilience to climate variability (Li et al. 2023). Such practices will be critical to sustaining agricultural productivity and adapting to climate change. Technological innovation is a key enabler of sustainable agriculture and land use. Digital technologies integration (e.g., precision agriculture, information technology, remote sensing) can increase resource use efficiency and minimise losses, thus leading to decreased emissions (Li 2023). Moreover, climate-resilient crop development and spread technologies can ensure food security, while simultaneously reducing the impact of agricultural production on the environment (Chen 2023). Nonetheless, the successful deployment of these technologies calls for supportive policy frameworks and investments in research and development that allow farmers to access and afford these technologies, especially in developing regions.

Sustainable Farming Practices

Climate change is causing twin threats of food insecurity and environmental unsustainability, at which sustainable farming practices would be the solution. Such practices involve various methods employed to become the most efficient agriculture production, but with little or no adverse effects on the ecosystem. Sustainable agriculture focuses on a myriad of techniques that build soil health, conserve water, preserve biodiversity, and reduce greenhouse gases by integrating ecological principles into production systems. One of the important sustainable agriculture practices is the use of ecological principles in the design and management of farming systems. Crop rotation, intercropping and cover crops are some of the agroecological practices that improve soil fertility and soil structure and reduce dependency on synthetic fertilisers and pesticides (Gather et al. 2019). Agriculture, which is like planting trees and crops in lines, as seen in agroforestry, can offer more than food, such as soil quality improvement, carbon sequestration and biodiversity (Gebru et al. 2019). Nebere et al. (2021) highlighted that agroforestry is not only a diverse way of production but also significantly contributes to the socio-economic sustainability of farmers through timber, as well as non-timber products (Nebere et al. 2021). Conservation agriculture is one of the related strategies, a knowledge-intensive practice grounded on these principles of minimum soil disturbance, soil coverage and crop rotation, which have the potential to improve soil health and land productivity. Several studies showed that this leads to better water retaining properties, a reduction of several soil erosion processes and increased resilience to climate variability (Sinaga et al. 2022).

One example is the increased use of conservation tillage, which can result in substantial decreases in carbon emissions from traditional plowing techniques. The academic research proves that applying these reins can boost the sustainability levels of land management and be an indicator to agricultural productivity (Zeweld et al. 2018). Integrated pest management (IPM) is another vital part of sustainable agriculture. IPM is a sustainable approach to managing pests by combining biological, cultural and chemical practices. IPM decreases reliance on chemical pesticides and protects the beneficials and maintains the ecosystem health. Using integrated pest management (IPM) techniques seems to help farmers control pests without harming the environment significantly (Skill et al. 2022). Innovative farming practices become sustainable solutions if they are not only environmentally friendly but also economically viable. This includes restricted access to resources, lack of knowledge and training, and socio-economic challenges facing smallholder farmers (Oyenuga and Omale 2025). For example, socio-psychological factors are essential barriers to the adoption of sustainable land management practices in resource-poor areas (Zeweld et al. 2018). At the same time, when they play a role, sustainable agriculture seldom has corresponding policies, or they are hardly put in place. Education and capacity building of the farmers through adequate technology transfer are therefore essential to overcome these challenges. Governments, NGOs and agricultural institutes can all play a vital role in ensuring that information and resources are accessible to farmers to adopt sustainable practices. Moreover, maintenance of supportive policies that would create incentives to engage in sustainable practices (e.g., subsidy for organic farming, research and development funding) are amongst other key features that can promote the practice of sustainable agriculture (Charatsari et al. 2022).

Agroforestry and Carbon Sequestration in Agricultural Lands

Agroforestry, which refers to the incorporation of trees and shrubs into agricultural landscapes, has emerged as an effective approach for improving carbon sequestration in conjunction with on-farm productivity. This diversified approach delivers climate change mitigation, but also improves biodiversity, soil health and resilience to climate variability. Agroforestry systems play an important role in mitigating greenhouse gas emissions and fostering sustainable land use, by sequestering atmospheric carbon dioxide (CO_2) into plant biomass and in soil through photosynthesis. One of the primary carbon-sequestration pathways for agroforestry and other activities is increased biomass on farmland. The vast amounts of carbon sequestered from trees and bushes (Agroforestry systems) with a potential captured carbon of up to 0.09 to 7.29 tons per ha/year, depending not only on the practice used in agroforestry systems but also on the ecological characteristics of the area of study (Király 2023).

Poplar-based integrated agroforestry systems in Punjab, India, for example, exhibited a high potential for carbon sequestration, with biomass production and soil organic carbon (SOC) content substantially higher than that achieved with conventional agricultural practices (Chauhan et al. 2015). Soil organic carbon plays a critical role as a carbon sink, improves soil fertility and structure, and promotes sustainable agricultural practices. Agroforestry systems also enhance soil aggregation and nutrient availability, which contribute to carbon storage. The role of agroforestry towards enhancing soil properties, enhancing the SOC pool and thereby increasing carbon density in soil has been reported (Loria et al. 2016).

The effectiveness of combining organic and inorganic fertilizers in these integrated systems can further enhance crop yields and the potential of carbon sequestration in soils, thus demonstrating the complementary functions of agroforestry as a land use practice for agricultural productivity and sustainable ecosystem functions. The systems also reduce greenhouse gas emissions (through no-till farming and appropriate fertilization regimes) while increasing carbon storage in biomass and soils. In southern Brazil, no-till agroforestry systems have been shown to bolster organic carbon and nitrogen content of the soil as well as microbial biomass, ultimately contributing to mitigation of greenhouse gas emissions (Torres et al. 2017). These management practices not only enhance carbon sequestration but also foster healthier ecosystems and more resilient agricultural systems. Agroforestry contains tremendous potential for carbon sequestration, the challenges lie in widespread adoption. Barriers, especially those associated with land tenure, access to inputs, and limited knowledge about agroforestry practices, can restrict its implementation, mainly on smallholder farmers. However, supportive policies and investments in research and development must be made to encourage adoption and improve the effectiveness of agroforestry to sequester carbon (Bogale and Bekele 2023).

Climate-Smart Agriculture Technologies

Climate-smart agriculture (CSA) refers to the approaches and technologies which improve agricultural productivity while also reducing greenhouse gas emissions and improving the capacity of the population to adapt to climate change. As the world's population continues to rise, and the ever-worsening climate crisis means we must provide innovative solutions for agriculture. CSA's three primary goals are to: (1) sustainably increase food production, (2) increase adaptive capacity to climate change, and (3) reduce or remove greenhouse gas emissions, where possible (Chuang et al. 2020). Climate-smart agriculture (CSA) is a commitment to using cutting-edge agricultural technologies to improve the efficiency of resource use and productivity. Precision agriculture will integrate technologies like GPS, remote sensing or data analytics to log crop farming, targeting a fine-level field of management. Mirzabaev indicated that data can help farmers prevent waste and reduce the impact on the environment by entering the land with more accurate inputs (water, fertilizers, pesticides). Studies show that precision agriculture can increase crop productivity and greatly reduce greenhouse gas emissions, so it seems a good option in terms of both adaptation and mitigation to climate change (Mirzabaev 2017). Improved crop varieties that are drought- and heat- and pest-tolerant are another critical aspect of climate-smart agriculture. Climate-resilient seeds are vital as they can ensure food security in the areas adversely impacted. According to a 2015 study, in Zambia, farmers that adopted varieties of improved seeds measured greater yields despite adverse climatic conditions (Arslan et al. 2015). Although these technologies are potentially effective in reducing the impacts of climate change, their effectiveness depends on the regional climate, indicating a need for regional-specific adoption of technologies (Arslan et al. 2015).

Another powerful climate-smart practice with potential is agroforestry: the intentional planting of trees in agricultural landscapes. It not only sequesters more carbon but also builds soil health and increases biodiversity and provides an alternative source of income through timber and non-timber forest products for the farmers that manage these lands (Toma et al. 2016). Agroforestry systems were observed to sequester a considerable amount of carbon and this approach was identified as an efficient option for climate change mitigation, improving agricultural yield as well (Dawid and Workalemahu 2021). While climate-smart agriculture technologies deliver immense advantages, there remain numerous obstacles to their adoption. Farmers may not implement this practice because they do not have access to information, financial assistance and/or technical support. As Toma et al. (2016) asserted, education and information transfer play a significant role in farmers' technology adoption. Well, this adds to the creation of centers of excellence which focus on training of trainers (Omale et al. 2024) as well as effective and sustainable knowledge dissemination that all help to strengthen the connections and linkages between research and education (Fasanmi et al. 2024), extension services, farmers and even enable the joint creation of knowledge through other formats and processes more suited to the local needs (Oyedele et al. 2024).

The socio-economic determinants of farmers' willingness to adopt climate-smart practices include land tenure security and market access. This demands policies that promote inclusive access to resources and support, especially to smallholder farmers who have been identified as critical to the adoption of CSA technologies (Tujo and Lecturer 2019). Furthermore, the role of different stakeholders, such as government agencies, NGOs (non-governmental organizations), and the private sector is also crucial to foster an enabling environment for climate-smart agriculture (Nwokolo et al. 2024).

Reducing Methane and Nitrogen Emissions in Agriculture

Agriculture constitutes a large percentage of greenhouse gas emissions and the two most significant from agricultural activity are methane (CH_4) and nitrous oxide (N_2O), both potent climate forcers with global warming potential that is in multiples of carbon dioxide (CO_2). When it comes to methane, emission sources are primarily enteric fermentation in livestock, rice paddies and manure management, and sources of nitrous oxide are mainly linked to nitrogen fertilizers use and soil management practices. Reducing these emissions is one of the keys to addressing global warming and to moving toward sustainable farming practices.

Improving livestock management approaches is one of the most potent tools for tackling methane emissions. Methane is produced in considerable amounts during the enteric fermentation (a natural process of digestion of ruminants). Studies have demonstrated that such a habit change or consumption of supplements (e.g., seaweed or tannins) can significantly reduce the amount of methane produced during digestion (Li 2024). For analogy, accordingly, useful from the ocean waters have suggested adding certain kinds of seaweed to cattle rations, which reduced methane emissions more than five times (Li 2024). More simply, as the adoption of better manure management practices (e.g., anaerobic digestion systems) capture this methane to be used as an energy source, this reduces emissions from manure storage (Shcherbak et al. 2014).

Reducing emissions from rice, a large fraction of global methane emissions, is effective using alternate wetting and drying (AWD) irrigation practices. AWD is a management practice that intermittently drains rice fields to decrease long-term anaerobic conditions favorable to methanogenesis (Ilić et al. 2022). Studies indicated that AWD could reduce methane emissions by around 30–50% compared to traditional continuous flooding systems (Lamothe 2024). In addition, a higher resilience to climate extremes through the adoption of improved rice varieties with higher tolerance to drought and flooding can maximize productivity with lower emissions (Huang et al. 2012). Nitrous oxide emissions, by contrast, can be traced mainly to nitrogen fertilizer use. Overapplication of nitrogen fertilisers is linked to greenhouse gas emissions, nutrient runoff and land degradation. It is important to mitigate emissions, and for optimising nitrogen use efficiency. For instance, smart practices, like precision agriculture, utilize updated technology to implement fertilizers in detail depending on the need of the soil and the plant (Ho et al. 2015). Precision nitrogen management can decrease N_2O emissions (20–30%) while maintaining/increasing crop yield (Wicki 2019). Mixed cropping, such as cover cropping and crop rotation, can increase soil health and capture nitrogen losses. Cover crops can enhance soil organic matter and, in the case of legumes, fix atmospheric nitrogen which makes nitrogen more available to later-grown crops while decreasing the need for synthetic fertilizers (Pang 2024). Planting cover crops in cropping systems would significantly reduce nitrous oxide emissions (Huang et al. 2020), according to past work. But these approaches are not without significant challenges, which make their widespread uptake hard. Farmers can be limited in their potential to implement emission-reducing practices by the socio-economic barriers, access to resources and knowledge gaps (Huang 2023). Sustainable practices guided by effective policy frameworks and supportive systems (Kumar et al. 2024; Nwokolo et al. 2023a) are also crucial to ensuring reductions in greenhouse gas emissions from agriculture (Oyenuga and Jeresa 2025; Nwokolo et al. 2023b).

Case Studies of Successful Land Use Practices

At the backdrop of climate change, food security, biodiversity, and environmental degradation challenges, sustainable land use practices are critical to solving them. Examples from different countries around the world show multiple case studies where it's been practised and proved its impact by improving agricultural productivity while sustaining ecological balance.

The second case study on smallholder potato farmers was conducted in Gilgil Sub-County, Kenya, where Climate-smart agriculture (CSA) practices were implemented. Waaswa et al. (2021) explored pathways for information diffusion of climate smart agriculture and found information diffusion of new crop and livestock technologies was highest when farmers interacted with knowledgeable neighbors. This not only allowed for two-way dialogue; it created a network of support, to hold each other accountable, and motivate the execution of sustainable solutions. The results underscore that social networks facilitate climate-smart practices, arguing that keeping community-level connections in mind can significantly undermine the effectiveness of agricultural strategies.

In Zambia, Manda et al. (2015) analyzed the current adoption of sustainable agricultural practices, as well as the impacts of such practices on maize yields and incomes. It was discovered that practices such as crop rotation, intercropping with legumes and conservation tillage had a positive influence on soil fertility and crop yield. Not only did they become more food secure, they also buried carbon in the soil by using greener soil management systems. The results flaunt the potential new sustainable practices have to improve both livelihoods and those same practices to address climate change.

There are several climate smart agriculture practices adopted in the Gangetic plains of Bihar, India (Aryal et al. 2018). Not only did CSA practices such as better irrigation methods and integrated crop-livestock systems significantly increase crop and livestock productivity; they also built resilience to climate variability. The study further demonstrated how CSA, through the application of diverse and locally appropriate agricultural practices, can solve the unique challenges that the farmers face.

In California, Gryze et al. (2010) emphasized the potential of alternative management practices, such as winter cover cropping and conservation tillage, to mitigate greenhouse gas emissions from agricultural systems. According to the study, these not only reduced emissions, but also enhanced soil health and increased water retention, leading to much more sustainable farming systems. This case, however, shows the holistic approach in land management considering environmental aspects with economic aspects.

Experiences of Ethiopian farmers illustrate the role of agroforestry in sustainable land use practices where Nkonya et al. (2017) asserted that the integration of trees within agricultural landscapes can contribute towards increased soil fertility, better water management and higher carbon sequestration potential. Agroforestry systems offer multiple ecosystem services by generating additional income through timber and non-timber products, and this may also involve biodiversity conservation and ecosystem resilience. The study of this case reveals the multiple advantages of agroforestry systems as sustainable practices of land use.

As these case studies illustrate, there are cases of success, but there are also challenges to widespread uptake of sustainable land use practices. Implementation can be difficult due to limited access to resources, lack of knowledge, and socio-economic barriers. For instance, as Chavula and Turyasingura highlight, access to land and security of land tenure are essential to incentivizing smallholder farmers to adopt climate-smart agriculture technologies. Further enabling policies and investment in education and capacity building must accelerate the transition to sustainable land use practices.

Implications

Adopting sustainable agriculture and land use practices for emission reduction could have profound and multifaceted implications. First, these practices can make a major dent in climate change by cutting greenhouse gas emissions for farming. As an example, some technologies of climate-smart agriculture, including precision agriculture and agroforestry, have the potential to achieve significant reductions in methane and nitrous oxide emissions, which will be key for national and global climate targets (Hou 2023; Qi et al. 2023). These strategies combined help sequester more carbon in soils and out of the atmosphere into vegetation than is emitted while improving agricultural productivity to help solve problems of food security in a changing climate. In addition, sustainable agriculture can help make farmers more resilient to climate-change impacts. Optimizing agricultural practices, as well as diversifying cropping systems, can improve the adaptive capacity of farmers and reduce their vulnerability to climate shocks (Shu et al. 2023). It is crucial for sustaining livelihoods and for food security, especially in areas that are particularly vulnerable to climate variability, in addition to preserving soil health and increasing biodiversity.

Soil organic matter is increased by cover cropping, crop rotation and reduced tillage practices (Marpaung 2024), and enhancing the activity of beneficial soil organisms, which are critical for nutrient cycling and pest management. Such relevance reaches beyond the immediate advantages of the waste cycle, a positive correlation can also be drawn towards a boost in agricultural output and sustainability, meanwhile benefiting both the earth and the farmer. There are many available options for the effective adoption of sustainable practices, but supportive policy frameworks and investment in research and development are crucial. The importance of government support is acknowledged as a necessary factor for farmers to acquire resources and incentives for the adoption of low-carbon agricultural technologies (Ye 2024). Addressing socio-economic constraints including land tenure, security (Oyedele 2025), and access to credit is also essential for enabling smallholders to adopt sustainable practices.

Conclusion, Limitations and Future Scope

Environmental sustainability and climate-vulnerable greenhouse gas emissions are heavily influenced by sustainable agriculture and land use practices. Innovative technologies and practices adopted by farmers offer solutions for enhancing productivity and for climate change mitigation. With case studies and evidence provided, this chapter shows how these sustainable practices are the key to balancing emission reduction targets along with food security and resilience. Sustainable agriculture is not just a responsibility of farmers; it needs collective actions from multiple stakeholders (governments, NGOs, and private sector organizations) to enable an ecosystem to adopt sustainable agriculture practices. Education, capacity expansion, and research will finance farmers to execute sustainable practices effectively. At a time when the world is fighting against climate change, ensuring we promote sustainable agriculture and land use will be vital in creating a robust and sustainable future food system for generations to come.

References

Arslan A, McCarthy N, Lipper L, Asfaw S, Cattaneo A, Kokwe M (2015) Climate smart agriculture? Assessing the adaptation implications in Zambia. J Agric Econ 66(3):753–780. https://doi.org/10.1111/1477-9552.12107

Aryal J, Jat M, Sapkota T, Khatri-Chhetri A, Kassie M, Rahut D, Maharjan S (2018) Adoption of multiple climate-smart agricultural practices in the gangetic plains of Bihar, India. Int J Climate Change Strateg Manage 10(3):407–427. https://doi.org/10.1108/ijccsm-02-2017-0025

Bogale G, Bekele S (2023) Sustainability of agroforestry practices and their resilience to climate change adaptation and mitigation in sub-saharan Africa: a review. Ekológia (Bratislava) 42(2):179–192. https://doi.org/10.2478/eko-2023-0021

Charatsari C, Lioutas E, Papadaki-Klavdianou A, Koutsouris A, Michailidis A (2022) Experiential, social, connectivist, or transformative learning? Farm advisors and the construction of agroecological knowledge. Sustainability 14(4):2426. https://doi.org/10.3390/su14042426

Chauhan S, Sharma R, Singh B, Sharma S (2015) Biomass production, carbon sequestration and economics of on-farm poplar plantations in Punjab, India. J Appl Nat Sci 7(1):452–458. https://doi.org/10.31018/jans.v7i1.631

Chavula P, Turyasingura B (2022) Land tenurial system influence among smallholder farmers' climate smart agriculture technologies adoption, sub-Saharan Africa: a review paper. Int J Food Sci Agric 6(1):8–16. https://doi.org/10.26855/ijfsa.2022.03.003

Chen P (2023) Sustainable agricultural management: how to achieve carbon neutrality in agriculture – evidence from China agricultural sustainable development plan. Sustain Dev 32(3):2846–2857. https://doi.org/10.1002/sd.2821

Chuang J, Wang J, Liou Y (2020) Farmers' knowledge, attitude, and adoption of smart agriculture technology in Taiwan. Int J Environ Res Public Health 17(19):7236. https://doi.org/10.3390/ijerph17197236

Dawid I, Workalemahu S (2021) Review on windbreaks agroforestry as a climate-smart agriculture practice. Am J Agric Forestry 9(6):342. https://doi.org/10.11648/j.ajaf.20210906.12

Fasanmi OO, Olusanya BB, Oyenuga MO (2024) Entrepreneurial training as a retirement succorance for Nigerian Retirees. Twist 19(4):137–147. https://doi.org/10.5281/zenodo.10049652#321

Gebru B, Wang S, Kim S, Lee W (2019) Socio-ecological niche and factors affecting agroforestry practice adoption in different agroecologies of southern Tigray, Ethiopia. Sustainability 11(13):3729. https://doi.org/10.3390/su11133729

Gryze S, Wolf A, Kaffka S, Mitchell J, Rolston D, Temple S, Six J (2010) Simulating greenhouse gas budgets of four California cropping systems under conventional and alternative management. Ecol Appl 20(7):1805–1819. https://doi.org/10.1890/09-0772.1

Ho A, Reim A, Kim S, Meima-Franke M, Termorshuizen A, Boer W, Bodelier P (2015) Unexpected stimulation of soil methane uptake as emergent property of agricultural soils following bio-based residue application. Glob Change Biol 21(10):3864–3879. https://doi.org/10.1111/gcb.12974
Hou S (2023) Regional heterogeneity of agricultural carbon emission reduction potential in China. Environ Eng Res 29(4):230399. https://doi.org/10.4491/eer.2023.399
Huang M (2023) Effect of fertilization on methane and nitrous oxide emissions and global warming potential on agricultural land in China: a meta-analysis. Agriculture 14(1):34. https://doi.org/10.3390/agriculture14010034
Huang J, Xiong R, Li F, Li T, Shen W (2020) Estimation of interannual trends of ammonia emissions from agriculture in Jiangsu Province from 2000 to 2017. Atmospheric Oceanic Sci Lett 13(3):268–273. https://doi.org/10.1080/16742834.2020.1736499
Huang X, Song Y, Li M, Li J, Huo Q, Cai X, Zhang H (2012) A high-resolution ammonia emission inventory in China. Global Biogeochem Cycles 26(1). https://doi.org/10.1029/2011gb004161
Ilić I, Jovanović S, Petrović-Ranđelović M (2022) Relationship between climate change and agriculture at the EU level. Econ Themes 60(3):323–342. https://doi.org/10.2478/ethemes-2022-0018
Király É (2023) Carbon sequestration in the aboveground living biomass of windbreaks—climate change mitigation by means of agroforestry in Hungary. Forests 15(1):63. https://doi.org/10.3390/f15010063
Kumar K, Singh R, Sharma A, Lehri BPS (2024) Climate change adaptation strategies for achieving net-zero economy. In: Singh R, Crowther D (eds) Transition towards a sustainable future. Approaches to global sustainability, markets, and governance. Springer, Singapore. https://doi.org/10.1007/978-981-97-5756-5_8
Lamb A, Green R, Bateman I, Broadmeadow M, Bruce T, Burney J, Balmford A (2016) The potential for land sparing to offset greenhouse gas emissions from agriculture. Nat Clim Chang 6(5):488–492. https://doi.org/10.1038/nclimate2910
Lamothe A (2024) 300-Year nh3 inventory in europe: uncovering combustion's recent share. https://doi.org/10.21203/rs.3.rs-3550976/v1
Li T (2024) Effects of planting density and nitrogen application on soil greenhouse gas fluxes in the jujube–alfalfa intercropping system in arid areas. Agronomy 14(2):273. https://doi.org/10.3390/agronomy14020273
Li B, Wu L, Zhu W, Qiao C, Zhang J, He W (2023) Influence of crop residue management and soil tillage method on reducing the carbon footprint of winter wheat production in the salt-affected arable land in the North China Plain. Agronomy 13(4):1018. https://doi.org/10.3390/agronomy13041018
Li C (2023) Research on the carbon reduction effect of digital transformation of agriculture in China. Polish J Environ Stud. https://doi.org/10.15244/pjoes/170001
Loria N, Bhardwaj S, Ndungu C (2016) Impact of cropping systems on soil properties, nutrient availability and their carbon sequestration potential in Shiwalik Hills of Himachal Pradesh. J Appl Nat Sci 8(3):1479–1484. https://doi.org/10.31018/jans.v8i3.987
Manda J, Alene A, Gardebroek C, Kassie M, Tembo G (2015) Adoption and impacts of sustainable agricultural practices on maize yields and incomes: evidence from rural Zambia. J Agric Econ 67(1):130–153. https://doi.org/10.1111/1477-9552.12127
Marpaung N (2024) The effect of agricultural policy on sustainability and environmental protection: omnibus law case study. IOP Conf Ser Earth Environ Sci 1413(1):012007. https://doi.org/10.1088/1755-1315/1413/1/012007
Mengesha A, Mansberger R, Damyanovic D, Stoeglehner G (2019) Impact of land certification on sustainable land use practices: case of Gozamin district, Ethiopia. Sustainability 11(20):5551. https://doi.org/10.3390/su11205551
Mirzabaev A (2017) Improving the resilience of central Asian agriculture to weather variability and climate change. 477–495. https://doi.org/10.1007/978-3-319-61194-5_20

MN (2023) Determinants of adoption of climate smart agricultural technologies among farm households in southern Karnataka, India. Int J Environ Climate Change 13(11):4354–4366. https://doi.org/10.9734/ijecc/2023/v13i113616

Nair P (2022). Carbon footprints and land-use systems. Carbon Footprints 1(1):2. https://doi.org/10.20517/cf.2022.07

Nebere H, Tolossa D, Bantider A (2021) Analyzing factors affecting the sustainability of land management practices in Mecha Woreda. Northwestern Ethiopia. Sustainability 13(13):7007. https://doi.org/10.3390/su13137007

Nkonya E, Koo J, Kato E, Johnson T (2017) Climate risk management through sustainable land and water management in sub-Saharan Africa, 445–476. https://doi.org/10.1007/978-3-319-61194-5_19

Nwokolo SC, Singh R, Khan S, Kumar A, Luthra S (2023a) Africa's awakening to climate action. In: CSR, sustainability, ethics & governance, pp 299–310. https://doi.org/10.1007/978-3-031-44514-9_9

Nwokolo SC, Singh R, Khan S, Kumar A, Luthra S (2023b) Remedies to the challenges of renewable energy deployment in Africa. In: CSR, sustainability, ethics & governance, pp. 59–74. https://doi.org/10.1007/978-3-031-44514-9_3

Nwokolo SC, Singh R, Khan S, Kumar A (2024) Key players in net zero technology. In: Elsevier eBooks, pp. 193–215. https://doi.org/10.1016/b978-0-443-31486-5.00007-0

Omale SA, Oyenuga MO, Abdullahi D, Madu I (2024) Moderating role of entrepreneurial education on innovative work behaviour and resilience of SMEs. Innovations 77(2):1166–1185. e0ce0–1166–1185.11600.pdf

Oyedele MO, Sunday AO, Abuh AI (2024) Fostering technological-enhanced training and development for business survival and performance in the new normal. J Propulsion Technol 45(3):1858–1869. https://propulsiontechjournal.com/index.php/journal/article/view/7479

Oyedele OM (2025) The cobra effect: exploring the dark side of innovation and environmental sustainability in developing nations. In: Singh R, Joshi A, Filho WL, Khan S (eds) Zero carbon industry, eco-innovation and environmental sustainability. World Sustainability Series. Springer, Cham. https://doi.org/10.1007/978-3-031-80220-1_15

Oyenuga MO, Jeresa S (2025) Adaptation and resilience: private, public, and individual change makers. In: Singh R, Filho WL (eds) Climate neutrality through smart eco-innovation and environmental sustainability. Climate Change Management. Springer, Cham. https://doi.org/10.1007/978-3-031-83250-5_2

Oyenuga MO, Omale SA (2024) Is Africa jinxed? Exploring the challenges of technology access and adoption in Africa. African J Econom Sustain Develop 7(4):142–161. https://doi.org/10.52589/AJESDULN1LRNF

Pang Y (2024) Dongxiang wild rice rhizomicrobiomes inhibit photosynthetic methanogenic microbes: a potential solution for reducing methane emissions in northeastern China. https://doi.org/10.21203/rs.3.rs-5431321/v1

Qi Y, Liu H, Zhao J (2023) Prediction model and demonstration of regional agricultural carbon emissions based on isomap–aco–et: a case study of Guangdong Province, China. Scientific Reports 13(1). https://doi.org/10.1038/s41598-023-39996-5

Shcherbak I, Millar N, Robertson G (2014) Global metaanalysis of the nonlinear response of soil nitrous oxide (N_2O) emissions to fertilizer nitrogen. Proc Natl Acad Sci 111(25):9199–9204. https://doi.org/10.1073/pnas.1322434111

Shu Q, Su Y, Hong L, Li F, Zhao Y, Du C (2023) Study on the spatial structure and drivers of agricultural carbon emission efficiency in Belt and Road Initiative countries. Sustainability 15(13):10720. https://doi.org/10.3390/su151310720

Sinaga R, Lubis M, Nasution F, Sembiring E (2022) The challenges of implementing agroecology as a social movement by members of the Serikat Petani Indonesia. IOP Conf Series Earth Environ Sci 1115(1):012101. https://doi.org/10.1088/1755-1315/1115/1/012101

Skill K, Passero S, Farhangi M (2022) Cultivating agroecological networks during the pandemic in Argentina: a sociomaterial analysis. Land 11(10):1782. https://doi.org/10.3390/land11101782

Toma L, Barnes A, Sutherland L, Thomson S, Burnett F, Mathews K (2016) Impact of information transfer on farmers' uptake of innovative crop technologies: a structural equation model applied to survey data. J Technol Transf 43(4):864–881. https://doi.org/10.1007/s10961-016-9520-5
Torres C, Jacovine L, Neto S, Fraisse C, Soares C, Neto F, Lemes P (2017) Greenhouse gas emissions and carbon sequestration by agroforestry systems in southeastern Brazil. Scientific Reports 7(1). https://doi.org/10.1038/s41598-017-16821-4
Tujo T, Lecturer G (2019) A big data Analytics framework in climate-smart agriculture. Comput Eng Intelligent Systems. https://doi.org/10.7176/ceis/10-6-01
Waaswa A, Nkurumwa A, Kibe A, Kipkemoi N (2021) Communicating climate change adaptation strategies: climate-smart agriculture information dissemination pathways among smallholder potato farmers in Gilgil sub-county, Kenya. Heliyon 7(8):e07873. https://doi.org/10.1016/j.heliyon.2021.e07873
Wicki L (2019) Biological progress and the use of nitrogen by cereal varieties 52:403–409. https://doi.org/10.22616/esrd.2019.148
Ye J (2024) Assessment of agricultural carbon emissions reduction potential and optimisation pathways based on a framework of equity and efficiency principles: evidence from Fujian Province in China. Agric Econom (Zemědělská Ekonomika) 70(3):125–136. https://doi.org/10.17221/329/2023-agricecon
Zeweld W, Huylenbroeck G, Tesfay G, Azadi H, Speelman S (2018) Impacts of socio-psychological factors on actual adoption of sustainable land management practices in dryland and water stressed areas. Sustainability 10(9):2963. https://doi.org/10.3390/su10092963

Chapter 23
Low-Carbon Transport and Mobility Solutions

Abstract Carbon transport and mobility solutions are reduced to be the major principle of climate change impact mitigation, urbanization, and sustainable development. Greenhouse gas reduction strategies and technologies in transportation, as a major source of world carbon emissions, have been addressed here. Through case studies and best practices, the chapter identifies the contribution of public transport systems, electric vehicles, cycling infrastructure, and multimodal transport modes to sustainable mobility. The chapter also discusses the contribution of policy frameworks, technological innovation, and community engagement to low-carbon transport transition. Finally, this chapter seeks to develop a vision of how low-carbon transport modes can propel emission reduction plans and improve urban livability and resilience.

Introduction

The transport industry is among the largest sources of greenhouse gas emissions, contributing approximately 14% of total emissions, with road transport accounting for the largest share. With urban populations continuing to grow and mobility needs increasing, the pressure for sustainable transport increases. Low-carbon mobility and transport solutions include various measures that have the purpose of lowering emissions but ensuring accessibility and efficiency in the transport system. Low-carbon mobility and transport solutions are important in reaching climate goals, reducing air pollution, and enhancing public health. Public transport system upgrading is one of the key approaches to lowering transport sector emissions. Investment in clean and accessible public transport can substantially cut the use of private cars, and hence overall emissions. Evidence shows that cities with a well-established public transport system emit less per capita than cities with high car use. And the combination of public transport and active modes like walking and cycling can deliver a more sustainable urban transport system. Electric vehicles (EVs) form another indispensable element of low-carbon transport strategies.

M. 0. Oyenuga et al., *Smart Technologies for Climate Change and Net Zero Policies: Practical Approaches Towards Sustainability*, Climate Change Management,
https://doi.org/10.1007/978-3-031-92221-3_23

Widespread take-up of EVs could dramatically reduce on-road transport emissions when combined with renewables-based electricity. The widespread adoption of EVs requires expansion of enabling infrastructure such as charging points as well as enabling policies that incentivize their deployment. Evidence from various locations shows that whole-of-government action involving subsidization and tax benefit support in addition to public information campaigns can introduce electric transportation much more quickly than unsupported consumer decisions would. Multimodal modes of transport, where two or more modes are used, also play an important role to enable low-carbon mobility. Multimodal systems can be made more efficient and reduce greenhouse gas emissions by allowing easy transfers from one mode to another. Multimodal transport corridors must be optimized to minimize carbon savings while continuing to meet urban mobility requirements.

Not only does this solution make the transport system more sustainable, but it also provides a more integrated experience for the user. Although low-carbon transport modes have the potential to be positive, there are several impediments to their implementation. These include low funding levels, poorly developed infrastructure, and opposition to change. To overcome these impediments, it is necessary to have concerted action by governments, private sectors, and communities to develop an enabling environment for low-carbon transport. Additionally, public awareness and engagement are instrumental in establishing a culture of sustainability and persuading people to shift to low-carbon mobility (Joshi et al. 2024).

Electrification of Transport Systems

The electrification of transport systems is a revolutionary step toward promoting sustainability in urban transport and reducing greenhouse gas emissions. Urbanization and air pollution are all pressing issues the world struggles to tackle along with climate change (Kumar et al. 2024), so the movement from fossil fuel-based cars to electric cars has taken hold.

One of the top drivers of the trend towards electrification of the transport sector is the widely-recognised need to reduce emissions, which makes up a significant portion of total greenhouse gas emissions released globally.

The electrification of transport has been held up as one of the big ideas for decarbonizing the sector, and most governments have put forward proposals to ban internal combustion engine cars (Zheng et al. 2019). A transition to electric mobility not only helps fight climate change but also improves air quality and public health by reducing harmful emissions from conventional cars. Leading this charge are electric vehicles, battery electric vehicles (BEVs), and plug-in hybrid electric vehicles (PHEVs). Ruggieri et al. suggest that electrifying public transport fleets (e.g., buses and taxis) can serve as a stepping stone to greater electrification.

The integration of Electric buses in public transport systems can significantly reduce the emission level while providing the public transport service that is needed and reliable. The development of charging infrastructure is also a key factor in the mass adoption of electric vehicles. Transitioning to electric mobility and optimizing transport systems is possible with fast-charging infrastructure and smart grid technologies. As a way to decarbonise logistics and supply chains, we will also see more freight transport turning electric, alongside passenger vehicles. Electric trucks and vans could reduce the carbon footprint for goods movement by a huge margin, particularly in cities where air pollution is a real problem. "As is noted by Gan et al. (2023), a lifecycle examination of electric cars shows that they have the promise of huge emissions savings over fossil fuel vehicles in terms of greenhouse. Despite this, its success in delivering emissions reductions also depends on whether electricity for recharging comes from renewable or otherwise sources.

The shift towards renewable energy for power generation is needed to reap the maximum environmental dividends of electric mobility (Lester et al. 2020). Though the potential of electrification looks encouraging, a few challenges still need to be addressed in its application (Oyedele 2025).

The lack of adequate charging facilities, the high upfront costs of electric vehicles, and concerns regarding battery life and performance can hinder their adoption. Public opinion about electric vehicles also significantly influences their adoption; public acceptance of electric vehicles is on the rise, as awareness of their environmental benefits grows (Lin et al. 2023). However, remedies for issues such as range anxiety and charging station convenience are crucial for the encouragement of mass adoption. In addition, transport system electrification should be followed by favorable policy environments and research and development investment. Policy makers have an important role to encourage electric vehicle uptake through incentives such as tax credits, subsidies, and charging infrastructure investment. As emphasized by Oke (2024) wide-ranging policies driving electrification together with other sustainable transport policies, including investment in public transit and active transportation infrastructure, are crucial to achieve a drastic reduction of emissions.

Public Transportation: Reducing Emissions Through Mass Transit

Public transport systems have a significant function in lowering carbon emissions and achieving sustainable urban transportation. With increasing traffic congestion and air pollution as cities expand, the necessity of effective and low-carbon modes of transport gains momentum. Mass transport modes, such as buses, rail, and the subway, are a sound option to personal auto travel, markedly reducing per capita emissions and raising the general efficiency of urban travel. A primary advantage of using public transit is that it discourages people from driving so much. From Li et al., regions with abundant access to public transport have remarkably lower levels of car ownership relative to regions that lack good public transport access Li et al. (2015). The fewer vehicles reduce the traffic congestion as well as lower emissions from the burning of fossil fuels. In addition, public transport systems tend to be more energy-efficient compared to private cars, with research showing that public transport can be four times more energy-efficient than automobiles (Sobrino and Monzón 2013). Through promoting the utilization of public transport, cities can readily reduce their carbon footprints and improve air quality. Electrification of public transport also enhances their potential for emissions reduction. Electric trains and buses do not produce tailpipe emissions, promoting cleaner cities. Zhang discovers that adopting electric public transport can contribute to a significant reduction in carbon emissions, particularly when the power consumed is renewable (Zhang 2023).

This transition not only decreases climate change but also enhances public health by decreasing air pollutants linked with conventional diesel-powered cars. Beyond emission reduction, public transport infrastructure can enhance social equity by enabling inclusive mobility to all citizens irrespective of their inability to afford personal cars. By offering public transport, it becomes frequent and low-cost, and cities can enable mobility for low-income citizens, eventually closing gaps in access to work, education, and services. By incorporating bike-sharing and pedestrian facilities into public transport networks, accessibility and environmentally friendly behavior can be improved (Fangzhou et al. 2022). But the efficient implementation of public transport systems requires heavy investment in infrastructure and policy support. As Prabhu and Pai (2012) point out, promotion of public transport usage in Indian cities requires development of efficient bus systems and enacting policies discouraging the utilization of private cars.

Second, public education campaigns are necessary (Oyetunde et al. 2023) for transforming travel patterns and motivating citizens to use public transportation. As much as there are many advantages of public transportation, there are obstacles to the general adoption of mass transit systems. Limited budget, poor infrastructure, and issues of safety and convenience perceived by the public may compromise the success of mass transit systems. To respond to these issues, governments and city planners must make public transport infrastructure investment a top priority, enhance the quality of service, and involve communities in planning (Guerra 2011).

Non-Motorized Transport: Bicycles and Pedestrian Systems

Integrating non-motorized transport (NMT), namely cycling and walking, into urban areas is increasingly viewed as an essential component of a sustainable urban transport system. As cities are faced with the dual challenge of traffic congestion and climate change, fostering non-motorized transport (NMT) presents a viable solution for both reducing greenhouse gas emissions and improving urban livability, public health. They are also two of the least polluting forms of transport, with zero emissions during use. Cities that spend money on cycling infrastructure and policy have more bicycle use, and this is linked with lower air pollution and greenhouse gas emissions (Pucher et al. 2010). One such is the incorporation of cycling, particularly in cities like Copenhagen and Amsterdam as part of the overall urban transport system, through which great reductions have been achieved in car dependency and associated emissions. Cities have invested significant amounts of money in dedicated cycling lanes, bike-sharing programs and public information campaigns, demonstrating that enabling infrastructure plays a key role in getting citizens cycling from A to B. Other than decreasing emissions, non-motorized transport contributes to public health improvement. Cycling and walking encourage physical activity and therefore reduce obesity, cardiovascular diseases, etc. and other illnesses associated with sedentary lifestyles.

Beyond individual health benefits, increased cycling and walking also reduces the costs of healthcare for populations (Pucher and Buehler 2010). Moreover, with less accidents and injuries in cities with higher levels of NMT use, due to reduced motor traffic, this means cities are much safer. The effectiveness of non-motorized transport systems is contingent upon the construction of appropriate infrastructure. These include dedicated bike paths, pedestrian sidewalks, and fully protected crossings that allow pedestrians and cyclists to move safely through the urban environment. According to Leão and Urbano (2020), street connectivity and urban design influence walking behavior, where it is higher on more connected streets. The integration of NMT into public transport systems enables ease, vehicular access and sustainable travel in a better way. The push for non-motorised transport has not been without its challenges, however. Insufficient infrastructure, safety concerns, and cultural attitudes towards cycling and walking can limit their uptake. According to Hofmeister and Keitsch (2016), it is crucial to recognise social practices and cultural contexts of cycling to develop successful mobility solutions. Women and vulnerable groups' perception of public safety should also be addressed to ensure equal access to non-motorized means of transport (Leão and Urbano 2020). Comprehensive policies and dissemination through the community are required to surmount such barriers. Policies such as subsidies for the purchase of bicycles, investment in efficient cycling systems, public information, etc., to popularize the benefits of NMT could be initiated by governments. Furthermore, local stakeholders' involvement in planning and decision-making can help foster a sense of ownership, ensuring non-motorised transport measures´ uptake (Festa and Forciniti 2019).

Innovations in Green Aviation and Shipping

Shipping and aviation account for a considerable percentage of the world's greenhouse gas emissions. As the world continues to awaken to the realization that it needs to curb climate change, green innovation in shipping and aviation has become a sine qua non to the sustainability of shipping and aviation. Concerning some of the technologies applied in aviation transport, it is arguably the creation and application of Sustainable Aviation Fuels (SAFs).

SAFs are created from renewable materials such as biomass, waste products, and even captured atmospheric carbon. The use of SAFs will be in a position to reduce the aviation sector's carbon footprint significantly, consistent with international climate agreements, without loss of operational performance (Hasan et al. 2021). The International Air Transport Association (IATA) has also been ambitious in its goals for the aviation sector to reduce net emissions by 50% by 2050 as compared to levels in 2005, and SAFs take a pivotal role in their success (Hasan et al. 2021). Nonetheless, the universal utilization of SAFs is constrained due to its costly production and shortages, making research and further investments essential in producing more to reduce its unaffordability and scarcity (Champeecharoensuk 2023).

Another innovative approach in aviation is the electrification of aircraft. Electric and hybrid-electric propulsion systems are being developed to reduce reliance on fossil fuels and lower emissions. Wang (2024) highlights that electric aircraft have gained significant attention due to their environmental benefits, including reduced greenhouse gas emissions and noise pollution. Though commercially operated flights that are entirely electric are yet to come, technological innovations in battery efficiency and energy consumption are leading the way to their potential future feasibility.

The potential for electric aviation to transform short-haul flights and regional travel is substantial, offering a pathway to decarbonize a segment of the aviation industry that is often overlooked. In the shipping industry, technologies used to decrease emissions are also emerging. For instance, the International Maritime Organization (IMO) set ambitious goals in reducing greenhouse gas emissions related to shipping by at least 40% by 2030 from 2008 levels (Wang 2023). Alternative measures for the implementation of the vision include alternative fuel types, which are hydrogen and ammonia, whose release is substantially lower, even by one order of magnitude from the traditional fuels, which are today marine fuels (Czermański et al. 2020).

Another potential course to left, it said, aimed at making shipping more fuel-efficient and lower emissions, was through wind-assisted propulsion technologies such as sails and kites. These technologies harness wind energy to assist the engine power, resulting in lower fuel consumption and emissions (Yildiz et al. 2022). Shipping is also being decarbonized: Solar ships. But for all these promising green shipping and aviation innovations, there are also plenty of challenges. The many costs associated with developing and deploying new technologies, the bureaucracy and the burden of investing in infrastructure all serve to slow progress. Moreover, shaping policy and frameworks supporting the innovation as well as adoption of the emerging technologies is also dependent on the engagement of stakeholders, like governmental key players, industries, and research establishments during the transition to cleaner technologies and fuels (Lagouvardou et al. 2020; Oyenuga and Jeresa 2025; Nwokolo et al. 2024g).

Case Studies of Low-Carbon Mobility in Developing Regions

While the global community is tackling the issues of climate change and urbanization, developing countries are increasingly acknowledging the significance of low-carbon transport solutions. They not only cater to environmental considerations but also boost accessibility and overall quality of life for citizens (Nwokolo et al. 2023c). One of the best case studies is the launch of a Bus Rapid Transit (BRT) system in Bogotá, Colombia. TransMilenio, launched in 2000, has transformed urban travel within the city through the offer of a low-cost and effective mode of public transportation. Santos et al. (2010) report that the BRT system has made travel times plummet and use of public transport increase, displacing private car usage and resultant emissions. TransMilenio's success has seen the same or similar systems duplicated in other Latin American cities as well as beyond, showing how BRT systems can be used to encourage low-carbon transportation options in the developing world. Details of these contributions can be found in Mahadevia's urban strategy (India Udaipur total low carbon transport strategy with integrated land use and transport planning) where the exercise of deriving low-carbon transport scenarios (i.e. measures that reduce dependence on private motorised travel) with explicit consideration of Sustainable Development Goals (SDGs) was undertaken in the integrated land-use transport planning context (Mahadevia 2023). In addition to improving public transport infrastructure, the city advertised alternatives like cycling and long-distance walking.

Its integrated approach has led to lower emissions and better quality of life in cities. One classic instance occurred in the Philippines when the government had introduced its "Green Transport Program" which promotes electric vehicles (EVs) and non-motorized transport.

The program is intended to decrease the transport sector carbon footprint with an electric public transport bus, and to promote cycling infrastructure development (Bakker et al., 2017). Involvement is the basis of the successful implementation of low-carbon transport solutions, and the community and stakeholders support (on a local level) of the above-mentioned initiative confirm that. In South Africa, Johannesburg has taken it further to promote non-motorized transport by building dedicated cycling infrastructure and pedestrian facilities (Nwokolo et al. 2024b).

There are lessons to be learnt from the integration of walking and cycling into the transport system in the city of Johannesburg and Cape Town, South Africa (Aderibigbe and Gumbo 2022). More than that, the efforts of the city to provide a sustainable form of transport illustrates how non-motorized transport can be a complement to available public transport and achieve sustainable urban mobility. The Indonesian Makassar has also focused on developing non-motorized transport infrastructure to achieve sustainable mobility. Irafany et al. Emphasize sustainable urban design as the main component for enhancing the potential of pedestrian paths and cycling lanes (Irafany et al. 2023). The program is designed to keep residents walking and cycling their main mode of transport, reducing dependence on motorized transport and emissions. While there is success to be drawn on, as shown by the case studies above, challenges remain for developing countries in implementation of mass low-carbon mobility solutions. Limited finances, poor infrastructure, and social challenges can also be barriers (Oyenuga Omale 2024). The economic precariousness of small cities makes them more constrained regarding adopting sustainable transport policies compared to better-resourced and economically stable states and cities (Pojani and Stead 2015). To promote low-carbon mobility, these challenges must be overcome through government, NGOs, and community collaboration to create policies and infrastructure that enable low-carbon mobility.

Implications

The path to low-carbon transport and mobility systems has significant consequences for the environment, human health, and economies. First, greening transportation by decoupling it from dependence on fossil fuel via electric vehicles, mass transit, and non-motorized forms of travel can eliminate huge quantities of greenhouse gases. Besides, infrastructure for public transport and electric car charging infrastructure can generate jobs and provide local economic stimulus.

When integrated with renewable energy sources, the electrification of transport networks can significantly mitigate carbon emissions (Bamidele et al. 2023). Such a transition will be key to meeting global climate targets and limiting the impacts of climate change. Furthermore, low-carbon transport alternatives can enhance public health via air quality and noise reduction. Public transport and non-motorized transportation systems such as cycling and walking can help individuals to live a healthier life and reduce incidences of respiratory diseases caused by air pollution. Cities that prioritize public transport and active modes are shown to be at lower risk of rates of obesity and associated illness. This improves not just individual health but also reduces costs at the societal and all levels of government.

The impacts of low-carbon transport transition are significant at the economy level, too. Jobs in public transport infrastructure and recharging networks for electric cars can help drive local economies. The production of clean transport technology can spur innovation and new business into the clean energy economy (Alves et al. 2017). Besides that, improved public transport also helps to reduce traffic congestion, thereby increasing productivity through reduction in travelling time and associated costs by both businesses and individuals. But real-world implementation of low-carbon transport projects is fraught with challenges. Poor financial budgets, lackluster infrastructure, perceptions of public safety and the lack of convenience written in by individuals discourage the adoption of sustainable mobility solutions. Larger countries are also at an advantage, along with better-developed countries, if the smaller cities are having a tough time because they have less money and are more susceptible to instability. Governments, private enterprise and communities must act collectively to break these constraints, developing policies and infrastructure in favour of low-carbon mobility.

Conclusion, Limitations and Future Scope

Low-carbon mobility options and transport are needed to create a sustainable feeling of urban development and comply with climate change imperatives on all fronts. By investing in public transport infrastructure, promoting non-motorized modes of transport and integrating new and intelligent technologies, cities can save significant greenhouse gases while boosting public health and economic resilience. The solutions were identified in this chapter and presented with evidence and detailed case studies to show how to achieve livable cities and make them more sustainable. A wide range of stakeholders, including governments, industry and civil society actors, must work together to create the necessary supportive environment for sustainable transport practices to be adopted and shared between different actors. Unleashing the full potential of low-carbon transport solutions will require a focus on enabling infrastructure, policy and public adoption barriers. As part of the efforts to combat climate change and advance sustainable development, it is also important to prioritize low-carbon transport and mobility solutions for a more resilient and sustainable future for all.

References

Aderibigbe O, Gumbo T (2022) Variations in mode choice of residents prior and during COVID-19: an empirical evidence from Johannesburg, South Africa. Sustainability 14(24):16959. https://doi.org/10.3390/su142416959

Alves H, Silva L, Marinheiro R, Moura J (2017) Pmipv6 integrated with mih for flow mobility management: a real testbed with simultaneous multi-access in heterogeneous mobile networks. Wireless Pers Commun 98(1):1055–1082. https://doi.org/10.1007/s11277-017-4908-6

Bakker S, Contreras K, Kappiantari M, Tuấn N, Guillen M, Gunthawong G, Maarseveen M (2017) Low-carbon transport policy in four ASEAN countries: developments in Indonesia, the Philippines, Thailand and Vietnam. Sustainability 9(7):1217. https://doi.org/10.3390/su9071217

Bamidele R, Öztüren A, Haktanir M, Ogunmokun O (2023) Realizing green airport performance through green management intransigence, airport reputation, biospheric value, and eco-design. Sustainability 15(3):2475. https://doi.org/10.3390/su15032475

Champeecharoensuk A (2023) Climate change mitigation in Thailand's domestic aviation: mitigation options analysis towards 2050. Energies 16(20):7199. https://doi.org/10.3390/en16207199

Czermański E, Pawłowska B, Oniszczuk-Jastrząbek A, Cirella G (2020) Decarbonization of maritime transport: analysis of external costs. Front Energy Res 8. https://doi.org/10.3389/fenrg.2020.00028

Fangzhou T, Binglang C, Lu Z, Ningyuan F (2022) Spatial and temporal characteristics of Bike-Sharing trips at street scale: a case study of Chengdu City. J Civil Eng Urban Plan 4(4). https://doi.org/10.23977/jceup.2022.040402

Festa D, Forciniti C (2019) Attitude towards bike use in Rende, a small town in southern Italy. Sustainability 11(9):2703. https://doi.org/10.3390/su11092703

Gan Y, Lü Z, He X, Wang M, Amer A (2023) Cradle-to-grave lifecycle analysis of greenhouse gas emissions of light-duty passenger vehicles in China: towards a carbon-neutral future. Sustainability 15(3):2627. https://doi.org/10.3390/su15032627

Guerra E (2011) Too much riding on climate change? Berkeley Plan J 23(1). https://doi.org/10.5070/bp32318385
Hasan A, Mamun A, Rahman S, Malik K, Amran M, Khondaker A, Al-Ismail A (2021) Climate change mitigation pathways for the aviation sector. Sustainability 13(7):3656. https://doi.org/10.3390/su13073656
Hofmeister T, Keitsch M (2016) Framing complexity in design through theories of social practice and structuration: a comparative case study of urban cycling, 9. https://doi.org/10.21606/drs.2016.47
Irafany S, Jinca M, Wunas S, Zamhuri M (2023) Development of non-motorized transport (NMT) facilities as sustainable and environmentally friendly transportation in the Losari Beach area of Makassar City, South Sulawesi, Indonesia. IOP Conf Series Earth Environ Sci 1134(1):012003. https://doi.org/10.1088/1755-1315/1134/1/012003
Joshi A, Singh R, Kumar V, Rani S (2024) Net zero strategy: exploring opportunities, confronting challenges for developing countries. In: Singh R, Crowther D (eds) Transition towards a sustainable future. Approaches to Global Sustainability, Markets, and Governance. Springer, Singapore. https://doi.org/10.1007/978-981-97-5756-5_11
Khan S (eds) Zero carbon industry, eco-innovation and environmental sustainability. World Sustainability Series. Springer, Cham. https://doi.org/10.1007/978-3-031-80220-1_15
Kumar K, Singh R, Sharma A, Lehri BPS (2024) Climate change adaptation strategies for achieving net-zero economy. In: Singh R, Crowther D (eds) Transition towards a sustainable future. Approaches to global sustainability, markets, and governance. Springer, Singapore. https://doi.org/10.1007/978-981-97-5756-5_8
Lagouvardou S, Psaraftis H, Zis T (2020) A literature survey on market-based measures for the decarbonization of shipping. Sustainability 12(10):3953. https://doi.org/10.3390/su12103953
Leão A, Urbano M (2020) Street connectivity and walking: an empirical study in londrina- pr. Semina Ciências Exatas E Tecnológicas 41(1):31. https://doi.org/10.5433/1679-0375.2020v41n1p31
Lester M, Bramstoft R, Münster M (2020) Analysis on electrofuels in future energy systems: a 2050 case study. Energy 199:117408. https://doi.org/10.1016/j.energy.2020.117408
Li J, Zhang P, Lo K, Guo M, Wang M (2015) Reducing carbon emissions from shopping trips: evidence from China. Energies 8(9):10043–10057. https://doi.org/10.3390/en80910043
Lin C, Sinclair T, Mitchell I (2023) Advancing transportation and urban mobility optimization: insights from a comprehensive review and analysis. https://doi.org/10.31219/osf.io/38wjd
Mahadevia D (2023) Including SDGs in low-carbon transport scenarios: a case of Udaipur. EC 4(2):2525. https://doi.org/10.54517/ec.v4i2.2525
Nwokolo SC, Singh R, Khan S, Kumar A, Luthra S (2023c) Africa's awakening to climate action. In: CSR, sustainability, ethics & governance, pp 299–310. https://doi.org/10.1007/978-3-031-44514-9_9
Nwokolo SC, Singh R, Khan S, Kumar A (2024b) Net zero technology and circular economy. In: Elsevier eBooks, pp 237–247. https://doi.org/10.1016/b978-0-443-31486-5.00009-4
Nwokolo SC, Singh R, Khan S, Kumar A (2024g) Key players in net zero technology. In: Elsevier eBooks, pp. 193–215. https://doi.org/10.1016/b978-0-443-31486-5.00007-0
Oke D (2024) Reducing economy-wide greenhouse gas emissions with electrofuels and biofuels as the grid decarbonizes. Energy Fuels 38(7):6048–6061. https://doi.org/10.1021/acs.energyfuels.3c04833
Oyedele OM (2025) The cobra effect: exploring the dark side of innovation and environmental sustainability in developing nations. In: Singh R, Joshi A, Filho WL, Ruggieri R, Ruggeri M, Vinci G, Poponi S (2021) Electric mobility in a smart city: European overview. Energies 14(2):315. https://doi.org/10.3390/en14020315
Oyenuga MO, Jeresa S (2025) Adaptation and resilience: private, public, and individual change makers. In: Singh R, Filho WL (eds) Climate neutrality through smart eco-innovation and environmental sustainability. Climate Change Management. Springer, Cham. https://doi.org/10.1007/978-3-031-83250-5_2

Oyenuga MO, Omale SA (2024) Is Africa jinxed? Exploring the challenges of technology access and adoption in Africa. African J Econom Sustain Develop 7(4):142–161. https://doi.org/10.52589/AJESDULN1LRNF

Oyetunde TO., Oyenuga MO, Adoga GJ (2023) Green management practices and organisational efficiency in higher institutions. Budapest International Research and Critics in Linguistics and Education (BirLE) Journal 6(3), 276–290. https://bircu-journal.com/index.php/birle/article/view/7716

Pojani D, Stead D (2015) Sustainable urban transport in the developing world: beyond megacities. Sustainability 7(6):7784–7805. https://doi.org/10.3390/su7067784

Prabhu A, Pai M (2012) Buses as low-carbon mobility solutions for urban India. Transport Res Record J Transport Res Board 2317(1):15–23. https://doi.org/10.3141/2317-03

Pucher J, Buehler R (2010) Walking and cycling for healthy cities. Built Environ 36(4):391–414. https://doi.org/10.2148/benv.36.4.391

Pucher J, Dill J, Handy S (2010) Infrastructure, programs, and policies to increase bicycling: an international review. Prev Med 50:S106–S125. https://doi.org/10.1016/j.ypmed.2009.07.028

Santos G, Behrendt H, Teytelboym A (2010) Part ii: policy instruments for sustainable road transport. Res Transp Econ 28(1):46–91. https://doi.org/10.1016/j.retrec.2010.03.002

Sobrino N, Monzón A (2013) Management of urban mobility to control climate change in cities in Spain. Transport Res Record J Transport Res Board 2375(1):55–61. https://doi.org/10.3141/2375-07

Wang S (2024) An energy efficiency optimization method for electric propulsion units during electric seaplanes' take-off phase. Aerospace 11(2):158. https://doi.org/10.3390/aerospace11020158

Wang J (2023) Review of EU maritime Transport greenhouse gas emissions Reduction Policies: Better prospects for policy makers in China? Int J Front Sociol 5(9). https://doi.org/10.25236/ijfs.2023.050920

Yildiz M, Kale U, Nagy A (2022) Analyzing well-to-pump emissions of electric and conventional jet fuel for aircraft propulsion. Aircr Eng Aerosp Technol 94(10):1605–1613. https://doi.org/10.1108/aeat-02-2021-0032

Zhang W (2023) Calculation of carbon emissions and study of the emission reduction path of conventional public transportation in Harbin City. Sustainability 15(22):16025. https://doi.org/10.3390/su152216025

Zheng Y, Li S, Xu S (2019) Transport oil product consumption and GHG emission reduction potential in China: an electric vehicle-based scenario analysis. PLoS ONE 14(9):e0222448. https://doi.org/10.1371/journal.pone.0222448

Chapter 24
Decarbonizing Industrial Processes

Abstract Decarbonisation of industrial processes is essential to mitigating climate change and potential sustainability targets. This chapter summarises different strategies and technologies that can be used to minimise greenhouse gas emissions from major industrial sectors such as manufacturing, energy and transport. Through case studies and best practices, the chapter gauges the contribution of renewable energy, energy efficiency and sustainable material for decarbonisation. It also discusses other factors affecting the transition to low-carbon production processes, including policy instruments, stakeholder involvement, and technological change development. Ultimately, this chapter seeks to add to the wider picture of decarbonization pathways and their role in the evolution of a more resilient and sustainable future.

Introduction

The industrial sector is one of the largest contributors to greenhouse gas emissions globally, accounting for approximately 21% of total emissions. As the world confronts the dual realities of climate change, energy and food security, decarbonisation of industry has never been such a priority. This transition not only means adopting practices that also enable greater use of resources, improve productivity and enable sustainable economic growth, but also those that reduce emissions. The process of decarbonization can refer to several processes, including implementation of renewable power, use of energy efficiency technology, transitioning to sustainable material production and use. One example is the integration into the industrial processes of solar, wind and biomass energy, which can particularly help to reduce dependency on fossil fuels and thus lower some carbon emissions. Sansavini (2024) also highlights the necessity of a systems perspective for understanding trade-offs between distinct decarbonization pathways to formulate sound policies. In addition, energy efficiency interventions, such as process optimization and the application of advanced manufacturing technologies, can result in high levels of energy savings and emission reductions. Decarbonization of industrial processes cannot be achieved without the contribution of sustainable materials.

M. O. Oyenuga et al., *Smart Technologies for Climate Change and Net Zero Policies: Practical Approaches Towards Sustainability*, Climate Change Management,
https://doi.org/10.1007/978-3-031-92221-3_24

The transition to circular economy principles that emphasize reuse, and recycling can also lead to a big drop in manufacturing process carbon emissions. As brought out by Axelson et al., incremental as well as radical business model innovation are required in bringing about an industrial transition for sustainability (Axelson et al. 2021). Besides driving down the emission levels, it also boosts the level of innovation and competitiveness within industries. But the actual fulfillment of decarbonization policy requires policy backstops and partnerships with multiple actors. Governments also have an essential function to introduce regulation and incentive programs that spur environmentally friendly production behaviors in the sector. As reported by Teixeira, those policies, for instance, encouraging investment in cleaner technologies and green energy, will push for a decarbonization shift (Teixeira 2023). In addition, involving businesses, communities, and research institutions in developing and implementing decarbonization plans is critical to building a culture of sustainability and innovation. Despite the advances made towards decarbonizing industrial activities, there are challenges. Access to limited financing, technological constraints, and change aversion may discourage the uptake of sustainable practices. As emphasized by Mba, it is essential to know the impediments to sustainable practices to inform development towards sustainability across different industries (Maundu 2020). In addition, there is a need for ongoing research and development to determine and adopt viable decarbonization routes specific to certain industries and settings.

Reducing Emissions in Manufacturing

Manufacturing is one of the major causes of greenhouse gas emissions globally, with an estimated contribution of 21% of the total.

As the fight against climate change heats up, emissions reductions in manufacturing are a perennial concern of policymakers, corporate actors, and scholars. Various methods for reducing emissions in the manufacturing sector, each focusing on innovation, efficiency, and greener practices, have been discussed above. Energy-efficient technology is also one of the best options for emissions elimination in manufacturing. Liu et al. show that by using energy-efficient approaches, we can greatly reduce carbon emissions and cut down on operational costs at the same time (Liu et al. 2018). For instance, high-end manufacturing technology from automation to robotics can improve device manufacture, optimise energy usage, and reduce waste. Furthermore, energy management systems enable businesses to track and measure energy consumption in real-time, highlighting areas where adjustments are possible and facilitating the implementation of energy-saving measures. One more primary center of how to reduce emissions in manufacturing is switching to renewable energy. Switching from fossil fuels to clean energy sources, like solar, wind and biomass, exponentially lowers the carbon footprint for manufacturers. Research by Xu points out that the utilization of renewable energy saves not just on emissions but also renders production processes more insensitive to price and supply uncertainty in the energy sector (Xu 2023). Also, policies favoring the deployment of renewable energy technologies can help accelerate the transition towards a low-carbon manufacturing sector. Circular economic adoption is equally crucial in helping save emissions in manufacturing. Through the emphasis on recycling, waste avoidance, and resource efficiency, manufacturing businesses can lower their impact on nature without compromising on productivity. The provided source by Caro et al., however, does not directly support the claim regarding circular economy practices to introduce radical declines in carbon emissions (Caro et al. 2013).

In addition, its contribution to emission reduction cannot be ignored. Proper coordination and cooperation between the supply chain members will result in best-in-class logistics, transportation emissions, and enhanced utilisation of resources. Li et al. (2022) reiterate that fairness issues involved in the operations of the supply chain can shape the decisions made by manufacturers on investing in emission reduction (Li et al. 2022). By facilitating open communication and cooperation among supply chain players, businesses can foster better efforts towards emission reduction among them. As attractive as the implementation of such practices may appear, however, there are barriers to the mass implementation of emissions reduction practices in manufacturing. Limited financing, technological constraints, and change resistance may stifle development. Understanding the determinants of manufacturers' behaviors for emissions reduction is essential in guiding practice and policy (Xu et al. 2023). Moreover, ongoing research and development are required to determine and apply effective emissions reduction strategies specific to industries and circumstances.

Innovations in Clean Energy for Heavy Industries

The industrial sector, which includes manufacturing, mining and construction, is responsible for a large portion of global greenhouse gas emissions. However, as global pressure is intensifying to action against climate change, the deployment of clean energy technologies in these sectors has become essential for decarbonizing the world and fostering sustainable development. Integration of renewable energy sources (wind, solar, biomass energy) into the industrial process is the most attractive innovation in terms of clean industrial energy. One of the key means to achieve sustainability and reduce carbon footprints is through the use of renewable energy solutions in production (Usman 2024). Solar panels, for instance, can be placed on factory rooftops to supply clean energy for their operation, and wind turbines can power energy-intensive production processes in exchange for their generation of electricity. Shift to renewables (low carbon) energy replaces a fossil energy dependency and grants energy security, immunity to price spikes in the legacy energy market. Other breakthroughs are associated with innovations on energy-efficient products and processes (Nwokolo et al. 2023b). Advanced production processes like automation can help reduce energy consumption and waste for the manufacturing industry. For example, Industry 4.0 technology not only utilizes big data analytics but also utilizes the Internet of Things (IoT) to allow producers to monitor their energy consumption in real time, providing a maximum opportunity for improvement (Calabrese et al. 2020). This is not just a cost advantage but also contributes to emission reduction (Apata et al. 2024) since adopting energy-efficient practices is capable of significantly enhancing the sustainability performance of an organization while reducing operating costs (Ghobakhloo et al. 2021).

Further, the application of hydrogen as a clean fuel is picking up momentum in heavy industries. Hydrogen can be made from renewable feedstocks and be utilized as a fuel for high-temperature processes, including steelmaking. The shift is highly applicable to challenging-to-decarbonize industries because hydrogen combustion has only water vapor as a byproduct. Additionally, the development of carbon capture, utilization, and storage (CCUS) technologies is also essential in curbing emissions in heavy industries. CCUS involves capturing carbon dioxide emissions from industrial processes and recycling it to serve another function or sequestering it underground so that it is not released into the atmosphere. Tiwari (2024) indicates the potential of CCUS technologies to significantly lower emissions in energy-intensive industries such as cement and steel production. Introduction of CCUS must be facilitated by a tremendous investment and policy aid to enhance take-up. Regardless of the supportive clean energy innovation, some weaknesses hinder their widescale application across heavy industries. Technological inefficiencies (Oyenuga and Omale 2024; Nwokolo et al. 2024; Nwokolo et al. 2023a, b, c), hefty initial capital expenses, and requisite skilled workforce (Singh et al. 2024; Oyedele et al. 2024) may make it hard to switch over to cleaner technology. These challenges can be addressed only by a multi-dimensional approach that includes technological innovation, along with social and economic factors (Awan et al. 2020). Furthermore, the policymaker's and incentive agency's role in the government is crucial to persuading heavy industries to shift to clean energy technology.

Circular Economy in Industrial Processes

The principle of the circular economy (CE) is a revolutionary method of industrial processes, which seeks to develop a closed-loop system with the least amount of waste, the highest efficiency in the use of resources, and sustainability. Contrary to the conventional linear economy, based on the take-make-dispose approach, the circular economy focuses on the ongoing use of resources and minimizing the ecological footprint of industrial processes. Essentially, the circular economy relies on several fundamental principles: reducing the use of resources, recycling products, recycling waste, and recovering energy. Placing these principles at the forefront, industries can reduce their use of virgin materials and reduce greenhouse gas emissions by a considerable margin. Geissdoerfer et al. (2017) argue that besides solving environmental issues, intuitive prior solution to the circular economy stimulates economy by generating new economic opportunities and improving competitiveness. For example, circularity-oriented companies reduce costs related to raw materials and waste management, which results in better-profitability. Nonetheless, the building sector can adopt circular principles like employing recycled materials, designing to disassemble, and having waste management systems in place to minimize its impact on the environment. For instance, construction experts have demonstrated that the application of circular economy principles may enable more sustainable construction processes and improved resource efficiency (Torgautov et al. 2021). For instance, using recovered materials to remanufacture old buildings, in addition to saving more material, also avoids emissions from producing virgin material. Although circular economy is promising, various challenges stall industries from employing it on large scales; (Kumar et al. 2024).

One of the major challenges is limited willingness and awareness of circular principles of the industry players. Small and medium-sized enterprises (SMEs) face many barriers and are one of the drivers of obstacles to the adoption of circular economy business models(Rizos et al. 2016). Additionally, circular practices may be stifled by regulatory frameworks and market conditions, making it less predictable for businesses to transform. To tackle these issues, we must promote cooperation between governments, companies, and research institutions. Through the design of incentives, funding, and research and development support, policymakers can also play a key role in creating an enabling environment for circular economic activity. Educational and training programs can also provide the skills and knowledge needed for industry practitioners to effectively adopt circular practice.

Case Studies of Decarbonizing Industries in Developing Countries

Decarbonization of industrial activities is at the heart of climate change mitigation, and enabling sustainable development, particularly for developing countries, where industrial development typically leads to a rise in greenhouse gas emissions using the Carbon Capture and Storage (CCS) technology. Lau's study assesses the viability of CCS in the ASEAN nations, highlighting the fact that ASEAN nations have adequate storage potential in oil and gas fields as well as in saline aquifers to store anthropogenic CO_2 emissions from industry and power generation for decades (Lau 2022). The study highlights the CCS investment as an alternative to carbonizing fossil fuel-fired power and heavy industries that prevail in the majority of developing countries. With the sequestration of CO_2 emissions before their release into the atmosphere, CCS has the potential to be a national climate enabler while allowing ongoing industrial production.

As steel is responsible for a significant share of the country's carbon emissions, the steel industry has been at the forefront of decarbonisation efforts in China. Jin et al. conducted an empirical study on technological progress in the energy area and the relationship between technological development and carbon emission (Jin et al. 2017). The study indicated that enhanced energy efficiency and cleaner production technology have led to substantial emission reductions by the steel sector. For example, the production using electric arc furnaces that recycle steel and consume less energy than traditional blast furnaces has caused less carbon emissions but equal production. The illustration depicts how technological innovation can spearhead decarbonization in high-energy-used sectors. The other illustration of notes is decarbonization of the cement industry, which is notoriously challenging to decarbonize due to high process emissions. Ju et al. (2021) studied Japan's national mitigation scenarios and emphasized the necessity of transitioning to substitute raw materials and fuels and encouraging energy efficiency in cement production. The study identifies that transition to low-carbon technologies, like the use of waste products as substitute raw materials to conventional materials, has the potential to decrease emissions substantially in the cement industry.

This methodology is most suited to urbanized and infrastructure-growing developing nations. In the context of energy-guzzling sectors, Aina's research presents the issues and opportunities introduced by developing countries towards decarbonizing (Aina 2024). The research calls for a comprehensive approach that includes national features and pioneering policies to initiate decarbonization. Employing a random forest distribution model, the research identifies strong drivers of emissions reduction and advocates for the value of real-time monitoring and forecasting in the design of policy.

Additionally, an industrial symbiosis case in the Russian ferrous metallurgy industry proves the possible co-operation on decarbonisation. Novikova et al. (2023) describe the experience of utilisation of by-product materials in one industry as a material resource in another with enormous potential in diminishing emissions and augmenting the resource efficiency. The recycling route not only cuts waste, but it also catalyzes innovations as well as a culture of partnership among sectors and can demonstrate possibilities in decarbonization based on mutual benefits. While there are encouraging developments in the decarbonization of developing nations' industries, more effort is needed. Limited access to funding, technological challenges, and capacity building can hinder momentum. Overcoming these hurdles involves an integrated approach that brings together technological advancement with social and economic factors (Nyangarika et al. 2022). In addition, the intervention of government policies and incentives plays a fundamental role in ensuring the uptake of clean technologies and sustainable practices across industrial sectors.

Challenges to Decarbonizing the Industrial Sector

Decarbonization of the industrial process is an essential aspect of combating climate change and sustainability endeavors on a worldwide level. Decarbonizing, though, comes with several challenges in the developed nations as much industrial expansion occurs together with increasing levels of greenhouse gases. An awareness of such issues will allow proper policies to support low-carbon industry culture to be effectively crafted. A major challenge is the use of fossil fuels in energy-intensive sectors like steel, cement, and chemical production. These industries are marked by high energy consumption and process emissions that are hard to abate. Franco (2023) points out that the inherent nature of operations in hard-to-abate industries tends to demand high-temperature heat sources, and thus it is difficult to identify alternative sources of energy to fossil fuels. Shifting to low-carbon energy sources like hydrogen or biomass demands huge investment in new technologies and infrastructure, which may be a hindrance for most companies, especially in developing countries (Oyedele et al. 2023). Another challenge is the economic cost of shifting to low-carbon practices. Most sectors have high capital costs involved with the implementation of new technologies that can discourage investments. Estevan et al. (2016) explain how environmental policies for the European Union encouraged the post-carbon industry to become counterproductive as they cause cash shortages for existing industries, giving rise to a dilemma for companies that are yet to be transitioned (Estevan et al. 2016). This serves to underscore the importance of effective financial support structures (Fasanmi et al. 2025) that can promote the transition without undermining the economic sustainability of the current industries.

The complexity of supply chains (Omale et al. 2024a, b) is another obstacle to decarbonization. For example, Bajželj et al. (2013) highlight that not only are emissions produced during product manufacturing, but emissions are also accrued throughout the entire lifecycle of the product, from supply chain upstream processes and transportation. And so, with this interconnectedness, it is difficult, if not impossible, to have effective emissions reduction mechanisms that don't consider the large-scale supply chain dynamics. Firms must work with suppliers and logistics suppliers to identify opportunities for emissions cuts, something that may be complicated by a variation in regulatory environments and market conditions.

Besides, the heterogeneity of industries makes it difficult to establish universal decarbonization pathways. Yuan points out that the varying industrial processes and products make it challenging to comprehend effective decarbonization strategies (Yuan 2023). Every sector could have industry-specific solutions considering operating features, regulation, and demand in the marketplace. This heterogeneity requires continued research and innovation to discover the best practices and scalable solutions applicable across sectors. Additionally, an absence of skills and know-how in applying low-carbon technology can act as a hindrance. As emphasized by Gai, poor nations tend to experience a skills deficit that constrains their capacity for successful adoption and operation of new technologies (Gai 2024). Education and training program investment (Omale et al. 2024a, b) is necessary to provide the workforce with the required competencies to facilitate the shift towards a low-carbon economy. Finally, public opinion and acceptance of decarbonization measures can also be challenging. Sectors are likely to face opposition from stakeholders who fear the economic effects of shifting towards low-carbon business. The decision-making process must involve engaging communities and stakeholders to gain support and ensure the success of the benefits of decarbonization through proper communication (Oyenuga and Jeresa 2025).

Implications

Industrial process decarbonization is fundamental to achieving carbon neutrality not only for environmental sustainability but also for economic and social justice. As industries move to low-carbon processes, they will be able to reduce their greenhouse gas emissions by a significant amount, reinforcing international climate goals as well as making them more resilient to climate threats. Cleaner technologies and processes can help improve energy efficiency, cost of operation, and competitiveness in an environment of uncertainty. For instance, carbon capture technology and renewable energy can help industries minimize their carbon footprint without an impact on productivity (Nwokolo et al. 2023c). Decarbonization can also drive innovation and new business models. When companies attempt to minimize their emissions, they will likely spend on R&D, which will trigger new technological innovation that not only benefits the environment but also the economy. Transitioning into a circular economy, which is all about efficiency of resources and minimizing waste, can unleash fresh opportunities for companies to innovate and diversify products (Oyedele 2025). This transformation is also able to create employment into new sustainable industries such as renewable energy, energy efficiency, and waste management. Decarbonization comes with its own set of challenges, however. Sectors are faced with daunting barriers in the guise of significant up-front capital costs, technology limitations, and policy barriers. The shift of European industries toward use of renewable energy is reliant on the levels of high investment and a favorable regime of policy for facilitating ease in the process.

Furthermore, the confidence of the skilled workforce and experience in using low-carbon technology may be a challenge, especially in developing countries where access to training and resources may be limited. Breaking these barriers requires a collaborative effort from governments, companies, and society to create an enabling environment that promotes sustainable actions.

Conclusion, Limitations and Future Scope

Decarbonization of industrial processes is an important component of both climate change mitigation and sustainable development. Industries can help the planet by adopting new technologies, energy efficiency and renewable energy, they can reduce greenhouse gas emissions and be economically viable and resilient. The case studies showcased in the chapter, along with the evidence, point to the success of decarbonization in different sectors and what industries can do to cash in on a low-carbon world. Achieving this transition will require collaboration between stakeholders, including governments, industry and research organisations. To facilitate the implementation of low-carbon practices in industrial processes, it is critical to adopt policies, invest in research and development, and build capacity. Seeking to fight against climate change, the decarbonisation of industrial processes will continue as an essential aspect to achieving a sustainable and resilient future. Such a type of transformation will not only be environment friendly with low carbon emissions but will also provide economic and social fairness in the world of interconnectivity.

References

Aina O (2024) Decarbonization, opportunities and dilemma for developing countries. Am J Comput Eng 7(1):1–12. https://doi.org/10.47672/ajce.1761

Apata AO, Oyenuga MO, Abeoba M, Ugom A (2024) Internet of Things (IoT) solutions for smart transportation infrastructure and fleet management. J Propulsion Technol 45(4):1492–1505. https://propulsiontechjournal.com/index.php/journal/article/view/8325

Awan U, Arnold M, Gölgeci İ (2020) Enhancing green product and process innovation: towards an integrative framework of knowledge acquisition and environmental investment. Bus Strateg Environ 30(2):1283–1295. https://doi.org/10.1002/bse.2684

Axelson M, Oberthür S, Nilsson L (2021) Emission reduction strategies in the EU steel industry: implications for business model innovation. J Ind Ecol 25(2):390–402. https://doi.org/10.1111/jiec.13124

Bajželj B, Allwood J, Cullen J (2013) Designing climate change mitigation plans that add up. Environ Sci Technol 47(14):8062–8069. https://doi.org/10.1021/es400399h

Calabrese A, Dora M, Ghiron N, Tiburzi L (2020) Industry 4.0 transformation process: how to start, where to aim, what to be aware of. Prod Plan Control 33(5):492–512. https://doi.org/10.1080/09537287.2020.1830315

Caro F, Corbett C, Tan T, Zuidwijk R (2013) Double counting in supply chain carbon footprinting. Manuf Serv Oper Manag 15(4):545–558. https://doi.org/10.1287/msom.2013.0443

Estevan D, Mezquita A, Ferrer S, Monfort E (2016) Unwanted effects of the European Union environmental policy to promote a post-carbon industry. The case of energy in the European ceramic tile sector. J Clean Prod 117:41–49. https://doi.org/10.1016/j.jclepro.2016.01.021

Fasanmi OO, Adeniyi BC, Oyenuga MO (2025) Exploring the role of innovation on economic performance of manufacturing firm in Nigeria. J Econom Finance Manage Stud 8(1):694–706. https://doi.org/10.47191/jefms/v8-i1-67

Franco A (2023) Routes for hydrogen introduction in the industrial hard-to-abate sectors for promoting energy transition. Energies 16(16):6098. https://doi.org/10.3390/en16166098

Gai G (2024) Research on China's environmental law and low-carbon emission economy trade surplus under the background of carbon neutrality. IETI Trans Data Analysis Forecasting (Itdaf) 2(1):23–32. https://doi.org/10.3991/itdaf.v2i1.49197

Geissdoerfer M, Savaget P, Bocken N, Hultink E (2017) The circular economy—a new sustainability paradigm? J Clean Prod 143:757–768. https://doi.org/10.1016/j.jclepro.2016.12.048

Ghobakhloo M, Iranmanesh M, Grybauskas A, Vilkas M, Petraitė M (2021) Industry 4.0, innovation, and sustainable development: a systematic review and a roadmap to sustainable innovation. Bus Strat Environ 30(8):4237–4257. https://doi.org/10.1002/bse.2867

Jin L, Duan K, Shi C, Ju X (2017) The impact of technological progress in the energy sector on carbon emissions: an empirical analysis from China. Int J Environ Res Public Health 14(12):1505. https://doi.org/10.3390/ijerph14121505

Ju Y, Sugiyama M, Kato E, Matsuo Y, Oshiro K, Herran D (2021) Industrial decarbonization under Japan's national mitigation scenarios: a multi-model analysis. Sustain Sci 16(2):411–427. https://doi.org/10.1007/s11625-021-00905-2

Kumar K, Singh R, Sharma A, Lehri BPS (2024) Climate change adaptation strategies for achieving Net-zero economy. In: Singh R, Crowther D (eds) Transition towards a sustainable future.

approaches to global sustainability, markets, and governance. Springer, Singapore. https://doi.org/10.1007/978-981-97-5756-5_8
Lau H (2022) Evaluation of decarbonization technologies for ASEAN countries via an integrated assessment tool. Sustainability 14(10):5827. https://doi.org/10.3390/su14105827
Li C, Guo T, Chen Y (2022) Robust emission reduction strategies under cap-and-trade and demand uncertainty. Sustainability 14(20):13445. https://doi.org/10.3390/su142013445
Liu Y, Huang C, Song Q, Li G, Xiong Y (2018) Carbon emissions reduction and transfer in supply chains under a cap-and-trade system with emissions-sensitive demand. Syst Sci Control Eng 6(2):37–44. https://doi.org/10.1080/21642583.2018.1509398
Maundu J (2020) impact of sustainable business practices in petroleum industry: a case study of national oil corporation of Kenya. Strat J Bus Chang Manag 7(3). https://doi.org/10.61426/sjbcm.v7i3.1680
Novikova O, Samarina V, Paytaeva K (2023) Russian experience of ferrous metallurgy decarbonization in the context of global trends. Bio Web Conf 63:03012. https://doi.org/10.1051/bioconf/20236303012
Nwokolo SC, Singh R, Khan S, Kumar A, Luthra S (2023a) Technological pathways to net-zero goals in Africa. In: Africa's path to net-zero. CSR, sustainability, ethics & governance. Springer, Cham. https://doi.org/10.1007/978-3-031-44514-9_5
Nwokolo SC, Singh R, Khan S, Kumar A, Luthra S (2023b) Decarbonizing hard-to-abate sectors in Africa. In: CSR, sustainability, ethics & governance, pp 211–236. https://doi.org/10.1007/978-3-031-44514-9_6
Nwokolo SC, Singh R, Khan S, Kumar A, Luthra S (2023c) Remedies to the challenges of renewable energy deployment in Africa. In: CSR, sustainability, ethics & governance, pp 59–74. https://doi.org/10.1007/978-3-031-44514-9_3
Nwokolo SC, Singh R, Khan S, Kumar A (2024) Threats to the rapidity of sustainability transitions posed by technological changes. In: Elsevier eBooks, pp 39–75. https://doi.org/10.1016/b978-0-443-31486-5.00002-1
Nyangarika A, Mikhaylov A, Muyeen S, Yadykin V, Mottaeva A, Pryadko I, Shvandar K (2022) Energy stability and decarbonization in developing countries: random forest approach for forecasting of crude oil trade flows and macro indicators. Front Environ Sci 10. https://doi.org/10.3389/fenvs.2022.1031343
Omale SA, Oyenuga MO, Abdullahi D, Madu I (2024) Moderating role of entrepreneurial education on innovative work behaviour and resilience of SMEs. Innovations 77(2):1166–1185 e0ce0-1166-1185.11600.pdf
Omale SA, Yusuf SO, Oyenuga MO, Ikemefuna M, Ojo SS, Momodu ID (2024) Organizational learning in the post-COVID-19 era: a prerequisite for stakeholder satisfactions. Twist 19(3):265–272. https://twistjournal.net/twist/article/view/339
Oyedele MO, Marcus GO, Ahungwa AI (2023) Do consumers care about green marketing practices? Insight from a developing nation. Budapest Int Res Crit Inst J (BIRCIJournal) 6(3):(1424–1436). https://bircu-journal.com/index.php/birci/article/view/7665
Oyedele MO, Sunday AO, Abuh AI (2024) Fostering technological-enhanced training and development for business survival and performance in the new normal. J Propulsion Technol 45(3):1858–1869. https://propulsiontechjournal.com/index.php/journal/article/view/7479
Oyedele OM (2025) The cobra effect: exploring the dark side of innovation and environmental sustainability in developing nations. In: Singh R, Joshi A, Filho WL, Khan S (eds) Zero carbon industry, eco-innovation and environmental sustainability. World sustainability series. Springer, Cham. https://doi.org/10.1007/978-3-031-80220-1_15
Oyenuga MO, Jeresa S (2025) Adaptation and resilience: private, public, and individual change makers. In: Singh R, Filho WL (eds) Climate neutrality through smart eco-innovation and environmental sustainability. Climate Change Management. Springer, Cham. https://doi.org/10.1007/978-3-031-83250-5_2

Oyenuga MO, Omale SA (2024) Is Africa jinxed? Exploring the challenges of technology access and adoption in Africa. African J Econom Sustain Develop 7(4):142–161. https://doi.org/10.52589/AJESDULN1LRNF

Rizos V, Behrens A, Gaast W, Hofman E, Ιωάννου A, Kafyeke T, Topi C (2016) Implementation of circular economy business models by small and medium-sized enterprises (SMEs): barriers and enablers. Sustainability 8(11):1212. https://doi.org/10.3390/su8111212

Sansavini G (2024) Integrated decarbonization of hard-to-abate industry utilizing biomass relieves burden on power sector. https://doi.org/10.21203/rs.3.rs-5458903/v1

Singh R, Crowther D, Ravesangar K (2024) Harnessing human Capital for a Sustainable Future: HR's pivotal role toward net zero policies. In: Approaches to global sustainability, markets, and governance, pp 27–47. https://doi.org/10.1007/978-981-97-5756-5_2

Teixeira R (2023) Exploring sustainable human manufacturing: industry 4.0, manufacturing, and green marketing for a balanced future. Concilium 23(15):772–790. https://doi.org/10.53660/clm-1784-23m18

Tiwari P (2024) Advanced fungal biotechnologies in accomplishing sustainable development goals (SDGs): what do we know and what comes next? J Fungi 10(7):506. https://doi.org/10.3390/jof10070506

Torgautov B, Zhanabayev A, Tleuken A, Türkyılmaz A, Mustafa M, Karaca F (2021) Circular economy: challenges and opportunities in the construction sector of Kazakhstan. Buildings 11(11):501. https://doi.org/10.3390/buildings11110501

Usman F (2024) Integrating renewable energy solutions in the manufacturing industry: challenges and opportunities: a review. Eng Sci Technol J 5(3):674–703. https://doi.org/10.51594/estj.v5i3.865

Xu Y (2023) Research on supply chain emission reduction decisions considering loss aversion under the influence of a lag effect. Sustainability 15(17):13092. https://doi.org/10.3390/su151713092

Xu C, Zhang Y, Yang Y, Gao H (2023) Carbon peak scenario simulation of manufacturing carbon emissions in northeast China: perspective of structure optimization. Energies 16(13):5227. https://doi.org/10.3390/en16135227

Yuan M (2023) Accelerating the net zero transition in Asia and the Pacific: low-carbon hydrogen for industrial decarbonization. https://doi.org/10.56506/ghcg4604

Chapter 25
Green Technologies for Water Management

Abstract As water scarcity increases, the need for sustainable solutions adds to this idea that green technologies could be a fix, and this is particularly true in the case of water management technologies. With water resources increasingly threatened by urbanisation and climate change, new sustainable principles need to be established to restore a balance between sufficient water for human usage and the health of the ecosystems that have become critical to water delivery. Green technology in this chapter is defined by smart metering, rainwater harvesting and irrigation systems that save water. By demonstrating how these technologies are being taken up by urban water supply systems, the chapter acknowledges that they can improve water efficiency, lessen environmental impacts, and bolster the resilience of water systems. Simply put, it demands a shift to green technology as an essential component of water sustainability in an era defined by evolving environmental threats.

Introduction

Water is a precious resource that sustains life, powers economic development, and sustains critical ecosystems. However, increasing populations, industrial needs, and climate change have put unprecedented stress on the exploitable freshwater (Bolson et al. 2018). Resource base of the world, requiring serious measures and novel strategies towards efficient management of water which, alongside green technology as sustainable decision-making designed to minimize negative environmental impact and achieve part of the resource productivity (Oyetunde et al. 2023; Oyedele et al. 2023) has come into play as viable responses to combating extreme water challenges (Siddique et al. 2023). Integrating these technologies into water management infrastructures is crucial not only to increase efficiency, and resilience, but also to support sustainable engagement to adjust to shifting environmental contexts. For instance, urban water systems have to grapple with the interconnected challenges of an aging infrastructure, pollution and rising demand for fresh water.

But these interdependencies exceed traditional dynamics for a comprehensive understanding of waterfreeness, resulting in water overexploitation and water scarcity

M. O. Oyenuga et al., *Smart Technologies for Climate Change and Net Zero Policies: Practical Approaches Towards Sustainability*, Climate Change Management,
https://doi.org/10.1007/978-3-031-92221-3_25

(e.g., Garay et al. 2018). Introducing green technologies such as recent Internet of Things (IoT) technologies and smart irrigation and water recycling technologies can change the water infrastructure and services to be cleaner and more responsive (Alshami et al. 2024). For example, such efforts, in the form of smart metering, may offer immediate information about consumer water use behavior, and will encourage consumers to make better decisions when it comes to water use, thereby promoting behaviors that can save water (Aivazidou et al. 2021). Moreover, the current paradigm shifts towards sustainable development considers the integration of ecological priorities into the management of water in practice. Water sustainability goals cannot be met without green technology, which is becoming increasingly important to stakeholders as they understand the value of protecting the environment (Aroonsrimorakot and Laiphrakpam 2023). Examples of such benefits can be better groundwater recharge from rainwater harvesting systems, reduced water losses from advanced leakage detection technologies, and more efficient agricultural water use via precision irrigation systems (Smith et al. 2023; Neupane and Guo 2019). But there are also tremendous challenges in applying green technologies in water management. Such impediments in the form of budgetary restrictions, regulatory issues, and low levels of public knowledge can hamper the journey of integrating such new-age solutions within established structures (Bolson et al. 2018; Steiner et al. 2021). It is hence necessary to understand such impediments and outline channels through which technology can be deployed effectively towards achieving better sustainable water management techniques that are effective in countering existing as well as anticipated future water requirements.

Sustainable Water Usage in Agriculture

Sustainable water use for agriculture is pivotal to ensure long-term food safety and environmental well-being in an increasingly water-scarce and climate change-tight world. As agriculture alone uses approximately 70% of the global freshwater, it is now imperative that effective water management practices are incorporated to ensure optimum agricultural productivity and ensure the water resource sustainability (Rosa et al. 2020a, b). One of the greatest steps forward in encouraging the application of sustainable water in agriculture is through embracing sophisticated irrigation technologies. Such as traditional processes for irrigation; e.g. flood irrigation that can waste water through evaporation and runoff. Improved irrigation systems (sprinklers, drip irrigation, etc.) can maximize water provided to plants with minimal loss (Sankari et al. 2024). One example is the impact of dripping irrigation that uses 50% less water than traditional processes (Sankari et al. 2024). Additionally, the processes may enhance the efficiency of water use and therefore increase the productivity of agriculture per unit of water consumed (Piao et al. 2010). New tools like rainwater harvesting and soil moisture sensors can also help sustainability. Rosa et al. (2020a, b), rainwater harvesting systems can be used to supplement irrigation requirements by collecting and storing rainwater that consequently decreases reliance on groundwater and sustainability of resources. Soil moisture sensors that alert farmers about the condition of their crops in real time also ensure that decisions regarding irrigation schedules are guided by real crop demand, instead of over-irrigating the crops; this coupled with below-ground sensors ensures that water is conserved (Watto and Mugera 2014). Soil health and water holding capacity improvement measures (agroecological measures) are another important aspect as to the sustainable agricultural water use. Crop rotation, cover cropping, and conservation tillage could improve soil structure and organic composition for infiltration and water holding (Wada and Bierkens 2014).

As these practices contribute to saving water, they also contribute to biodiversity enhancement and climate change resilience of these systems to keep the agricultural systems productive despite environmental fluctuations (Kumbhar et al. 2014). Another component of sustainable water management is climate-smart agriculture. This is consistent with the adoption of farming practices and technologies to buffer climate change and simultaneously improve productivity and reduce greenhouse gas emissions (Sankari et al. 2024). Innovative crops that require less water, along with new irrigation management strategies, are helping farmers adapt to unpredictable climatic conditions, all aspects of climate-smart agriculture that allow continued efficiencies in water use.

Stakeholder engagement is also all-important for effective water management practices in agriculture. Another important driver to adopt collaborative water management practices, knowledge sharing, and policy influence that favours sustainable water use are water user associations and farmer groups (Maru et al. 2023) Governance participation through community engagement (Nwokolo et al. 2024a; Oyenuga and Jeresa 2025) has led to higher coordination in water resource management, where stakeholders' needs are addressed in sustainable agriculture.

Water Conservation Technologies

Water conservation technology has a major role to play in the development of water conservation strategies in agriculture and urbanized systems whose demand for water continues to grow. But, as climate change, population and industrialization continue to put pressure on water resources, new technologies need to be adopted to improve water efficiency and promote long-term sustainability. The most water-saving technology is drip and sprinkler irrigation, even known as precision irrigation, where Low loss of evaporation and runoff water is applied directly to the root zones of crops (Lamm et al. 2017). For example, drip irrigation (which can reduce water use by 50% compared to traditional flood irrigation methods) provides a more effective way to maintain agricultural output without exhausting water potential (Warner et al. 2020). Moreover, soil moisture sensors are used to improve the irrigation scheduling through soil water monitoring: water application is carried out, when needed (Lamm et al. 2017). Other revolutionary technologies such as rainwater harvesting are also making tremendous contributions around sustainable water utilization. Harvesting rainwater, collecting it for irrigation and sanitary purposes, thus reducing conventional water sources for the latter. Implementing rainwater harvesting systems not only relieves stress on freshwater resources; it can also save money for farmers and households that depend on traditional water sources. The Internet of Things (IoT) (Apata et al. 2024) smart water management systems, are also highly promising to increase water conservation. Developing software for IoT devices and sensors can at least track in real time the way water is consumed, leaks occurring, and system inefficiencies.

This provides information that can be used to optimise the supply of water and improve the management of resources in general (Cheong et al. 2015). To illustrate, smart water grids can deliver an automatic water supply system management where water is adequately consumed without waste (Cheong et al. 2015). This type of adaptation (Nwokolo et al. 2023a, b) contributes to the preservation of water and reduces the operating cost of municipal water supply services. War and Peace Water, therefore, can be improved not only through technological innovation, but also through sustainable use policies that guarantee the encouragement of the production of water-saving technologies. As an example, policy frameworks which place users into incentive-based regimes for utilising efficient methods of irrigating will create the conditions for farmers (Fasanmi et al. 2024; Oyedele et al. 2024) investing into new technology (Steiner et al. 2021).

Training programs for farmers will also promote awareness (Oyenuga and Omale 2024) and understanding of available water-saving technologies. Through promoting an innovation and adaptation culture (Oyenuga and Jeresa 2025), the programs can be able to promote the adoption of sustainable interventions that support water conservation initiatives (Warner et al. 2020). Last but not least, water conservation is very relevant in urban areas. Urban regions can utilize integrated stormwater management systems that collect and treat runoff for reuse, reducing the effects of flooding while conserving dwindling groundwater resources. Incorporating green infrastructure (Oyetunde et al. 2023; Oyedele et al. 2023), such as permeable pavement and green roofs, can improve water management in the urban region through facilitated onsite rainwater infiltration and lighter loads on conventional drainage systems (Watto and Mugera 2014).

Reducing Emissions in the Water Supply Chain

The water supply chain contributes a major share to greenhouse gas emissions through the energy consumption in sourcing, treating, distributing, and disposing of water. With the desire for sustainability during climate change, emission reduction related to the water supply chain has become vital. Various methods can be implemented at various levels of the water supply chain to minimize greenhouse gas emissions and enhance the sustainability of water management.

1. **Energy Efficiency in Water Treatment and Distribution:** Energy consumption within water treatment plants and pumping of water through distribution networks is one of the biggest emission contributors to the water supply chain. There have been reports of significant savings of energy with upgrades to efficient equipment, e.g., variable speed pumps and efficient motors (Smith et al. 2015). In addition, utilizing renewable energy (Nwokolo et al. 2023b), e.g., solar or wind energy, to power water treatment and distribution can drastically lower the carbon footprint of water supply services. For example, decentralized water treatment systems based on solar energy offer a less emissions-intensive option while improving water supply reliability in local areas (Smith et al. 2015).
2. **Infrastructure Improvement:** Old and inefficient infrastructure may result in large water losses through leakage, which not only wastes a valuable resource but also boosts energy use related to the treatment and distribution of extra water supplies. Evidence suggests a statistically significant relationship between water leakage and greenhouse gas emissions from urban water supply systems (Smith et al. 2015). Hence, investing in infrastructure modernization, such as leak detection technologies and replacing aged pipes, can result in both water savings and lower emissions. The use of smart water management systems with IoT technology enables real-time leak monitoring and facilitates optimizing maintenance activities (Sathaye et al. 2010).
3. **Innovative Water Conservation Strategies:** Conservation strategies like rainwater harvesting systems and the utilization of greywater for irrigation and non-potable uses can decrease treated water demand to a considerable extent, subsequently minimizing the energy needed for water operations and resultant emissions along the water supply chain (Suryadi and Susanti 2024). Additionally, water-saving fixtures and raising community awareness in conservation methods can enable consumers to help drive emission reductions.
4. **Transportation and Logistics Optimization:** Transportation logistics of the water supply chain can also be a source of greenhouse gas emissions. Improved routing and the adoption of cooperative transportation plans can minimize emissions through optimized delivery operations (Danloup et al. 2015). Regional distribution centers and bundling deliveries can also minimize the effects of transportation on overall emissions (Li et al. 2017).
5. **Integration of Development Policies:** There is a need for a systemic framework aspect, specifically long-term policies that encourage sustainable behavior throughout the water supply chain. Incentives from regulators, such as carbon taxes or cap-and-trade systems, can also lead water utilities to use lower-emission technology, while incentivizing water conservation behavior (Das and Jharkharia 2018). It could also enhance transparency, because increased disclosure can force other stakeholders to adopt emissions-reduction best practices.

Case Studies of Green Water Management Technology

With increasing efforts to combat global water scarcity, the need for sustainable water management could not be greater. Case studies provide examples of relevant green water management technology (GWMT) that can be applied efficiently and effectively in water-related problems without lock-in of any environmental issues.

1. Urban Rainwater Harvesting: RWH (Rainwater harvesting) is an established green water management technology with extensive documentation where rainwater is collected and stored in tanks for various applications such as irrigation, groundwater recharge, and drinking water supply. In this context, the implementation of RWH systems was done in Bangalore, India, also, to mitigate the water shortage in the city due to rapid urbanization and growing population. The task involved installation of rainwater harvesting (RWH) systems in government and private buildings. This led to a considerable decrease in reliance on municipal water supplies, which meant that residents would spend less money buying water, as well as an estimated 15% decrease in the total demand for water within the city (Das and Jharkharia 2018). This case illustrates the pressing need for rainwater harvesting, a decentralized solution, to improve urban water security.
2. Smart Irrigation Solutions in Agriculture: In agriculture, smart irrigation technology is proving to be a boon to conserve water and enhance production. For instance, farmers in California's Central Valley adopted precision irrigation technology, using drip irrigation practices in conjunction with soil moisture sensors. As a result, this allowed the watering to take place only when necessary, thus maximizing the use of water during periods of growth. Consequently, participants saved up to 40% water, while achieving the same level of crop production (Karaman et al. 2024). This case illustrates how smart irrigation can lay the foundation for green agriculture and lower the environmental impact of farming.
3. Constructed Wetlands for Wastewater Treatment: That's another innovative way to manage water. As part of the project to improve wastewater management and neighbouring biodiversity, the team built a constructed wetland system in the township of Narowal in Pakistan. It treated both agricultural runoff and municipal sewage, thereby minimising the pollutant load discharged in the nearby water bodies. Considering the provided space in the constructed wetlands, several fauna and flora species could establish their habitats and enhance local biodiversity (Ibrahim et al. 2021). This example shows what engineered natural systems can do for wastewater treatment issues at the same time they are applied for ecological restoration.
4. Impermeable Pavements in Urban Planning: It means the use of pavements in the urban areas helps for a two in one solution for stormwater management and groundwater recharge. This has reduced surface runoff and flooding in public spaces during rainstorms. By way of example, the lot is blanketed in permeable pavements, which enable rainwater to seep through the surface to replenish the groundwater table and reduce strain on drainage systems. Results from the monitoring identified a substantial reduction of stormwater runoff, that in turn

yielded enhanced water quality of the upstream water bodies (Jing et al, 2024). This is an example of how green infrastructure offers sustainable urban water management.

5. Integrated Water Resource Management (IWRM) in South Africa: Lastly, the approach of integrated water resource management (IWRM) projects in South Africa is a comprehensive strategy for managing water resources that takes into account environmental, economic and social interests. It has also meant looking at integrated water resource management (IWRM) measures towards strengthening water conservation efforts across-the-board (agriculture, industry, urbanization). Resource and knowledge sharing has been advocated by cross-sector stakeholders as a way to improve management of water access, and reduce competition over scarce resources. Therefore, sustainable water systems could tackle the issues of climate variability and population growth (Irum et al. 2018).

The Role of Policy in Ensuring Sustainable Water Practices

In the global context of growing water scarcity and climate change, policy may be an important driver of progress toward sustainable water practice. A sound water management policy can support the uptake of the techniques and technologies needed to promote sustainable use of water resources. By improving regulations for the control of water pollution, initiating conservation practices, and pursuing complete water resource management systems, governments and institutions can reach a sustainable climate that promotes water utilization (Sahoo and Goswami 2024). Stringent wastewater discharge standards, for example, can compel industries toward cleaner production practices and, in the case that it can be recycled again in the environment, advanced water treatment technologies, which will make the water reusable before it is reinjected back into the environment (Sahoo and Goswami 2024).

Moreover, steps that influence economics such as water pricing and agriculture-use improvement have great potential to push the agriculture and industry practices towards less resource uses and more efficient water management. Not only does the application of such economic measures make better use of resources, but at the same time it encourages the development of innovation focused on saving water (Makanda et al. 2022).

Furthermore, IWRM policies also focus on prioritization of integrated water, land, and related resources development and management. They achieve this by motivating various actors, such as governments, businesses, and communities to create integrated strategies that apply to specific environments (Makanda et al. 2022). For instance, regulation that encourages community-based governance of water resources has been effective in most contexts because it fosters accountability and stewardship at the local level, resulting in more sustainable practices (Toàn et al. 2023). As cited in global climate change policy (Kumar et al. 2024) and sustainability guidelines (Oyedele 2025), policy drives sustainable water management.

The use of a process policy is an innovation driver through the desire to achieve certain targets (Nwokolo et al. 2024b) such as the reduction of greenhouse gas emissions associated with water supply and treatment activities (Nwokolo et al. 2024b), which encouraged the implementation of sustainable green technologies. Within the context of case studies, the implementation of such policy increases efficiency in both water use and resilience to climate change impacts (Li and Qian 2018), this is not only required but a matter of ensuring a water secure future. Effective policies to stimulate sustainable water management rest on good governance institutions that uphold transparency, accountability and stakeholder participation. Policies alone won't do it; we need education and outreach to instil values of sustainable water use in the general population. Combining elements of technological uptake, stakeholder collaboration, and regulatory systems, these policies are positioned to facilitate an integrated approach to sustainable water management which fosters environmental integrity and socio-economic growth (Santos et al. 2023).

Implications

Regulation which recognizes the participation of communities in the management of water resources has had success in many contexts by fostering a sense of responsibility and ownership at the local level, leading to more sustainable practices (Toàn et al. 2023). Policy as a sustainable water management driver is also advanced about climate change policy (Kumar et al. 2024) and sustainability principles (Oyedele 2025). The implications of these technologies are critical in terms of environmental sustainability, economic affordability, and social justice. These technologies will be designed to make better use of water, reduce wastage and address pressure on availability, resulting in a more resilient environment. However, cities are also subject to the processes of urbanization and climate change and such green technologies as rainwater harvesting systems, permeable pavements and smart irrigation can all be effective solutions designed to minimize the impacts of flooding, desertification and the deterioration of water quality (Redzhepov et al. 2024). Moreover, the transition to green infrastructure is also in line with broader sustainable development goals that provide sustainable water management solutions for communities while at the same time improving ecosystems.

By reducing the dependence on centralized supply water systems through local treatment and recycling plants, for example, cities can reduce infrastructure and operational costs (Khasanova 2024).

Innovative technologies in water management are expected to deliver energy savings and thus help reduce greenhouse gas emissions from pumping, treatment, and distribution (Smith et al. 2015). Furthermore, the promotion of novel water management practices could spark local economic development through local job opportunities from this technology's development, installation, and maintenance. Furthermore, implementing green technologies in water management empowers people by increasing the capacity of communities to effectively handle resources.

Educating and involving stakeholders in decision-making about sustainable use of water makes them more concerned and responsible for the conservation of water resources (Nassani et al. 2023). These participatory mechanisms not only deliver better environmental outcomes; they create social capital, which strengthens communities' climate-resilient ability to cope with water challenges. It should be highlighted that the impact of the adoption of green technology is not limited to water management alone. The technologies are climate adaptable and can sequester, which makes cities more resilient overall. As an example, green roofs mitigate urban heat islands, while simultaneously stormwater management, thus addressing multitude of environmental problems simultaneously (Abuseif 2023).

Conclusion, Limitations and Future Scope

The use of green technologies in water management is a revolutionary prospect to address the pressing issues that arise due to water scarcity, urbanization, and climate change. Maximizing the utilization of water, reducing emissions, and promoting sustainable practice make these technologies enhance the effectiveness of water supply systems and enhance overall environmental and social goals (Li and Yin 2014). As the context for water management also continues to develop, governments, stakeholders, and communities are responsible for cooperating to support the implementation of green technologies backed up by policy facilitating research, innovation, and social engagement. Their effects run over into economic, social, and environmental fields and thus make them necessary approaches to keep water resources sustainable. Going forward, what can be gleaned from successful green technology implementations will provide a guiding principle to policymakers and practitioners everywhere, leading toward a more secure and resilient water management future. As water challenges mount globally, embracing green technologies is not merely embracing an urgency but also a responsibility to future generations, which will ultimately translate into healthier ecosystems and better living conditions for everyone.

References

Abuseif M (2023) Exploring influencing factors and innovative solutions for sustainable water management on green roofs: a systematic quantitative review. Architecture 3(2):294–327. https://doi.org/10.3390/architecture3020017

Aivazidou E, Banias G, Lampridi M, Vasileiadis G, Anagnostis A, Papageorgiou E, Bochtis D (2021) Smart technologies for sustainable water management: an urban analysis. Sustainability 13(24):13940. https://doi.org/10.3390/su132413940

Alshami A, Ali E, Elsayed M, Eltoukhy A, Zayed T (2024) IOT innovations in sustainable water and wastewater management and water quality monitoring: a comprehensive review of advancements, implications, and future directions. IEEE Access 12:58427–58453. https://doi.org/10.1109/access.2024.3392573

Apata AO, Oyenuga MO, Abeoba M, Ugom A (2024) Internet of Things (IoT) solutions for smart transportation infrastructure and fleet management. J Propulsion Technol 45(4):1492–1505. https://propulsiontechjournal.com/index.php/journal/article/view/8325

Aroonsrimorakot S, Laiphrakpam M (2023) Green innovation to save water and energy for environmental sustainability in agricultural farming: a review. Agric Nat Res 57(4). https://doi.org/10.34044/j.anres.2023.57.4.18

Bai H, Tang K, Zhao X, Yu Z (2021) Water policy and regional economic development: evidence from Henan Province, China. Water Policy 23(2):397–416. https://doi.org/10.2166/wp.2021.167

Bolson J, Sukop M, Arabi M, Pivo G, Lanier A (2018) A stakeholder-science based approach using the national urban water innovation network as a test bed for understanding urban water sustainability challenges in the U.S. Water Resour Res 54(5):3453–3471. https://doi.org/10.1029/2017wr021191

Cheong S, Choi G, Lee H (2015) Barriers and solutions to smart water grid development. Environ Manage 57(3):509–515. https://doi.org/10.1007/s00267-015-0637-3

Danloup N, Mirzabeiki V, Allaoui H, Gonçalves G, Julien D, Mena C (2015) Reducing transportation greenhouse gas emissions with collaborative distribution. Manag Res Rev 38(10):1049–1067. https://doi.org/10.1108/mrr-11-2014-0262

Das C, Jharkharia S (2018) Low carbon supply chain: a state-of-the-art literature review. J Manuf Technol Manag 29(2):398–428. https://doi.org/10.1108/jmtm-09-2017-0188

Fasanmi, O.O; Olusanya, B.B and Oyenuga M.O (2024). Entrepreneurial training as a retirement succorance for Nigerian Retirees. Twist 19(4):137–147. https://doi.org/10.5281/zenodo.10049652#321

Garay L, Font X, Corrons A (2018) Sustainability-oriented innovation in tourism: an analysis based on the decomposed theory of planned behavior. J Travel Res 58(4):622–636. https://doi.org/10.1177/0047287518771215

Ibrahim N, Razak R, Wahab M, Osman A, Rahman S (2021) Supply chain risks and roles of the strategy of green supply chain management practices. Int J Acad Res Business Social Sci 11(7). https://doi.org/10.6007/ijarbss/v11-i7/10322

Irum S, Qureshi M, Ashfaq M, Sami A, Bhatti M, Umar A (2018) A review of green supply chain management practices in Asian countries. Int J Eng Technol 7(2(29):1094. https://doi.org/10.14419/ijet.v7i2.29.14319

Jing K, Yee H, Shafiei M, Ismail R (2024) Impacts of green site management practices on energy and water consumption efficiency in the Malaysian construction industry. Int J Manage Stud 31(1):61–88. https://doi.org/10.32890/ijms2024.31.1.3

Karaman A, Ellili N, Uyar A (2024) Do sustainable supply chain practices mitigate carbon emissions? The role of supplier environmental, social and governance training. Bus Strateg Environ 33(8):8126–8148. https://doi.org/10.1002/bse.3931

Khasanova M (2024) Green technologies are an integral element in environmental protection. Bio Web Conf 82:06008. https://doi.org/10.1051/bioconf/20248206008

Kumar K, Singh R, Sharma A, Lehri BPS (2024) Climate change adaptation strategies for achieving net-zero economy. In: Singh R, Crowther D (eds) Transition towards a sustainable future. approaches to global sustainability, markets, and governance. Springer, Singapore. https://doi.org/10.1007/978-981-97-5756-5_8

Kumbhar V, Choudhury S, Sen A, Singh T (2014) Assessment of irrigation and agriculture potential using geospatial techniques: a case study of "Bhima-Ujjani" project. Proc Soc Behav Sci 157:277–284. https://doi.org/10.1016/j.sbspro.2014.11.030

Lamm A, Warner L, Taylor M, Martin E, White S, Fisher P (2017) Diffusing water conservation and treatment technologies to nursery and greenhouse growers. J Int Agric Ext Educ 24(1):105–119. https://doi.org/10.5191/jiaee.2017.24110

Li S, Sun S, Wang Y, Xia H (2017) Dynamic game on carbon emission reduction in intermodal supply chain. Destech Trans Environ Energy Earth Sci (seee). https://doi.org/10.12783/dteees/seee2016/6567

Li P, Qian H (2018) Water resources research to support a sustainable China. Int J Water Resour Dev 34(3):327–336. https://doi.org/10.1080/07900627.2018.1452723

Li J, Yin W (2014) Effect of green water management on river base flow. Appl Mech Mater 692:90–96. https://doi.org/10.4028/www.scientific.net/amm.692.90

Liu X, Zhao A (2024) Advancing sustainability in construction and environmental management, about innovative materials, technologies, and policy frameworks. Appl Comput Eng 66(1):107–112. https://doi.org/10.54254/2755-2721/66/20240923

Makanda K, Nzama S, Kanyerere T (2022) Assessing the role of water resources protection practice for sustainable water resources management: a review. Water 14(19):3153. https://doi.org/10.3390/w14193153

Maru H, Haileslassie A, Zeleke T (2023) Impacts of small-scale irrigation on farmers' livelihood: evidence from the drought-prone areas of the upper awash sub-basin, Ethiopia. Heliyon 9(5):e16354. https://doi.org/10.1016/j.heliyon.2023.e16354

Nassani A, Hussain H, Condrea E, Grigorescu A, Yousaf Z, Haffar M (2023) Zero waste management: investigation of green technology, the green supply chain, and the moderating role of CSR intentions. Sustainability 15(5):4169. https://doi.org/10.3390/su15054169

Neupane J, Guo W (2019) Agronomic basis and strategies for precision water management: a review. Agronomy 9(2):87. https://doi.org/10.3390/agronomy9020087

Nwokolo SC, Singh R, Khan S, Kumar A, Luthra S (2023a) Technological pathways to net-zero goals in Africa. In: Africa's path to net-zero. CSR, sustainability, ethics and governance. Springer, Cham. https://doi.org/10.1007/978-3-031-44514-9_5

Nwokolo SC, Singh R, Khan S, Kumar A, Luthra S (2023b) Remedies to the challenges of renewable energy deployment in Africa. In: CSR, sustainability, ethics & governance, pp 59–74. https://doi.org/10.1007/978-3-031-44514-9_3

Nwokolo SC, Singh R, Khan S, Kumar A (2024a) Key players in net zero technology. In: Elsevier eBooks, pp 193–215. https://doi.org/10.1016/b978-0-443-31486-5.00007-0

Nwokolo SC, Singh R, Khan S, Kumar A (2024b) Net zero technology and circular economy. In: Elsevier eBooks, pp 237–247. https://doi.org/10.1016/b978-0-443-31486-5.00009-4

Oyedele MO, Marcus GO, Ahungwa AI (2023) Do consumers care about green marketing practices? Insight from a developing nation. Bp Int Res CritS Inst J (BIRCIJournal) 6(3):1424–1436. https://bircu-journal.com/index.php/birci/article/view/7665

Oyedele MO, Sunday AO, Abuh AI (2024) Fostering technological-enhanced training and development for business survival and performance in the new normal. J Propulsion Technol 45(3):1858–1869. https://propulsiontechjournal.com/index.php/journal/article/view/7479

Oyedele OM (2025) The cobra effect: exploring the dark side of innovation and environmental sustainability in developing nations. In: Singh R, Joshi A, Filho WL, Khan S (eds) Zero carbon industry, eco-innovation and environmental sustainability. World Sustainability Series. Springer, Cham. https://doi.org/10.1007/978-3-031-80220-1_15

Oyenuga MO, Jeresa S (2025) Adaptation and resilience: private, public, and individual change makers. In: Singh R, Filho WL (eds) Climate neutrality through smart eco-innovation and environmental sustainability. Climate Change Management. Springer, Cham. https://doi.org/10.1007/978-3-031-83250-5_2

Oyenuga MO, Omale SA (2024) Is Africa jinxed? Exploring the challenges of technology access and adoption in Africa. African J Econom Sustain Develop 7(4):142–161. https://doi.org/10.52589/AJESDULN1LRNF

Oyetunde TO, Oyenuga MO, Adoga GJ (2023) Green management practices and organisational efficiency in higher institutions. Bp Int Res CritS Inst J (BirLE) 6(3):276–290. https://bircu-journal.com/index.php/birle/article/view/7716

Piao S, Ciais P, Huang Y, Shen Z, Peng S, Li J, Fang J (2010) The impacts of climate change on water resources and agriculture in China. Nature 467(7311):43–51. https://doi.org/10.1038/nature09364
Redzhepov M, Nazarova N, Cherkina V, Romanishina T (2024) Green innovations in construction: how environmental technologies shape the economy of sustainable cities. E3s Web Conf 515:03004. https://doi.org/10.1051/e3sconf/202451503004
Rosa L, Chiarelli D, Sangiorgio M, Beltran-Peña A, Rulli M, D'Odorico P, Fung I (2020b) Potential for sustainable irrigation expansion in a 3 °c warmer climate. Proc Natl Acad Sci 117(47):29526–29534. https://doi.org/10.1073/pnas.2017796117
Rosa L, Chiarelli D, Rulli M, Dell'Angelo J, D'Odorico P (2020) Global agricultural economic water scarcity. Sci Adv 6(18). https://doi.org/10.1126/sciadv.aaz6031
Sahoo S, Goswami S (2024) Theoretical framework for assessing the economic and environmental impact of water pollution: a detailed study on sustainable development of India. J Future Sustain 4(1):23–34. https://doi.org/10.5267/j.jfs.2024.1.003
Sankari M, KarpuraDheepan G, Sridevi N, Hemalatha C (2024) Innovations in irrigation: water conservation in agriculture. https://doi.org/10.47716/978-93-92090-50-9
Santos E, Carvalho M, Martins S (2023) Sustainable water management: understanding the socioeconomic and cultural dimensions. Sustainability 15(17):13074. https://doi.org/10.3390/su151713074
Sathaye N, Horvath A, Madanat S (2010) Unintended impacts of increased truck loads on pavement supply-chain emissions. Transport Res A Policy Practice 44(1):1–15. https://doi.org/10.1016/j.tra.2009.09.002
Siddique M, Ahmed S, Rani K (2023) Green technology for water scarcity: an entrepreneurial business may bring transformation in water management practices. Irrig Drain 72(4):1149–1164. https://doi.org/10.1002/ird.2832
Smith K, Liu S, Chang T (2015) Contribution of urban water supply to greenhouse gas emissions in China. J Ind Ecol 20(4):792–802. https://doi.org/10.1111/jiec.12290
Smith D, Sikka A, Dirwai T, Mabhaudhi T (2023) Research and innovation in agricultural water management for a water-secure world. Irrig Drain 72(5):1245–1259. https://doi.org/10.1002/ird.2872
Steiner J, Devlin D, Perkins S, Aguilar J, Golden B, Santos E, Unruh M (2021) Policy, technology, and management options for water conservation in the Ogallala Aquifer in Kansas, USA. Water 13(23):3406. https://doi.org/10.3390/w13233406
Suryadi E, Susanti I (2024) Indonesian local learning ecosystem in energy transition era ensure wealth stay in local economies. Int J Supply Chain Manage 13(4):16–20. https://doi.org/10.59160/ijscm.v13i4.6247
Toàn T, Hanh D, Thu D (2023) Management models and the sustainability of rural water supply systems: an analytical investigation in Ha Nam Province, Vietnam. Sustainability 15(12):9212. https://doi.org/10.3390/su15129212
Wada Y, Bierkens M (2014) Sustainability of global water use: past reconstruction and future projections. Environ Res Lett 9(10):104003. https://doi.org/10.1088/1748-9326/9/10/104003
Warner L, Lamm A, White S, Fisher P, Beattie P (2020) A new perspective on adoption: delivering water conservation extension programming to nursery and greenhouse growers. J Agric Educ 61(1):172–189. https://doi.org/10.5032/jae.2020.01172
Watto M, Mugera A (2014) Efficiency of irrigation water application in sugarcane cultivation in Pakistan. J Sci Food Agric 95(9):1860–1867. https://doi.org/10.1002/jsfa.6887

Chapter 26
The Future of Net Zero and Digital Sustainability in Developing Regions

Abstract This chapter focuses on how the world emerges from the global economic and ecological challenges, it assesses how net zero ambitions and digital sustainability could intertwine in emerging economies, especially considering the role of digital technologies to foster space for sustainable interventions, but also as a key towards climate change mitigation. How digital tools are being woven into sustainability initiatives presents unique opportunities and challenges for countries targeting net zero emissions. This chapter is a critical analysis of successful projects applying digital technologies (to drive energy efficiency, resource utilization, and stakeholder engagement) within the areas of water, agriculture, and urban planning. It discusses how smart technologies, data analytics, and IoT applications are not only promoting decision-making processes but also modernizing operations that will eventually help achieve net zero targets. The chapter outlines the importance of promoting inclusive access to such technologies across the developing world, so that all societies may benefit from their digital sustainability efforts. Drawing on case studies, the chapter proffers practical recommendations for policymakers and practitioners regarding how to best reconcile net zero ambitions with digital innovation in the interest of resilient and sustainable development in the Global South.

Introduction

The existential threat posed by climate change, and the far-reaching implications of such a crisis, have created a broad international push towards net zero by mid-century (Kumar et al. 2024). The goal, based on the 2015 Paris Agreement, is to limit global warming to significantly less than 2°. This means lowering emissions that generate the greenhouse gases across all sectors in an economy, from disease energy to transportation, agriculture, and water (DeAngelo et al. 2021). This thus makes developing regions (which are often characterised by a range of challenges such as low resources, poor infrastructure and heightened climate vulnerability (Oyenuga and Omale 2024)) particularly keen on exploring innovative approaches

M. 0. Oyenuga et al., *Smart Technologies for Climate Change and Net Zero Policies: Practical Approaches Towards Sustainability*, Climate Change Management,
https://doi.org/10.1007/978-3-031-92221-3_26

to achieve green and net zero aligned developmental trajectories. Digital technologies are increasingly viewed as integral facilitating tools for sustainability, represent possible avenues to scale efficiency, improve resource management, and increase stakeholder engagement (Nwokolo et al. 2024a, b, c; Nwokolo et al. 2024b).

The rise of Internet of Things (IoT), data analytics, smart transportation and infrastructure can change the way societies interact with their natural resources (Apata et al. 2024), supporting information for decision processes as well as implementation of sustainable practices (Lenz 2021). For example, in the water sector, digital solutions like smart metering systems can manage water use better, which can cause less waste in distribution and improved accessibility to water supplies ("Closing in on net zero" 2024). Micro-irrigation technology, for example, can enable precision farming practices, ensuring that water, fertilizers, and pesticides are applied judiciously and thus minimizing environmental ramifications (Pandiangan et al. 2023). Although the benefits of integrating digital technologies into sustainability processes are immense and staggering, developing areas also face challenges that need to be addressed to take advantage of these resources. Issues related to access to technology, digital literacy, data privacy, and infrastructure disparity can exacerbate inequalities and restrict equality of access to digital solutions (Shang and Zhang 2022). This is and must be true for digital sustainability initiatives as well, which should be inclusive and open to all societal segments for individuals and communities to participate in and contribute to the delivery against the net zero agenda (Aytaç and Tran 2022).

Scaling up Green Technologies

Green technologies, across various sectors, from energy to materials recycling, will need to be scaled up to meet sustainable development goals and climate change challenges. Advanced green technologies, in particular, can help achieve significant improvements by integrating cleaner production processes, renewable energy systems, and sustainable water management practices to reduce environmental impacts and enable economic growth. Yet, scaling these technologies will be an event with multi-faceted solutions/approach from policy-support, financial investment, stakeholder engagement, and capacity building to mainstream adoption. From public policy frameworks that encourage both innovation and adoption, to the uptake of effective technology solutions, perhaps the single most important incentive to scale up green technologies comes from public policy.

It implies that pollution avoidance, which is the emphasis of green technologies that can improve emissions reductions and resource efficiency in China's iron and steel industry (Xi et al. 2019) can have a considerable impact by enhancing these efficiencies through waste prevention via green processes and practices (Oyedele et al. 2023). Developing the green technologies requires that there are enough good conditions (including regulatory frameworks, financial incentives, etc.). Notably, green credit policies can offer loans to companies that invest in new technologies, thereby boosting their investment in green business operations (Du et al. 2023). The future is even more promising because, in addition to traditional tools, the digitalization has enormous scaling potential (Nwokolo et al. 2025). Data-human transmission perspective, in addition, the digital economy promotes the innovation of green technology, and the demand for labor and economy resilience (Sun et al. 2023).

The scale of data analytics and IoT applications will enable organizations to utilize their resources and track their environmental performance in much more detail and will help drive the shift towards sustainability. This incorporation of technology provides organizations with conclusions that determine environmental objectives alongside operational efficiency.

Particularly, collaboration among stakeholders is critical to the successful scaling of green technologies (Nwokolo et al. 2024a; Oyenuga and Jeresa 2025). This means working with companies, governments, NGOs and communities across green tech development and sustainable practices implementation. For example, social network analyses of green technology diffusion in agriculture show that high costs of diffusion stop farmers from adopting sustainable production practices, hence reducing demand in the market (Wang et al. 2020). Lowering cost barriers and enhancing communication between stakeholders can produce critical gains in the uptake and diffusion of green technologies. Moreover, it is vital to address the structural challenges that prevent the ubiquitous adoption of green technologies (Oyetunde et al. 2023).

In many regions without infrastructure and the ability to launch these technologies, this can result in missed opportunities for sustainable development. Thus, focusing on educating and training activities to build local capacity and generate awareness of the potential benefits of green technologies will form a good platform to scale initiatives (Joshi et al. 2024). Qualitative evaluations of partial projects help generate lessons and best practices that other contexts can replicate and adapt (Darko and Chan 2016). Lastly, scaling up of green technology presents numerous opportunities for reducing environmental impact and promoting sustainable development. The development of resilient economies and the establishment of a new sustainable future will require a multidisciplinary approach; integrating sound policy instruments and progressive digital tools with stakeholder mobilization and capacity-building projects.

Opportunities for International Collaboration

International collaboration has a market in the pursuit of sustainable development goals, especially with water management and climate resilience. Having cross-national cooperation can help to address problems on a regional level, in the fight against unprecedented challenges on the environment that the regions are facing, and support each other to solve these issues through best practices as well as to foster innovative solutions to overcome common problems. One way this can happen is through the idea of collaborative frameworks to pool resources, knowledge, and expertise and create synergies to increase effectiveness, which can lead to better management of critical resources such as managing water. International collaboration in water management is often underpinned by best practice and technology sharing. Even with the reality of this day and age, countries can still be fortunate to learn from the experiences, trials and tribulations of other countries like theirs. As an example, partnerships being formed under the Belt and Road Initiative (BRI) have enabled cross-border supply chain cooperation, boosting water management technologies throughout participating countries. It can improve understanding of local conditions while also using advanced technologies that have been built in contexts of greater industrialization (Cui et al. 2020). Thus, cross-border networks can serve as a platform where knowledge, capabilities, and technologies can be exchanged, which ultimately lead to efficiencies and sustainable practices (Liu et al. 2021).

Moreover, international partnerships might also strengthen the research and development, especially in the case of green technologies. Collaborative efforts between universities, research institutes and industry can foster innovation and speed up the transfer of advanced technologies. Research has shown how R&D partnerships are correlated with the development of green innovation, as effective partnerships lead to better products and processes, hence the sustainability of operations across borders (Sánchez-Sellero and Bataineh 2021). Sharing its spirit of cooperation to further focus on sharing knowledge to develop collaboration among them, it holds discussions in waters resources and environmental protection, more when the matters are complex. Additionally, tackling transboundary challenges require cooperation (Oyedele 2025). Water management in cross- and trans-boundary river basins often calls for integrated approaches supported by data sharing and joint governance frameworks. At the national level, Kenya and Tanzania have embarked on cross-border water management strategies to tackle issues resulting from shared water resources, highlighting a requirement for coordinated policies and joint action to promote sustainable use. They manage conflicts and secure common resources while supporting ecological sustainability such collaborations (Pakura 2020) Moreover, global health challenges, exemplified by the COVID-19 pandemic, also highlight the need for collaboration to develop resilient water supply chains. It also allows for coordination of resources, information, and planning among stakeholders, mitigating the threats to health that the improper management of water poses. Improving access to clean water to strengthen global health security requires integrated strategies which include capacity building and resource sharing (Sami and Chun 2024).

Emerging Technologies for Net Zero

As nations seek their climate goals, made more urgent by the challenge of climate change, emerging technologies are crucial for accelerating the transition to a net-zero emissions future. Central to this shift are innovations that align with the use of renewable energy sources, the increase of energy efficiency, and sustainability of practices in sectors such as transportation, tourism (Alam et al. 2024; Oyenuga and Akwen 2025) and urbanization. Green hydrogen is shining one of the brightest spots on the path to net zero. The electrolysis of water powered by electricity from renewables can produce green hydrogen, which is a clean energy vector that could allow decarbonisation of hard-to-abate sectors such as aviation and heavy industry (Shirizadeh et al. 2023). Realizing the full potential of green hydrogen will require greater technology scalability, infrastructure buildout and awareness of the critical mineral inputs that could limit its supply (Greenwald et al. 2024). Solar and wind energy integrated into the energy grid is also a key point to attain net-zero emissions, alongside the green hydrogen production (Nwokolo et al. 2023).

Theoretically, research indicates that solar and wind technologies could supply global energy demands nearly 100-fold over (Sun 2024)! The improved efficiency and cost-competitiveness of these technologies are leading to their wider deployment, which will be crucial in implementing energy-transition strategies in developed and developing contexts alike. Around even a higher case of energy storage solutions can be enabled by solar energy which can give energy to changing demands, which can reduce the consumption of fossil fuels and related greenhouse gas emissions (Sun 2024).

Carbon capture, utilization and storage (CCUS) technologies are also a critical frontier in the journey to net-zero emissions. CCUS is particularly viable in sectors that are hard to decarbonize, including cement production and fossil fuel power generation (Nagireddi et al. 2023). These innovations are applied to the emission sources themselves and will capture emissions before they enter the atmosphere, reducing their environmental footprint and facilitating a transition towards cleaner industries. Additionally, the digitalisation of industries enhances the prospects of attaining net zero emissions. As big data, IoT and artificial intelligence (AI) are integrated into the energy systems and supply chains, they not only improve efficiencies but also help make real-time decisions that will ultimately reduce energy use and waste (Xue et al. 2022). The use of predictive analytics enables organizations to optimize their energy consumption and improve resource allocation in the context of net zero. (Mehmood et al. 2024). This includes net-zero energy buildings (NZEBs), representing an opportunity for urban infrastructure to enable net-zero ambitions. A net-zero building should generate as much energy as it consumes in its operational lifetime, which relies on an integrated concept of energy-efficient design, renewable energy technology, and smart controlling systems (Isaacs et al. 2022). Evidence of the scalability of sustainable building practices can also be found in the significant energy savings and emissions reductions achieved through retrofitting existing buildings (Ananwattanaporn et al. 2021). Furthermore, the success and scaling of these emerging technologies relies heavily on cross-sector and cross-geography collaboration. Therefore, it is crucial to understand what "king" means locally to accelerate processes related to green transitions, and collaborative governments, businesses, and communities can accelerate turnaround of innovation and knowledge (Froy et al. 2022).

So, policies must make this happen by establishing conditions that generate more R&D investments, getting clean technologies to scale and making financing available that facilitate net-zero transitions. With the advancements in clean energy technology, with CCUS deployed, with energy efficiency scalable through digital transformation and with collaboration at its core, society will be able to face the challenges built around climate change while accelerating economic growth. These technologies are also scalable, and scalarisation of these is a prime environmental challenge but also an opportunity for resilient communities and a functioning and palatable planet for future generations.

The Road Ahead for Sustainable Development

Climate change and sustainable development will be challenges that transcend movements; hence international collaboration will be a key approach to designing functioning solutions. The links between the global economy and its environmental systems mean that countries must share what is left of the resources and the know-how and technology to address difficult problems of sustainable development. Working together like this can lead to knowledge on both sides being shared, equity in other parts of the world being achieved by and innovation from different learning experiences. One of the biggest opportunities for international cooperation is with technology transfer, especially in renewable energy. Nations with the advanced clean energy technologies, like wind, solar and hydropower, can share their innovations and best practices with developing countries interested in modernizing their energy infrastructure. Partnerships between developed and developing countries enable the transmission of technology and the capacity to implement projects effectively. Alternatively, implementation of financial schemes such as earmarked grants and concessional loans that support renewable energy projects can help concede the expenses of such arrangements to developing countries (Akpan and Olanrewaju 2023). Collaboration, internationally, in education and research, is also essential (Oyedele and Iember 2021). Interdisciplinary sustainability programs can encourage collaboration across the disciplines for both students and researchers, mentoring tomorrow's global workforce to solve sustainability challenges. A particular example could be academic programs and international research projects for networking, which can help networks to innovation systems to address the knowledge gap through promoting their regional innovations (Annan-Diab and Molinari 2017).

In addition, transnational research can assist countries in developing evidence-driven policies to address pressing sustainability needs such as water scarcity and agricultural resilience. The SDGs offer an inclusive global framework for cooperation, addressing all facets of our interconnected challenges, from poverty to climate action. They will be complemented by increasing synergies across countries to reach the same objective. GSK is also a signatory of various collaborative initiatives that address access to clean water and sanitation, which are key public health outcomes that reduce inequality (Obersteiner et al. 2016). But, along with international commitment, SDGs are also the master key for coordination, thus magnifying targeted development, expertise and resources which are still kept harnessed. Multinational enterprises also have an emerging potential for internalizing sustainability strategies at relatively lower costs throughout their global supply chains. Sustainability of production and supply-chain resilience in response to environmental risk can be encouraged through international collaboration. From the other end, transnational corporations can encourage solutions region-specialized by which achieving sustainable development goals (SDGs) gets promoted sustainably and their competitive position can be improved through corporate responsibility activities (Alkan and Kamaşak 2023). Finally, technological transfer, education initiatives, collaborative research and the framework of the SDGs are part of one way that countries can relate to these topics. These are derived from the collaborative, multi-sector approaches that are necessary to drive innovation, generate climate change resilience, and build pathways to a people- and planet-positive future.

Implications

Transition to net zero emissions and digital sustainability poses significant implications for developing regions economically, environmentally, and socially. First, digital technology inclusion sustainable centric programs ensure greater resource efficiency, which includes managing land, water and energy resources sustainably. Developing regions can effectively utilize resources, reduce waste, and diminish the environmental footprint moving towards industries such as smart factories, IoT, and big data analytics (Feroz et al. 2021). This is especially relevant now because developing countries often depend on growth to protect the environment, and this approach will put them in a better position to contribute to local and global sustainability targets. A coalition of government, private sector, and civil society to implement digital sustainability initiatives creates jobs and builds a robust economy. Investing in green technologies and digital infrastructures could promote the local economy, as well as create green jobs, whilst making sure that nobody is left behind throughout this transition process in developing regions (Oyedele et al. 2023). One example is that digital literacy efforts can enable communities to find their voice in emerging technologies, advancing progress toward social equity and development objectives.

The relationship between economic growth and social equity underscores the necessity for especially the distribution of both the benefits and responsibilities of digital transformation and sustainable development among different stakeholders. Another critical implication of this transition is that emerging markets could become leaders in sustainable innovation (Oyedele 2025). Adopting net zero pathways and digital sustainability early in their respective development journeys will allow these regions to be at the forefront of the global green economy. A significant example is the increasing interest of international companies in incoming sustainable foreign direct investment (FDI), the notion that the more a country has proven sustainable efforts, the more they attract investment (Hintošová and Bódy 2023).

However, the realization of these implications depends on a strong policy framework that promotes innovation, advances the use of technologies and promotes international collaboration. Policymakers need to make certain that regulations are conducive to sustainable practices and digital inclusion, thus establishing the landscape that enables sustainability stakeholders and other stakeholders to live and grow (Zhang et al. 2022). This can be through public–private partnerships that mobilize resources and expertise, and through knowledge-sharing initiatives that tailor initiatives to the unique challenges and opportunities of different regions.

Conclusion, Limitations and Future Scope

The journey to net-zero emissions and digital sustainability in developing countries is fraught with challenges and opportunities that necessitate strategic intervention and cooperation. As noted in this chapter, the confluence of digital technologies with sustainability work has transformative potential to make resources more efficient, spur economic growth, and enhance social equity. By embracing innovation such as renewable energy systems, cutting-edge water management technology, and digital platforms that enable the optimal allocation of resources, the developing world can move towards a more sustainable world. The transition has implications far beyond environmental consideration, such as the overall socio-economic dividend that can benefit society and spur inclusive growth. By taking advantage of these possibilities, emerging economies can address domestic challenges in the near term and play a leading role in the global green economy. But doing so will necessitate collective action through policies, dialogues and forums across national borders. So early investments in infrastructure, training, and institution-building will be critical to building momentum and ensuring real reform. Knowledge sharing and cooperation will be instrumental in tearing down barriers and achieving common goals, the role of international cooperation cannot be understated. After all, at a time when the world struggles to come to terms with the very real consequences of climate change and the need for sustainable action, the new economies must seize an opportunity and pave the way to a net-zero world by converging a new digital transformation with sustainable development. This coordinated approach not only promotes environmental sustainability but also enhances overall quality of life so that future generations inherit a world which is not only fair but also capable of being prosperous in the face of an evolving climate.

References

Akpan J, Olanrewaju O (2023) Sustainable energy development: history and recent advances. Energies 16(20):7049. https://doi.org/10.3390/en16207049

Alam MN, Turi JA, Bhuiyan AB, Kharusi SA, Oyenuga M, Zulkifli N, Iqbal J (2024b). Factors influencing intention for reusing virtual reality (VR) at theme parks: the mediating role of visitors' satisfaction. Cogent Social Sciences, 10(1). https://doi.org/10.1080/23311886.2023.2298898

Alkan D, Kamaşak R (2023) The practice of "sustainable development goals washing" in developing countries. https://doi.org/10.33422/6th.conferenceme.2023.03.150
Ananwattanaporn S, Patcharoen T, Bunjongjit S, Ngaopitakkul A (2021) Retrofitted existing residential building design in the energy and economic aspect according to Thailand building energy code. Appl Sci 11(4):1398. https://doi.org/10.3390/app11041398
Annan-Diab F, Molinari C (2017) Interdisciplinarity: practical approach to advancing education for sustainability and for the sustainable development goals. Int J Manage Educ 15(2):73–83. https://doi.org/10.1016/j.ijme.2017.03.006
Apata AO, Oyenuga MO, Abeoba M, Ugom A (2024) Internet of Things (IoT) solutions for smart transportation infrastructure and fleet management. J Propulsion Technol 45(4):1492–1505. https://propulsiontechjournal.com/index.php/journal/article/view/8325
Aytaç S, Tran C (2022) Towards net zero energy library buildings. Issues Sci Technol Libr (100). https://doi.org/10.29173/istl2701
Closing in on net zero (2024) In: OECD Net Zero+ Policy Papers. https://doi.org/10.1787/bbe55705-en
Cui L, Gao M, Sarkis J, Lei Z, Kusi-Sarpong S (2020) Modeling cross-border supply chain collaboration: the case of the belt and road initiative. Int Trans Oper Res 30(3):1187–1215. https://doi.org/10.1111/itor.12874
Darko A, Chan A (2016) Review of barriers to green building adoption. Sustain Dev 25(3):167–179. https://doi.org/10.1002/sd.1651
DeAngelo J, Azevedo I, Bistline J, Clarke L, Görland S, Kannengießer S (2021) A matter of time? Sustainability and digital media use. Digit Policy Regul Govance 23(3):248–261. https://doi.org/10.1108/dprg-11-2020-0160
Du Q, Pan H, Liang S, Liu X (2023) Can green credit policies accelerate the realization of the dual carbon goal in China? examination based on an endogenous financial cycle model. Int J Environ Res Public Health 20(5):4508. https://doi.org/10.3390/ijerph20054508
Feroz A, Zo H, Chiravuri A (2021) Digital transformation and environmental sustainability: a review and research agenda. Sustainability 13(3):1530. https://doi.org/10.3390/su13031530
Froy F, Heroy S, Uyarra E, O'Clery N (2022) What drives the creation of green jobs, products and technologies in cities and regions? insights from recent research on green industrial transitions. Local Econ J Local Econ Policy Unit 37(7):584–601. https://doi.org/10.1177/02690942231170135
Greenwald J, Zhao M, Wicks D (2024) Critical mineral demands may limit scaling of green hydrogen production. Front Geochem 1. https://doi.org/10.3389/fgeoc.2023.1328384
Hintošová A, Bódy G (2023) Sustainable FDI in the digital economy. Sustainability 15(14):10794. https://doi.org/10.3390/su151410794
Isaacs M, Garcia-Navarro J, Ong W, Jiménez-Calvo P (2022) Is photocatalysis the next technology to produce green hydrogen to enable the net zero emissions goal? Global Challenges 7(3). https://doi.org/10.1002/gch2.202200165
Joshi A, Singh R, Kumar V, Rani S (2024) Net zero strategy: exploring opportunities, confronting challenges for developing countries. In: Singh R, Crowther D (eds) Transition towards a sustainable future. approaches to global sustainability, markets, and governance. Springer, Singapore. https://doi.org/10.1007/978-981-97-5756-5_11
Kumar K, Singh R, Sharma A, Lehri BPS (2024) Climate change adaptation strategies for achieving net-zero economy. In: Singh R, Crowther D (eds) Transition towards a sustainable future. approaches to global sustainability, markets, and governance. Springer, Singapore. https://doi.org/10.1007/978-981-97-5756-5_8
Lenz S (2021) Is digitalization a problem solver or a fire accelerator? Situating digital technologies in sustainability discourses. Soc Sci Inf 60(2):188–208. https://doi.org/10.1177/05390184211012179
Liu Y, Ma J, Song H, Qian Z, Xiao L (2021) Chinese universities' cross-border research collaboration in the social sciences and its impact. Sustainability 13(18):10378. https://doi.org/10.3390/su131810378

Luderer G, Byers E, Davis S (2021) Energy systems in scenarios at net-zero CO_2 emissions. Nat Commun 12(1). https://doi.org/10.1038/s41467-021-26356-y
Mehmood K, Kautish P, Mangla S, Ali A, Kazançoğlu Y (2024) Navigating a net-zero economy future: antecedents and consequences of net-zero economy-based green innovation. Bus Strateg Environ 33(5):4175–4197. https://doi.org/10.1002/bse.3685
Nagireddi S, Agarwal J, Vedapuri D (2023) Carbon dioxide capture, utilization, and sequestration: current status, challenges, and future prospects for global decarbonization. Acs Engineering Au 4(1):22–48. https://doi.org/10.1021/acsengineeringau.3c00049
Nwokolo SC, Singh R, Khan S, Kumar A, Luthra S (2023) Technological pathways to net-zero goals in Africa. In: Africa's path to net-zero. CSR, Sustainability, Ethics & Governance. Springer, Cham. https://doi.org/10.1007/978-3-031-44514-9_5
Nwokolo SC, Singh R, Khan S, Kumar A (2024a) Harnessing Industry 4.0 for Africa's net zero economy through technological pathways. In: Elsevier eBooks, pp 249–282. https://doi.org/10.1016/b978-0-443-31486-5.00010-0
Nwokolo SC, Singh R, Khan S, Kumar A (2024b) Net zero technology and circular economy. In: Elsevier eBooks, pp 237–247. https://doi.org/10.1016/b978-0-443-31486-5.00009-4
Nwokolo SC, Singh R, Khan S, Kumar A (2024c) Key players in net zero technology. In: Elsevier eBooks, pp 193–215. https://doi.org/10.1016/b978-0-443-31486-5.00007-0
Nwokolo SC, Singh R, Khan S, Kumar A (2025) Role of digitalization and connectivity for achieving a net zero economy in Africa. In: Technological pathways for Africa's net-zero economy, pp 175–192. https://doi.org/10.1016/B978-0-443-31486-5.00006-9
Obersteiner M, Walsh S, Frank S, Havlík P, Cantele M, Liu J, Vuuren D (2016) Assessing the land resource–food price nexus of the sustainable development goals. Sci Adv 2(9). https://doi.org/10.1126/sciadv.1501499
Oyedele MO, Marcus GO, Ahungwa AI (2023) Do consumers care about green marketing practices? Insight from a developing nation. Bp Int Res CritS Inst J (BIRCI Journal) 6(3):(1424–1436). https://bircu-journal.com/index.php/birci/article/view/7665
Oyedele OM, Iember AA (2021) Covid-19 and the future of higher education. IEEE Technol Policy Ethics 6(4):1–3. https://doi.org/10.1109/ntpe.2021.9778140
Oyedele OM (2025) The cobra effect: exploring the dark side of innovation and environmental sustainability in developing nations. In: Singh R, Joshi A, Filho WL, Khan S (eds) Zero carbon industry, eco-innovation and environmental sustainability. World Sustainability Series. Springer, Cham. https://doi.org/10.1007/978-3-031-80220-1_15
Oyenuga MO, Akwen PI (2025) Tourism and intercultural understanding in developing nations. Afr J Econ Sustain Dev 8(1):1–14. https://doi.org/10.52589/ajchrt-okuxnmsq
Oyenuga MO, Jeresa S (2025) Adaptation and resilience: private, public, and individual change makers. In: Singh R, Filho WL (eds) Climate neutrality through smart eco-innovation and environmental sustainability. Climate Change Management. Springer, Cham. https://doi.org/10.1007/978-3-031-83250-5_2
Oyenuga MO, Omale SA (2024) Is Africa jinxed? Exploring the challenges of technology access and adoption in Africa. Afr J Econ Sustain Dev 7(4):142–161. https://doi.org/10.52589/AJESDULN1LRNF
Oyetunde TO, Oyenuga MO, Adoga GJ (2023) Green management practices and organisational efficiency in higher institutions. Bp Int Res CritS Linguist Educ (BirLE) J 6(3):276–290. https://bircu-journal.com/index.php/birle/article/view/7716
Pakura S (2020) Open innovation as a driver for new organisations: a qualitative analysis of green-tech start-ups. Int J Entrep Ventur 12(1):109. https://doi.org/10.1504/ijev.2020.105135
Pandiangan M, Kurniawan T, Aurelia A, Rasya T (2023) Enhancing airport environmental sustainability through net zero energy building principles. IOP Conf Ser Earth Environ Sci 1268(1):012044. https://doi.org/10.1088/1755-1315/1268/1/012044
Sami D, Chun S (2024) Strengthening health security at ground border crossings: key components for improved emergency preparedness and response—a scoping review. Healthcare 12(19):1968. https://doi.org/10.3390/healthcare12191968

Sánchez-Sellero P, Bataineh MJ (2021) How R&D cooperation, R&D expenditures, public funds and R&D intensity affect green innovation? Technol Anal Strat Manag 34(9):1095–1108. https://doi.org/10.1080/09537325.2021.1947490

Shang Z, Zhang L (2022) The sustainable digitalization in the manufacturing industry: a bibliometric analysis and research trend. Mob Inf Syst 2022:1–11. https://doi.org/10.1155/2022/1451705

Shirizadeh B, Ailleret A, Guillon A, Bovari E, Khatib N, Douguet S, Trüby J (2023) Towards a resilient and cost-competitive clean hydrogen economy: the future is green. Energy Environ Sci 16(12):6094–6109. https://doi.org/10.1039/d3ee02283h

Sun Z, Liu J, Tansuchat R (2023) China's digital economy and enterprise labor demand: the mediating effects of green technology innovation. Sustainability 15(15):11682. https://doi.org/10.3390/su151511682

Sun T (2024) Research on integrating hydrogen energy storage with solar and wind power for net-zero energy buildings. Appl Comput Eng 70(1):150–155. https://doi.org/10.54254/2755-2721/70/20241008

Wang W, Wang J, Liu K, Wu Y (2020) Overcoming barriers to agriculture green technology diffusion through stakeholders in China: a social network analysis. Int J Environ Res Public Health 17(19):6976. https://doi.org/10.3390/ijerph17196976

Xi Q, Wang X, Xu Y, Wei Y (2019) Exploring driving forces of green growth: empirical analysis on China's iron and steel industry. Sustainability 11(4):1122. https://doi.org/10.3390/su11041122

Xue L, Zhang Q, Zhang X, Li C (2022) Can digital transformation promote green technology innovation? Sustainability 14(12):7497. https://doi.org/10.3390/su14127497

Zhang Y, Su L, Jin H, Yu-nan Y (2022) The impact of globalization on renewable energy development in the countries along the belt and road based on the moderating effect of the digital economy. Sustainability 14(10):6031. https://doi.org/10.3390/su14106031

Chapter 27
Net Zero Technology and Smart Agriculture

Abstract With the unremitting global climate change, the global agriculture is facing more unprecedented challenges, so it is essential to adopt climate-smart agriculture (CSA) by combining advanced technologies to obtain high-consciousness, high-resilience, and sustainable agriculture systems. This chapter discusses the convergence of net-zero technology and smart agricultural practices, focusing on various technological complexities including precision farming, smart irrigation, and IoT powered solutions that collectively account for minimizing the impacts of Climate Change on sustainability. In addition, it highlights the necessity of coordination among stakeholders, including policymakers, farmers, and researchers, for such technologies to achieve their potential in favorable institutional settings. By detailing examples across agricultural backgrounds, the chapter illustrates that technology to achieve net-zero emission can transform farming practices and what such large-scale examples mean for national and global climate targets, Academics, researchers, and technologists necessarily turn to the very model whose flaws they aim to fix: they advocate for sustainable agriculture to be integrated into technological approaches to build food systems capable of persisting under the pressures of climate change.

Introduction

Integrating net-zero technologies in smart agriculture is a key transformation in the pursuit of sustainability and climate resilience. With escalating global concerns regarding climate change, the agricultural sector, which is a major source of greenhouse gas (GHG) emissions, comes under mounting scrutiny and pressure to implement practices that reduce its environmental impact (Singh et al. 2025a, b, c, d). Net-zero ambitions must inevitably apply across many sectors, including agriculture using generic levers, including innovation, policies, behavioral change (Olanrewaju et al. 2024). Ambitious targets such as these will require the implementation of novel solutions that include the combination of bioenergy with carbon capture and storage, better soil management, and other regenerative activities that enhance carbon sequestration (Rosa and Gabrielli 2023). Additionally, smart agriculture relies

M. O. Oyenuga et al., *Smart Technologies for Climate Change and Net Zero Policies: Practical Approaches Towards Sustainability*, Climate Change Management,
https://doi.org/10.1007/978-3-031-92221-3_27

on both digital technologies and climate-smart development practices to achieve this effect, but is also expected to add resilience in global economies, mitigation of emissions and adaptation (Dev et al. 2023; Oyenuga and Omale 2024). Digitalization has potential, enabling processes like precision agriculture and mechanized production, resulting in more resource-efficient systems (Vărzaru 2024; Mizik 2021).

A clear example of this integration can be the deployment of smart irrigation technologies, which include innovative approaches for the management and monitoring of water usage to optimize water extraction to tackle weather conditions changes and subsequently increase the food security level (Bwambale et al. 2022; Durmuş et al. 2024;). They need to be complemented with CSA(Climate Smart Agriculture) as it provides a common ground between agricultural productivity and environmental sustainability. Climate-Smart Agriculture (CSA) includes diverse strategies targeting enhanced agricultural productivity while addressing climate adaptation and mitigation (Kurgat et al. 2020; Daniel et al. 2023). For example, strategies of mulching, integrated crop-livestock systems and agroforestry are promoted to improve soil health and resilience whilst at the same time minimizing emissions, (Legesse et al. 2019; Balo and Mahata 2022). The relationship between innovative technologies and socio-economic measures must be integrated; improving farmers' training and resources to implement these practices is vital to converting the potential of smart agriculture into a reality (Azadi et al. 2021; Phiri et al. 2022).

Understanding Net Zero Technology in Agriculture

Net-zero technology in agriculture is a critical evolution within the food production industry as we are all forced to get increasingly serious about sustainability (Dewasiri and Singh 2025). Its vision is no net contribution to climate change, through the balance between emissions generated with land use and emissions created out of atmosphere (Rosa and Gabrielli (2023) Effecting such a balancing act is needed in the wake that agriculture represents around 14% of global GHG emissions, which primarily include methane (CH_4) and nitrous oxide (N_2O) released during crop production and livestock management (Arjoo and Shreya 2022).With increasing pressures to cut emissions consistent with international climate commitments including the Paris Agreement, transformative and holistic approaches combining technological and sustainable ways of doing things are urgently needed (Allen et al. 2022; Luderer et al. 2018). Net-zero agriculture practices are characterized by a multifaceted set of mitigative and generative strategies that both increase carbon sequestration and reduce the exhaustion of resources and reliance on carbon-intensive input. Technological innovations (e.g., precision agriculture) where data analytics and automation can optimize farming practices for lesser chemical fertilizers and water without compromising emissions (Gichuki et al. 2023).

These augmentative agroecological practices, including but not limited to cover cropping, reduced tillage or organic farming, enable the bolstering of soil health and stable carbon storage through time (Reay 2020). These practices may create systems that can alleviate food insecurity, promote resilience and reinforce the ecosystem services essential for agronomy (Lamb 2024). The other important component of net-zero strategies of agriculture is the emphasis on soil as a carbon sink. Soil organic matter enhancing practices, including rotational grazing, reforestation, and organic amendments can have large emission reductions (Hutton et al. 2022). The ability of soils to sequester carbon can be one of the major means by which agriculture can contribute towards reaching net zero ambitions. Yet, soil management disruption as an identified risk factor causing highly variable GHG emissions needed to be managed tightly to preserve soil integrity and function (Hutton et al. 2022; Tresise et al. 2021). Tooling for net-zero agriculture offers enormous promise, though there are hurdles in realizing it. These are restraints in finance, training, technical assistance, and norms that do not and should not promote sustainable behavior (Gichuki et al. 2023; Arsic et al. 2023). Farmers moving toward low-emission technologies, for example, may find themselves needing to make up-front investments in new equipment or practices, which can be prohibitively expensive (Arjoo and Shreya 2022). Hence mechanisms are required to support such a transition in agricultural practices towards net-zero i.e., through R&D in the respective domain, financial facilitation, and public–private collaborations (Arjoo and Shreya 2022; Booth 2021).

Smart Agriculture: Innovations and Applications

Smart agriculture is a growing industry that employs technological advancement and innovative techniques to enhance the efficiency of modern agriculture practices and responsibilities to tackle climate change challenges. Smart agriculture will be essential for the world to overcome problems like a growing population and the efficient use of resources now and in the future. This integrated approach attempts to transform the traditional methodology of agriculture into a productive, pro-active system with the potential to react rapidly, proactively, and flexibly to changing, environmental factors using advanced techniques, such as the Internet of Things (IoT), precision agriculture, data analytics and artificial intelligence (Strong et al. 2022; Singh et al. 2025a, b, c, d). IoT devices, though used in transportation (Apata et al. 2024) are also among the game-changing innovations in smart agriculture by enabling real-time monitoring of environmental conditions and crop status. These devices may collect vast amounts of information on soil moisture, temperature and nutrient levels, drastically improving the decision-making processes for farmers (Strong et al. 2022) Sensors, for example, can communicate with irrigation systems indicating when and how much water should be supplied which subsequently saves water resources while optimizing crop yields (Barasa et al. 2021). Specifically, their implementation allows to increase effectiveness and decrease greenhouse gases emissions coming from traditional agricultural systems (Rosa and Gabrielli 2023). Furthermore, agriculture is the biggest economic activity in developing nations especially in Africa, and it's a revolution due to precision farming. Utilizing GPS technology and remote sensing allows farmers to apply fertilizers, pesticides, and water judiciously to reduce waste and to only apply inputs when and where they are needed (Karanisa et al. 2022).

This practice has been demonstrated to enhance soil health and plant growth while reducing the environmental footprint, aligning with the larger goal of net zero emissions in food production systems (Reay 2020). Zero tillage for example, among other conservation agriculture practices, can help in improving the soil structure and biodiversity necessary for sustaining agricultural productivity against changing climate scenarios (Meena et al. 2023). Technologies of modern import, integrated with climate-smart agricultural practices, would be necessary for the establishment of innovative sustainable food systems. The implementation of smart greenhouses taking the internal environment into account, for instance, is demonstrated to improve the productivity of cultivated crops while reducing water usage and energy consumption (Karanisa et al. 2022; Embaby 2024). These systems facilitate year-round continuous production and act as a way of harnessing renewable energy resources and also represent progress in terms of having sustainable solutions in agriculture with net-zero targets (Costa et al. 2022; Singh et al. 2025a, b, c, d). For example, aerial and grounded drones can also be used in crop monitoring, which allows farmers to monitor the state of the field, enabling prompt action and better resource management (Barasa et al. 2021). But the move to smart agriculture is not without its challenges. Their barriers include lack of access to technology, limited knowledge of the farmers, and lack of financial means (Strong et al. 2022; Maltby 2024). Other study-based data indicates the adoption of such technologies could develop via literacy and confidence programs executed by education systems (i.e., farmer field schools, community training programs) to improve farmer education literacy on smart practices (Strong et al. 2022; Barasa et al. 2021; Oyedele et al. 2024; Fasanmi et al. 2024). Additionally, with clear policies and support mechanisms, these barriers must be alleviated for the full potential of smart agriculture to contribute to sustainable development (Barasa et al. 2021; Sakrabani et al. 2023).

Sustainable Practices for Net Zero Agriculture

As global demand for agriculture continues to grow, the agriculture sector is also under increasing pressure to cut greenhouse gas (GHG) emissions without compromising food security. However, the solution is not as straightforward, it will require a mix of adoption of technology, behavioural change throughout a system and redistribution of agricultural policies. One primary approach is to implement nature-based solutions, for example, improving hedgerows, agroforestry and afforestation. Not only do those practices sequester atmospheric CO_2, they enable diversity and resilience of processes in nature (Tresise et al. 2021). Hedgerows are a common feature of many agricultural landscapes, and have been shown to effectively sequester carbon and enhance biodiversity (Tresise et al. 2021). Similarly, restoring low-quality lands may also be able to optimise carbon sequestration and agricultural productivity, and is therefore a more pragmatic target for intervention in an agricultural regime (Flack et al. 2022).

Equally necessary in the transition towards sustainable agriculture are technological innovations. Breakthroughs like the adoption of blockchain technology, which ensures transparency and sustainability in the agricultural supply chain, are a positive direction. It minimizes waste, optimizes the use of resources, and allows to track the carbon footprints of all stages of food production (Saha et al. 2022; Yadav et al. 2024).

Additionally, the use of Robotics and Autonomous Systems (RAS) can improve efficiency by optimizing the use of fertilizers, seeds, and energy, directly in support of net zero (Pearson et al. 2022). Partnership among sectors that work together (Oyenuga and Jeresa 2025) to address existing barriers is a cornerstone of net-zero agriculture. Regional challenges, such as institutional capacities, limited finances and past practices (Costa et al. 2022; Sakrabani et al. 2023) should be identified (as important factors). For instance, Costa et al (2022) study states that agriculture practices need to be more region specific so they do not serve only as net-zero strategies in a superficial way. Furthermore, they can also greatly contribute to the large-scale adoption of transformative practices and technologies (Maltby 2024), with extension services advocating to and enabling farmers on aspects of where to adopt these sustainable methods, as well as increasing reach and exposure.

Digital and Technological Interventions for Net Zero Agriculture

Digital and technological interventions are critical enablers of net-zero agriculture, as they assist in better efficiency, reduce carbon emissions, and optimally use resources in agricultural practices. This continuous digitalization in the agricultural sector is allowing traditional production methods to be replaced by more efficient and sustainable systems. This itself distinctly indicates the account of digital intervention being an example of provision of drastic reductions in terms of the greenhouse gas (GHG) emissions. Precision agriculture means utilizing modern technologies (remote sensing, GPS, and IoT devices) and better monitoring and management of agricultural inputs and outputs (Bocean 2024). This technology reduces reliance on chemical fertilizers and pesticides, a significant contributor to greenhouse gas emissions (Li et al. 2024) by enabling producers to make maximal use of the resources at hand. It also promotes more effective decision making based on data, creating productivity, compared to traditional industry practice, while minimizing adverse environmental impacts (Wang and Mei 2024). Similarly, the digital economy is key to decarbonizing agriculture. A digital infrastructure enables (Singh et al. 2024) innovative solutions catering for the needs of new-age farmers and paves the way for investment in low-carbon technologies and practices thus spanning towards sustainability in farming (Zhu et al. 2023).

Digital inclusive finance (Oyenuga et al. 2020) enables farmers to obtain capital more efficiently so that they can introduce more carbon-reducing practices, thus contributing to the decline of carbon emission intensity in agricultural production (Hong et al. 2024). Moreover, the expansion of agricultural digital platforms that connect producers and consumers directly and streamline supply chains adds to the transparency and efficiency within agricultural systems (Wang et al. 2024). By improving coordination between various parties and optimizing logistics, these platforms reduce waste and inefficiencies (Wang and Mei 2024), resulting in a marked reduction in carbon emissions across the entire supply chain. In addition, they foster sustainable consumption behaviour, while educating consumers about the ecological consequences of their acquisitions and motivating eco-friendly purchasing (Bocean 2024; Oyedele et al. 2023; Oyenuga et al. 2023; Singh et al. 2025a, b, c, d).

Challenges and Future Prospects

The integration of high-end technologies, such as Artificial Intelligence (AI) and the Internet of Things (IoT), is a significant challenge for digital agriculture, even if it has the potential to enhance productivity and sustainability, high costs, limited infrastructure, and a lack of digital literacy in the marketplace for farmers continue to limit these technologies (Alkhafaji et al. 2024; Alazzai et al. 2024). Recent review work has pointed out that the hurdles for deploying AI and IoT (AIoT) on real farms are huge and will need considerable effort from the scientific community (Aliev et al. 2023). Adoption of AIoT in agriculture varies by region and the scale of operation on the farm. However, in these less developed areas, the disparity in the availability of technology and information may lead to a widening gap of inequality in the agricultural sector (Wang et al. 2024). Moreover, pest and disease management presents its challenges with vertical and smart farming practices. Such systems, for instance, have been demonstrated to call for new types of integrated pest management, since the application of conventional chemical controls may be impractical (Roberts et al. 2020). The demand for new biological control approaches, for maintaining the harmony of modern agricultural systems, emphasizes the need for ongoing research and development of pest controls (Roberts et al. 2020; Kabir et al. 2023). Future studies might also focus on how such a system could devise an effective sound control strategy by efficiently balancing community engagement with technology-based incentives.

A second big problem is the environmental costs of implementing technologies. Precision agriculture aims to reduce waste and optimize resources, but the sustainability of these technologies must be continuously evaluated (Karunathilake et al. 2023; Nath and Saikia 2021). For instance, the potential to increase yield while decreasing inputs through precision farming technologies must be managed in terms of the power expenditure compared to data management and processing which negates some of those benefits if it is not carefully managed, thereby constituting a cobra effect (Finco et al. 2023; Mukhamedova et al. 2022; Oyedele 2025). In this respect, applied solutions are always needed to direct the efficient energy implementation to the digital solutions to complement the environmental impacts within the whole agriculture system (Finco et al. 2023; Mukhamedova et al. 2022).

As we gaze ahead, the road for digital-first agriculture is a sunny one. The rapid progress of technologies like AI, IoT, and precision farming provides opportunities for optimizing resource allocation and resilience of crops under climate change (Alazzai et al. 2024), Deepasree et al. (2024). Moreover, numerous biotechnologies offer transformative tools for crop improvement that have led to the development of varieties in sectors that are even more resistant to pests and climate threats, thus further improving the stability of food production systems (Sampath et al. 2023). Furthermore, the emphasis on financing for the digital divide will extend the outreach of these transformative practices; thus, smallholders and less affording farmers will have unrestricted access to advanced agricultural technologies and production systems (Hong et al. 2024; Modjo et al. 2024).

Therefore, projections for this digitally integrated future of agriculture hinge on the collaboration of industry stakeholders, governments, the private sector, and farmers alike, that are forced to rethink their approach to leveraging both challenges and opportunities presented through this digital transformation process to enable sustainable growth in agriculture.

Implications

The interaction of digital and technology interventions in agriculture is important not only for environmental sustainability, economic viability, social equity and food security. The Net Zero challenge means different things to different people, but it can be framed from three vital perspectives: environmental stewardship, economic reformation, and social responsibility to the agricultural hierarchy.

Environmental Stewardship: Technologies like precision agriculture, AI, and IoT are changing the way they impact the environment. Such technologies empower better tracking and analytics of inflows and outflows, resulting in minimized wastage and optimized resource utilization. Which in turn helps immensely decrease the greenhouse gas release, promote soil health and encourage biodiversity (Bocean 2024; Nath and Saikia 2021). Technologies such as remote sensing for accurate irrigation and nutrient management contribute significantly to avoiding the excessive use of water and chemicals, which are two major instigators of environmental degradation in traditional farming systems (Alazzai et al. 2024; Finco et al. 2023).

Economic Reformation: Digital agriculture reform will completely change the economic architecture in which agricultural production takes place. For example, productivity improvements through data driven decision making or automation can also lead to higher yield and lower cost of labor for farmers while establishing channels for farmers in the digital space (Zhu et al. 2023; Wang et al. 2024). While the transformation is indispensable to the sustainability of agri-food systems, it will require a capitalisation in technology and infrastructure spending that can place a heavy financial burden, especially on smallholder farmers (Alazzai et al. 2024; Oyedele 2025). Hence, there exist an unprecedented opportunity to design enabling policies and finance mechanisms that make such transition possible for a larger economic elevation (Hong et al. 2024; Modjo et al. 2024).

Driving Social Equity: Bringing marginalized groups into the fold of the technological revolution offers a significant opportunity to improve social equity in agriculture. Event agricultural digital platforms and digital inclusive financial services are developing, which provide smallholder farmers with access to critical resources and markets that were hitherto not accessible (Hong et al. 2024; Modjo et al. 2024). Conversely, the information divide characterized by disparities in access to technology and information has the potential to increase existing disparity if not aggressively mitigated (Wang et al. 2024; Alkhafaji et al. 2024). To provide equitable access to education, technology and financial resources, all stakeholders in the agricultural value chain must be empowered.

Conclusions, Limitations and Future Scope

Digital and technological interventions offer great potential for reaching net-zero agriculture, leading to a transformational age in agriculture where sustainability can be reconciled with profitability and equity. The implications of such technologies go beyond just increasing productivity; they represent a transition towards a more sustainable and resilient agricultural system that can combat the challenges of global warming and food scarcity. But to achieve this potential will require systemic change and collaboration across everyone involved: the public and private sectors, NGOs, and the farming community. This innovation is spread evidence-based global policy priorities. In addition, ongoing research should be undertaken to combat problems associated with the environmental consequences of new technologies, allowing them to jive with the general aim of sustainability. So, the future is bright for agriculture, provided it can embrace these digital flows, but also adhere to fundamental tenets of sustainability and equitable economic development. With this integrated approach based on innovation and inclusivity, we can move decisively towards a future where agriculture is not only net-zero, but in which net-zero agriculture is a reality.

References

Alazzai W, Abood B, Al-Jawahry H, Obaid M (2024) Precision farming: the power of AI and IoT technologies. E3s Web Conf 491:04006. https://doi.org/10.1051/e3sconf/202449104006

Aliev R, Курбанова M, Samoylova A (2023) Transformative potential of digital agriculture for enhancing global food security. Bio Web Conf 76:05010. https://doi.org/10.1051/bioconf/20237605010

Alkhafaji MA, Ramadan GM, Jaffer Z, Jasim L (2024) Revolutionizing agriculture: the impact of AI and IoT. E3S Web Conf 491:01010. https://doi.org/10.1051/e3sconf/202449101010

Allen MR, Friedlingstein P, Girardin CA, Jenkins S, Malhi Y, Mitchell-Larson E, Peters GP, Rajamani L (2022) Net Zero: science, origins, and implications. Annu Rev Environ Resour 47(1):849–887. https://doi.org/10.1146/annurev-environ-112320-105050

Apata AO, Oyenuga MO, Abeoba M, Ugom A (2024) Internet of things (IoT) solutions for smart transportation infrastructure and fleet management. J Propuls Technol 45(4):1492–1505. https://propulsiontechjournal.com/index.php/journal/article/view/8325

Arjoo VK, Shreya S (2022) Climate-smart agriculture: need of the hour. JART Special (01):32–36. https://doi.org/10.56228/jart.2022.sp106

Arsic M, O'Sullivan C, Wasson R, Antille D, Clarke W (2023) Beyond waste-to-energy: bioenergy can drive sustainable Australian agriculture by integrating circular economy with net zero ambitions. Detritus (23):28–34. https://doi.org/10.31025/2611-4135/2023.17278

Azadi H, Moghaddam SM, Burkart S, Mahmoudi H, Van Passel S, Kurban A, Lopez-Carr D (2021) Rethinking resilient agriculture: from climate-smart agriculture to vulnerable-smart agriculture. J Clean Prod 319:128602. https://doi.org/10.1016/j.jclepro.2021.128602

Balo S, Mahata D (2022) A review of climate-smart agriculture is a new approach to the farming system. Int J Environ Clim Chang 2682–2692. https://doi.org/10.9734/ijecc/2022/v12i1131256

Barasa P, Botai C, Botai J, Mabhaudhi T (2021) A review of climate-smart agriculture research and applications in Africa. Agronomy 11(6):1255. https://doi.org/10.3390/agronomy11061255

Bocean C (2024) A longitudinal analysis of the impact of digital technologies on sustainable food production and consumption in the European Union. Foods 13(8):1281. https://doi.org/10.3390/foods13081281

Booth R (2021) Pathways, targets and temporalities: analysing English agriculture's net zero futures. Environ Plan E Nat Space 6(1):617–637. https://doi.org/10.1177/25148486211064962

Bwambale E, Abagale FK, Anornu GK (2022) Smart irrigation for climate change adaptation and improved food security. IntechOpen eBooks. https://doi.org/10.5772/intechopen.106628

Costa C, Wollenberg E, Benitez M, Newman R, Gardner N, Bellone F (2022) Roadmap for achieving net-zero emissions in global food systems by 2050. Scientif Rep 12(1). https://doi.org/10.1038/s41598-022-18601-1

Daniel E, Tiri G, Mamman M (2023) Evaluation of climate smart indicators influencing agricultural practices on sorghum enterprise in Katsina and Sokoto states. Fudma J Agric Agric Technol 9(1):129–136. https://doi.org/10.33003/jaat.2023.0901.17

Deepasree A, Singh S, Sharma N, Rout S, Saritha B, Gautam S (2024) A comprehensive review on the future of smart farming and its role in shaping food production. J Exp Agric Int 46(5):486–493. https://doi.org/10.9734/jeai/2024/v46i52401

Dev P, Khandelwal S, Yadav S, Arya V, Mali H (2023) Climate-based smart agriculture: need for food security and sustainability. Int J Environ Clim Chang 13(3):224–231. https://doi.org/10.9734/ijecc/2023/v13i31702

Dewasiri NJ, Singh R (2025) Advancing climate neutrality through sustainable production and consumption: pathways to a resilient future. In: Singh R, Filho WL (eds) Climate neutrality through smart eco-innovation and environmental sustainability. Climate Change Management. Springer, Cham. https://doi.org/10.1007/978-3-031-83250-5_1

Durmuş Y, Arslansoy G, Gürbüz T (2024) Impact of smart irrigation systems and water management on climate change. Bio Web Conferences 85:01057. https://doi.org/10.1051/bioconf/202485 01057

Embaby Z (2024). Innovative greenhouse to improve economic and environmental conditions. https://doi.org/10.5772/intechopen.113335

Fasanmi OO, Olusanya BB, Oyenuga MO (2024) Entrepreneurial training as a retirement succorance for nigerian retirees. Twist 19(4), 137–147. https://doi.org/10.5281/zenodo.10049652#321

Finco A, Bentivoglio D, Belletti M, Chiaraluce G, Fiorentini M, Ledda L, Orsini R (2023) Does precision technologies adoption contribute to the economic and agri-environmental sustainability of Mediterranean wheat production? An Italian case study. Agronomy 13(7):1818. https://doi.org/10.3390/agronomy13071818

Flack J, Lukáč M, Todman L (2022) Woodland planting on UK pasture land is not economically feasible, yet is more profitable than some traditional farming practices. CentL Eur J 68(2):61–71. https://doi.org/10.2478/forj-2022-0001

Gichuki C, Osewe M, Ndiritu S (2023) Dissemination of climate smart agricultural knowledge through farmer field schools (FFS): analyzing the application cas knowledge by smallholder farmers. Int J Dev Issues 22(3):399–417. https://doi.org/10.1108/ijdi-04-2023-0109

Hong H, Li-li S, Zhao L (2024) Exploring the impact of digital inclusive finance on agricultural carbon emissions: evidence from the mediation effect of capital deepening. Sustainability 16(7):3071. https://doi.org/10.3390/su16073071

Hutton P, Mikhailova E, Lin L, Hao Z, Zurqani H, Post C, Shepherd G (2022) Net-zero target and emissions from land conversions: a case study of Maryland's Climate Solutions Now Act. Geographies 3(1):40–59. https://doi.org/10.3390/geographies3010003

Kabir M, Reza M, Chowdhury M, Ali M, Samsuzzaman S, Ali M, Chung S (2023) Technological trends and engineering issues on vertical farms: a review. Horticulturae 9(11):1229. https://doi.org/10.3390/horticulturae9111229

Karanisa T, Achour Y, Ouammi A, Sayadi S (2022) Smart greenhouses as the path towards precision agriculture in the food-energy and water nexus: case study of Qatar. Environ Syst Decis 42(4):521–546. https://doi.org/10.1007/s10669-022-09862-2

Karunathilake E, Le A, Heo S, Chung Y, Mansoor S (2023) The path to smart farming: innovations and opportunities in precision agriculture. Agriculture 13(8):1593. https://doi.org/10.3390/agriculture13081593

Kurgat B, Lamanna C, Kimaro A, Namoi N, Manda L, Rosenstock T (2020) Adoption of climate-smart agriculture technologies in Tanzania. Front Sustain Food Syst 4. https://doi.org/10.3389/fsufs.2020.00055

Lamb W (2024) The size and composition of residual emissions in integrated assessment scenarios at net-zero CO2. Environ Res Lett 19(4):044029. https://doi.org/10.1088/1748-9326/ad31db

Legesse B, Ute J, Hule M (2019) Climate smart agriculture and its implementation challenges in Africa. Curr J Appl Sci Technol 1–13. https://doi.org/10.9734/cjast/2019/v38i430371

Li J, Sheng X, Zhang S, Wang Y (2024) Research on the impact of the digital economy and technological innovation on agricultural carbon emissions. Land 13(6):821. https://doi.org/10.3390/land13060821

Luderer G, Vrontisi Z, Bertram C, Edelenbosch O, Pietzcker R, Rogelj J, Kriegler E (2018) Residual fossil CO_2 emissions in 1.5–2 °c pathways. Nature Climate Change 8(7):626–633. https://doi.org/10.1038/s41558-018-0198-6

Maltby J (2024) Avoiding siloed approaches: integrating psychological insights into sustainable farming. PLoS ONE 19(10):e0301881. https://doi.org/10.1371/journal.pone.0301881

Meena K, Srivastava R, Kumar A, Singh S, Tiwari A, Singh N (2023) Economic impact of wheat sowing through zero tillage technique against broadcasting under hot sub-humid (moist) eco-region, Uttar Pradesh, India. Int J Environ Clim Chang 13(10):851–858. https://doi.org/10.9734/ijecc/2023/v13i102726

Mizik T (2021) Climate-smart agriculture on small-scale farms: a systematic literature review. Agronomy 11(6):1096. https://doi.org/10.3390/agronomy11061096

Modjo A, Tapi T, Safruddin S, Ansar M, Fitriani D (2024) Harnessing technology for agricultural sustainability: case studies and future directions. Join 1(3):270–280. https://doi.org/10.59613/n2nrg748

Mukhamedova K, Cherepkova N, Korotkov A, Dagasheva Z, Tvaronavičienė M (2022) Digitalisation of agricultural production for precision farming: a case study. Sustainability 14(22):14802. https://doi.org/10.3390/su142214802

Nath PK, Saikia CR (2021) Assessing the prospects of agricultural sustainability in central Brahmaputra floodplain environment of Darrang district, Assam. Int J Rural Manag 18(3):358–375. https://doi.org/10.1177/09730052211034352

Olanrewaju O, Oduro P, Babayeju O (2024) Exploring capital market innovations for net zero goals: a data-driven investment approach. Finance Account Res J 6(6):1091–1104. https://doi.org/10.51594/farj.v6i6.1244

Oyedele MO, Marcus GO, Ahungwa AI (2023) Do consumers care about green marketing practices? Insight from a developing nation. Budapest Int Res Critics Instit J (BIRCIJournal) 6(3):(1424–1436). https://bircu-journal.com/index.php/birci/article/view/7665

Oyedele MO, Sunday AO, Abuh AI (2024) Fostering technological-enhanced training and development for business survival and performance in the new normal. J Prop Technol 45(3):1858–1869. https://propulsiontechjournal.com/index.php/journal/article/view/7479

Oyedele OM (2025) The cobra effect: exploring the dark side of innovation and environmental sustainability in developing nations. In: Singh R, Joshi A, Filho WL, Khan S (eds) Zero carbon industry, eco-innovation and environmental sustainability. World Sustainability Series. Springer, Cham. https://doi.org/10.1007/978-3-031-80220-1_15

Oyenuga MO, Jeresa S (2025) Adaptation and resilience: private, public, and individual change makers. In: Singh R, Filho WL (eds) Climate neutrality through smart eco-innovation and environmental sustainability. Climate Change Management. Springer, Cham. https://doi.org/10.1007/978-3-031-83250-5_2

Oyenuga MO, Omale SA (2024) Is Africa jinxed? Exploring the challenges of technology access and adoption in Africa. Afr J Econ Sustain Dev 7(4):142–161. https://doi.org/10.52589/AJESDULN1LRNF

Oyenuga MO, Jelili B, Bulugbe K (2020) E-banking and customer satisfaction in the banking industry. Niger J Bus Adm 18(1):112–132 https://api-ir.unilag.edu.ng/server/api/core/bitstreams/53eb4271-c536-4c99-8910-64622549ba98/content#page=120

Oyenuga MO, Akhaine ME, Omale SA (2023) We need it: how digital marketing tactics influence the purchasing behaviour of Nigerian millennials. Konfrontasi J Cult Econ Soc Chang 10(2):91–105. https://doi.org/10.33258/konfrontasi2.v10i2.275

Pearson S, Camacho-Villa T, Valluru R, Gaju O, Mini C, Gould I, Sklar E (2022) Robotics and autonomous systems for net zero agriculture. Curr Robot Rep 3(2):57–64. https://doi.org/10.1007/s43154-022-00077-6

Phiri A, Charimbu M, Edewor S, Gaveta E (2022) Sustainable scaling of climate-smart agricultural technologies and practices in sub-Saharan Africa: the case of Kenya, Malawi, and Nigeria. Sustainability 14(22):14709. https://doi.org/10.3390/su142214709

Reay D (2020) Land use and agriculture: pitfalls and precautions on the road to net zero. Front Clim 2. https://doi.org/10.3389/fclim.2020.00004

Roberts J, Bruce T, Monaghan J, Pope T, Leather S, Beacham A (2020) Vertical farming systems bring new considerations for pest and disease management. Ann Appl Biol 176(3):226–232. https://doi.org/10.1111/aab.12587

Rosa L, Gabrielli P (2023) Achieving net-zero emissions in agriculture: a review. Environ Res Lett 18(6):063002. https://doi.org/10.1088/1748-9326/acd5e8

Saha A, Raut R, Yadav V, Majumdar A (2022) Blockchain is changing the outlook of the sustainable food supply chain to achieve net zero? Sustainability 14(24):16916. https://doi.org/10.3390/su142416916

Sakrabani R, Garnett K, Knox J, Rickson R, Pawlett M, Falagán N, Simmons R (2023) Towards net zero in agriculture: future challenges and opportunities for arable, livestock and protected cropping systems in the UK. Outlook Agric 52(2):116–125. https://doi.org/10.1177/00307270231178889

Sampath V, Rangarajan N, Sharanappa C, Deori M, Veeraragavan M, Ghodake B, Kaushal K (2023) Advancing crop improvement through crispr technology in precision agriculture trends-a review. Int J Environ Clim Chang 13(11):4683–4694. https://doi.org/10.9734/ijecc/2023/v13i113647

Singh R, Gupta A, Kandpal V (2024) ESG Excellence in the digital world. Fintech for ESG and the Circular Economy 193–204. https://doi.org/10.1002/9781394234035.ch10

Singh A, Singh R, Katragadda R (2025) Leveraging artificial intelligence for climate neutrality: human-centric mitigation strategies to combat global warming. In: Singh R, Filho WL (eds)

Climate neutrality through smart eco-innovation and environmental sustainability. Climate Change Management. Springer, Cham. https://doi.org/10.1007/978-3-031-83250-5_7

Singh R, Rani S, Filho WL (2025) Climate change and the path to climate neutrality: understanding causes, impacts, and action strategies. In: Singh R, Filho WL (eds) Climate neutrality through smart eco-innovation and environmental sustainability. Climate Change Management. Springer, Cham. https://doi.org/10.1007/978-3-031-83250-5_17

Singh R, Sharma A, Kumar K (2025) Navigating carbon neutrality: strategies for sustainable future and environmental challenges. In: Singh R, Filho WL (eds) Climate neutrality through smart eco-innovation and environmental sustainability. Climate Change Management. Springer, Cham. https://doi.org/10.1007/978-3-031-83250-5_5

Singh, R., Filho, W.L., Lohiya, L. (2025). Fostering Human Factors in Smart Eco-Innovation: A Comparative Study on Sustainable Technology Adoption in Developing Countries. In: Singh, R., Filho, W.L. (eds) Climate Neutrality Through Smart Eco-Innovation and Environmental Sustainability. Climate Change Management. Springer, Cham. https://doi.org/10.1007/978-3-031-83250-5_10

Strong R, Wynn J, Lindner J, Palmer K (2022) Evaluating Brazilian agriculturalists' IoT smart agriculture adoption barriers: understanding stakeholder salience prior to launching an innovation. Sensors 22(18):6833. https://doi.org/10.3390/s22186833

Tresise M, Reed M, Chapman P (2021) Effects of hedgerow enhancement as a net zero strategy on farmland biodiversity: a rapid review. Emerald Open Res 3:23. https://doi.org/10.35241/emeraldopenres.14307.1

Vărzaru A (2024) Assessing agricultural impact on greenhouse gases in the European Union: a climate-smart agriculture perspective. Agronomy 14(4):821. https://doi.org/10.3390/agronomy14040821

Wang W, Mei T (2024) Research on the effect of digital economy development on the carbon emission intensity of agriculture. Sustainability 16(4):1457. https://doi.org/10.3390/su16041457

Wang H, Wu Y, Zhu N (2024) Measurement and spatio-temporal heterogeneity analysis of coupling coordination between development of digital economy and agricultural carbon emission performance. PLoS ONE 19(6):e0305231. https://doi.org/10.1371/journal.pone.0305231

Yadav N, Luthra S, Garg D, Singh S (2024) Development of a framework for blockchain technology-based net zero strategies in the agriculture supply chain: a TISM and grey-dematel approach. Bus Strateg Environ 34(1):1357–1382. https://doi.org/10.1002/bse.4054

Zhu Y, Wang X, Gong Z (2023) Blessing or curse? The impact of digital technologies on carbon efficiency in the agricultural sector of China. Sustainability 15(21):15613. https://doi.org/10.3390/su152115613

Chapter 28
Smart City and Urban Planning for Climate Change Innovation

Abstract The necessity with which the world must confront climate change has forced cities across the world to jump at the urban design challenge with unprecedented creativity of sustainability and smart technology. In detail, this chapter examines the notion of smart cities and how integrated digital solutions can support climate change adaptation and mitigation. It explores approaches towards energy-efficient, carbon–neutral and much more resilient design of urban areas subject to extreme weather events. It also examines the challenges and opportunities associated with deploying such solutions in the urban context, highlighting the importance of collaborative coproduction governance and community participation.

Introduction

As urban areas grow, now comprising over half the global population, cities have played a pivotal role in combating climate change. Cities account for approximately 70% of global greenhouse gas emissions and are uniquely placed to lead the transition to sustainable urban living (Fankhauser et al, 2021). However, sustainable smart cities guided through sensed data and other technologies that holistically improve system efficiency may be part of the solution to both climate change mitigation (via its greenhouse gas emissions) and adaptation (through community resilience and preparedness to climatic changes) (Duan and Kim 2023). Smart urban planning is the use of urban digital infrastructure, including the Internet of Things (IoT) (Apata et al. 2024), building information modeling, and geographic information systems, to monitor and manage urban resources in real time (Li et al. 2024). With climate change being a significant threat to urban environments (Singh et al. 2025a, b, c, d, e), adaptive urban planning is a vital component of it. The growing number of heatwaves and severe weather occurrences can cause disastrous impacts on human health (Singh et al. 2025a, b, c, d, e) and structural safety integrity (Khanh et al. 2023).

Cities around the globe must face these challenges and implement ecosystems-based comprehensive planning frameworks (Seto et al. 2021) that help prioritize

M. 0. Oyenuga et al., *Smart Technologies for Climate Change and Net Zero Policies: Practical Approaches Towards Sustainability*, Climate Change Management,
https://doi.org/10.1007/978-3-031-92221-3_28

climate adaptation and reduction of carbon footprints. This is in line with ensuring urban sustainability worldwide, which is highlighted by the Sustainable Development Goals (SDGs) set by the United Nations in 2015 (Ziozias et al. 2023). In this chapter, we explored some innovative strategies that can be implemented by smart cities to move toward carbon neutrality. By doing so, we highlight essential roles of natural carbon sinks, city green spaces, and smart energy systems in supporting sustainable city resilience (Fan and Wei, 2022). They position the interaction of various stakeholder groups as integral to inclusive and equitable urban environments through community input incorporated into design decision-making processes (Bricker et al. 2024).

The Nexus Between Smart Cities and Climate Change

Climatic preconditions as the most pressing global issues have tied the idea of smart cities with the use of big data, studying the relationship between appropriate infrastructure management, as well as sustainable resource management, and the resilience of smart cities to climatic changes. As urban centers contend with worsening threats from the impacts of climate change including flooding, extreme heat and sea-level rise, the need for frameworks to make them adaptive and resilient grows more urgent. Smart cities employ technology and data analytics to manage resources more effectively and improve public services and urban living.

The firms offer various products and services that should be central to mitigating specific facets of climate change, and therefore they provide for the synergies between technological development and ecological sustainability. Since climate change has become a global threat, there are new methods of urban planning that have emerged, which can lead to increased resilience. For example, combining geographical information systems (GIS) that can help urban planners sense vulnerabilities to influence plans and policies that target the distinct climate threats faced by cities, and even map hotspots (e.g., areas at risk of flooding or heat stress (Musakwa 2017; Landauer et al. 2018). With the use of appropriate data, smart city projects encourage data-driven decision-making processes that enable cities to respond proactively, rather than reactively, to climate hazards, leading to advancements of urban climate policies (Yin et al. 2021; Reckien et al. 2018). In addition, implementing energy saving and energy efficiency measures to reduce these emissions is crucial, and this is where smart technologies step up (Singh et al. 2025a, b, c, d, e). An example of these is smart grids that facilitate the integration of renewable (green) energy source that targets a more efficient energy distribution while helping to reduce dependency on fossil fuels, which helps to lower overall carbon footprints (Бакаева et al. 2021).

Green infrastructure and urban densification strategies strengthened via the deployment of green management (Oyetunde et al. 2023) and digital technologies enabled eco–sensitive solutions (Singh et al. 2024) encourage carbon sequestration, as well as enhanced living experiences of the people (Noy 2022). The digitisation of urbanised areas transitions amongst smart and environmentally affirming cities come with their difficulties (Oyedele 2025; Singh et al. 2025a, b, c, d, e). The successful implementation of these projects will necessitate municipalities adopting adaptive governance frameworks that allow for the effective collaboration of different stakeholders, including government agencies, private sector partners, and community organizations (Reckien et al. 2018; Aylett 2015; Oyenuga and Jeresa 2025). Urban living laboratories involve processes of experimentation and collective learning that give cities the chance to co-create and test new ways of adapting to climate change at a large scale, which are inclusive (Prestwood et al. 2018). A key factor with the nexus between smart cities and climate change is social equity. Policymakers must work with the potential for technological developments to compound existing inequities through inclusive responses that support all urban stakeholders, and particularly the vulnerable communities most severely impacted by climate stressors (Yin, 2024; Pietrapertosa et al. 2019). This entails embedding a social dimension into climate interventions in cities and interventions to urban climate issues so climate resilience itself becomes not just an outcome, but a shared outcome (Oyenuga and Omale 2024).

Technological Innovations for Smart and Sustainable Cities

Smart cities with digital technologies and data-based management have emerged as the next evolutionary phase of urban development to counter climate change and drive sustainability. Smart cities use the latest advancements in information and communication technologies (ICT) approach to improve urban services, quality of life, and economic growth, while reducing environmental impacts. Innovation-driven solutions (Fasanmi et al. 2025) that would enable sustainable city growth are among the key pillars of smart city development. The Internet of Things (IoT) is one of the examples that allows real-time data collection and analysis to help cities to optimize the use of the resources and minimize energy consumption (Bibri and Krogstie 2020a, b; Singh et al. 2025a, b, c, d, e). An example of this is enrolling smart grid technologies that allow better management of the electricity supply and demand, resulting in lower greenhouse gas emissions (Parks 2018). In addition, big data analytics enable smart transportation systems to reduce traffic congestion, leading to lower vehicular emissions and better urban air quality (Kirimtat et al. 2020; Apata et al. 2024).

Citizen participation is also a core element of smart city projects. This system of local governance promotes more democratic urban governance practices, allowing for localised, needs-based, and adaptive solutions to the unique challenges faced by each community. With citizens treated as active participants, the public becomes more trusting in the process, while the practice of sustainability can be embedded in practice, with citizens being co-creators of their city's future (Fernández and Peek 2020). This manifests itself in practice in community-led civic tech, initiatives that help residents engage in local governance and community sustainability. And while there are always positive aspects to smart cities, the technological change either is widening existing fault lines or creating them anew. One criticism against digital solutions is that they may not address the complex social-spatial dynamics of urban areas.

As cities pedal toward smart growth, they must also focus on equity so that innovations work for all their residents, especially marginalized communities. Adenekan et al. (2024) opined that governance strategies play a vital role in addressing the social side of smart city initiatives and that the environmental impact of smart city technologies deserves investigation. Smart solutions, though, tend to be a part of the growing number of issues creating unsustainable urban environments, such as e-waste production and data centre energy consumption (Kuntsman 2020), sustainability in smart city policies should not be limited to how cities implement smart technologies but rather integrated into urban governance as comprehensive urban strategies for sustainable urban development (Azizalrahman and Hasyimi 2020).

Sustainable Urban Planning Strategies

Cities around the world are facing a myriad of challenges, such as climate change, population growth, urban sprawl and resource depletion, all of which require sustainable urban planning strategies. They are designed to produce cities that serve present needs at the same time as ensuring that future generations can also meet their needs. Through the integration of technology and data, stakeholder engagement, and resilient design principles, urbanists can create cities that promote social equity, economic prosperity, and environmental sustainability.

In Sustainable urban planning, one of the fundamental stones is the application of green technologies and clean energy solutions in urban design (Korol et al. 2018; Bollano 2024). This includes, but is not limited to, the application of renewable energy, like solar and wind energy, in conjunction with energy-efficient infrastructure. Using sustainable construction materials that can decrease the ecological footprint of the building itself is also a given (Dewasiri and Singh 2025). Smart buildings Systems related technology presents high technologies in architecture that support energy saving and low waste (Korol et al. 2018). Those engaged in eco-development principles, be it as architects or urban planners, create spaces that cultivate the health and well-being of their inhabitants and bridge a relationship with the environment. Smart cities use data-driven decision-making and technology to optimize urban management and resource allocation, which facilitates efficiency (Bibri and Krogstie 2020a, b). Smart mobility solutions that aim to analyze real-time data from citizens, vehicles, public transportation services, etc., to alleviate traffic congestion and emissions are one such example of smart city applications that contribute to better air quality (Yiğitcanlar 2020).

In addition, Internet of Things technologies play a key role in real-time monitoring and maintenance of urban infrastructure, helping cities to promptly respond to changing conditions (Almalki et al. 2021). Sustainable urban foot is not only about participatory governance and conscious design that address local needs, but also about community engagement. In due consideration of the role of residents in the decision-making process, urban development initiatives must be informed by the needs and aspirations of community residents (Agbali et al. 2019; Adenekan et al. 2024). This hands-on approach nurtures stewardship and accountability at the local level, leading to the integration of sustainable behavior within communities. Such participatory processes can take multiple forms (e.g., public consultations, participatory budgeting) and can lead to more egalitarian parameters of urban space (Choi and Song 2022). Urban planning also struggles, especially with equity and access to green spaces and resources. Socioeconomic processes can exacerbate disparities in the availability of essential services and amenities between urban areas (Colding et al. 2019).

Getting over these inequities takes policies aimed to help disenfranchised communities level the playing field, durability to avoid disparities in the climate justice transition, and limit the extent to which green initiatives disproportionately affect those involved in the earlier transition towards sustainability. An approach like this could be to also provide affordable housing opportunities, equitable access to green places, and community-oriented public transportation solutions (Sabry 2022; Alhosani et al. 2021). Additionally, the necessity of robust and resilient infrastructure becomes progressively pertinent as cities are faced with the multi-dimensional repercussions of climate change. Therefore, this incorporation of climate adaptation measures leads to the sustainability of urban setups and adding of the resilience of built environments against climate variability (Barbala et al. 2023). Proposed strategies would include the development of urban green spaces designed not simply for beautification but to act as natural flood barriers, as well as, increasing the number of green spaces would enhance urban biodiversity and tourism (oyenuga and Akwen 2025) or putting in place climate sensitive building codes that would mitigate risks from heat waves and other extreme weather events (Romanelli 2022).

Climate Adaptation and Mitigation Policies

Climate change poses a formidable challenge to urban areas and demands unique adaptations as well as mitigative actions in policy-making that enable sustainable development. With cities aiming to build resilience against such climate-related events, all while reducing greenhouse gas emissions, effective climate adaptation and mitigation policies will ultimately be paramount. Here we explore some strategies and approaches, principles and diverse examples of climate adaptation and mitigation policies that cities can adopt in pursuit of these twin goals. A key principle in climate adaptation policy is to recognise local vulnerabilities and to tailor solutions that enhance resilience to specific climate threats. An example of a pioneer actor city that is realizing the promise of adaptive urbanism by investing in new strategies such as floating neighborhoods and green infrastructure to control stormwater runoff and reduce flooding is Rotterdam, Netherlands (Esenarro et al. 2024). Such measures are types of adaptive responses that exemplify the success of combining physical resilience measures with urban planning to mitigate climate hazards. Policy makers must learn from past lessons and design concrete plans consisting of methods to minimize disruption to pandemics and other disruptive occurrences including natural disasters (Sharifi and Khavarian-Garmsir 2020).

Urban planners will better optimize their responses to uncertain climate conditions if they employ multidimensional approaches based on historic data as well as future predictions. Additionally, integrating smart technologies into urban policy frameworks allows for better climate mitigation and adaptation actions. The application of artificial intelligence and big data analytics, for instance, supports the near-instantaneous registration of conditions within the urban surroundings, during which the data from a specific area can be collected and presented to assist with real-time decision-making on energy use and resource use planning (Obracht-Prondznyska et al. 2022). This allows city management to better allocate resources and monitor the impact of sustainability efforts. Policies that combine climate peace and mitigation are especially helpful because they create positive interactions.

Furthermore, urban centers are progressively pursuing sustainable energy solutions in response to climate action plans through the adoption of sustainability frameworks. As such, transforming urban energy systems toward renewable sources will be crucial in decreasing fossil fuel dependence and emissions on the urban level. Policies aimed at energy efficiency and renewable energy technologies can be especially beneficial for climate change mitigation as they have been in several European countries (Maestosi et al. 2021; Nik et al. 2020). These paradigm shifts in urban districts toward energy positive forms not only boost independence towards energy sources but also represent a role model in effective climate action. Adaptation strategies need to engage the community and all stakeholders during the planning and implementation phases. Adopting more inclusive policies that foster citizen participation can enhance acceptance and effectiveness of climate initiatives across different demographic groups, allowing cities to enhance climate resilience (Huang-Lachmann 2019).

Case Studies of Smart Cities Driving Climate Innovation

This subsection discusses remarkable examples of smart cities that are successfully adopting climate innovations in their governance and urban planning practices, thus highlighting the promise of technology-based initiatives to support urban resilience and environmental stewardship.

San Diego, California

One city that serves as a key example of urban sustainability and citizen engagement through smart innovations is San Diego. No doubt, a lot of sensor-based solutions and Internet of Things (IoT) technologies have been adopted by the city. San Diego seeks to manage traffic efficiently and conserve energy using intelligent traffic control systems and real-time monitoring of environmental conditions (Agbali et al. 2019). The use of these technologies proved critical in enabling citizens to engage more actively in urban services and sustainability initiatives (Agbali et al. 2019).

Vienna, Austria

Vienna is an example of a city that balances smart city strategies and climate adaptation, for functional returns. The Smart City Wien Framework is a common initiative that brings together diverse sustainable development initiatives, such as energy efficiency or sustainable transportation systems (Fernández and Peek 2020). In bringing previously siloed initiatives under one umbrella strategy, Vienna has improved management of interlinked urban services, helped intersectoral collaboration. It embraces the integrative nature of urban systems in a multifaceted approach towards climate adaptation (Fernández and Peek 2020).

Stockholm and Malmö, Sweden

In Sweden, the Royal Seaport of Stockholm and the Western Harbor of Malmö are two pioneering eco-city development strategies that bring together notions of the smart city and sustainability in successful ways. Both places focus on renewable energy, green transport systems, and energy-efficient building planning, to develop low-carbon cities (Bibri and Krogstie 2020a, b). Utilizing technologies such as big data and smart infrastructure, these cities are revolutionizing resource management and creating opportunities for citizens to engage in sustainability programs, among others. Sustainability and smart agenda dualization is underlined by the urbanization opportunities to be pursued in line with the notion of eco-cities (Bibri and Krogstie 2020a, b).

Yokohama, Japan

One of the best examples of exploratory energy innovation programs that systematically tackle climate change is the Yokohama Smart City Project. In this regard, the reactive potentiality of urban elements to the transformation of static structures and the utilization of fluid urban materialities to measure for urban business sustainability (Suwa 2019) persists as the city becomes reacquainted with its form through the evolution of local characteristics to suit energy-conscious practices. Yokohama is reducing greenhouse gas emissions and enhancing the lives of residents through the UX of smart grids and innovative energy management systems (Suwa 2019). The case exemplifies the key role played by local governments in implementing systemic changes toward urban development aligned with environmental goals.

Dubai, UAE

Dubai is an example in smart tourism and sustainable practices for a city. For example, Dubai embraces the growing concept of the smart city and uses technology such as artificial intelligence and data analytics to improve the quality of services in tourism, urban mobility, and civic engagement (Khan et al. 2017). Sustainability in tourism examples indicate a solution for some human needs that index expression in smart technology and how this trend promotes environmental sustainability. Dubai is a prime example of how urban areas can utilize climate innovation to accommodate their residents and tourists by combining smart tourism and sustainability frameworks (Khan et al. 2017).

Alexandria, Egypt

Principles of sustainable urban development are incorporated in the framework for smart eco-cities in Alexandria) The literature on smart eco-cities mainly focuses on enhancing the interconnectedness of smart solutions and ecological approaches (Tarek 2023). In Alexandria, this effort is about finding an equilibrium between technology and environmental sustainability, providing other developing countries a practical reality of how smart cities can redefine, promote and support local solutions to local problems of urban resilience (Tarek 2023).

Implications

Smart technologies integrated in urban planning are a multiple funnel for driving climate innovation and urban sustainability. The case studies discussed above lead to several important implications for urban governance, policy making and community engagement.

Joint solutions for smart cities show that technology can be applied to strengthen the resilience of cities to climate change effects. Real-time monitoring systems are just a few of the innovations that allow cities to address environmental challenges, decrease vulnerabilities and respond to environmental challenges such as flooding extremes (Agbali et al. 2019; Fernández and Peek 2020). Also, big data analysis and IoT technologies provide informed insights that enable optimized resource allocation across different sectors. We can create evidence-based strategies that optimize efficiency and reduce emissions, confirming cities' pledges in sustainability (Bibri and Krogstie 2020a, b; Suwa 2019). Meanwhile, the success of any smart city initiative relies on collaboration among various stakeholders from government agencies and private sector partners to residents. Involvement of the community in designing and implementing climate policies instills a sense of ownership and promotes sustainable behaviour among citizens (Khan et al. 2017). Recognizing and boosting communities and people offers ground-breaking alternatives that fit local green settings.

Also, the intersection of climate adaptation and mitigation in smart cities demonstrates the importance of cohesive policy integration. As more cities pursue multi-pronged strategies simultaneously to address resilience and emissions, they are making comprehensive urban governance models that will address the complexities of climate change (Tarek 2023). Therefore, the dialogue around climate innovation must be balanced with a commitment to social equity. Similarly, as smart technologies are adopted in cities, questions about equity and access emerge: Will these innovations be equitably shared with marginalized communities? Strategies cannot be written in a way that perpetuates current inequalities whilst ensuring opportunities for all to participate in climate action (Khan et al. 2017).

Conclusion, Limitations, and Future Scope

Smart cities proactively stimulating climate innovation demonstrate that urban areas can harness technology to meet climate change challenges. However, by increasing resilience, driving evidence-based decisions, enriching multi-actor cooperation and ensuring equal access to smart answers, cities can enable themselves to be front runners in sustainability and the climate fight. What these examples illustrate is the degree to which integrated policies across sectors and communities can inform climate-sensitive alternative land use options for mayors and city governments, leading to climate-resilient urban areas that also provide means of mitigation. As climate breakdown looms, cities have to adapt to changing times because innovative ideas combined with collaborative efforts will save a greener tomorrow. Thus, smart cities can be an example of how to use technology not just simply as an instrument of efficiency but as an indispensable lever in the broader quest towards sustainability and climate change resilience.

References

Adenekan O, Ezeigweneme C, Chukwurah E (2024) The evolution of smart cities: integrating technology, governance, and sustainable development. Int J Appl Res Soc Sci 6(5):891–902. https://doi.org/10.51594/ijarss.v6i5.1131

Agbali M, Trillo C, Ibrahim IA, Arayıcı Y, Fernando T (2019) Are smart innovation ecosystems really seeking to meet citizens' needs? Insights from the stakeholders' vision on smart city strategy implementation. Smart Cities 2(2):307–327. https://doi.org/10.3390/smartcities2020019

Alhosani K, Kaied Y, Darwish A (2021) Ajman, an environmentally friendly city with its quality of life: review of sustainability challenges and achievements by Ajman Municipality and planning department. Renew Energy Environ Sustain 6:12. https://doi.org/10.1051/rees/2021009

Almalki F, Alsamhi S, Sahal R, Hassan J, Hawbani A, Rajput N, Breslin J (2021) Green IoT for eco-friendly and sustainable smart cities: future directions and opportunities. Mobile Netw Appl 28(1):178–202. https://doi.org/10.1007/s11036-021-01790-w
Apata AO, Oyenuga MO, Abeoba M, Ugom A (2024) Internet of Things (IoT) solutions for smart transportation infrastructure and fleet management. J Propul Technol 45(4):1492–1505. https://propulsiontechjournal.com/index.php/journal/article/view/8325
Aylett A (2015) Institutionalizing the urban governance of climate change adaptation: results of an international survey. Urban Climate 14:4–16. https://doi.org/10.1016/j.uclim.2015.06.005
Azizalrahman H, Hasyimi V (2020). Towards a generic framework for smart cities. https://doi.org/10.5772/intechopen.85820
Barbala A, Sporsem T, Stol K (2023) Building climate resilience in smart cities using open data services. https://doi.org/10.24251/hicss.2023.248
Bibri S, Krogstie J (2020) The emerging data–driven smart city and its innovative applied solutions for sustainability: the cases of London and Barcelona. Energy Informatics 3(1). https://doi.org/10.1186/s42162-020-00108-6
Bibri SE, Krogstie J (2020b) Smart eco-city strategies and solutions for sustainability: the cases of Royal Seaport, Stockholm, and Western Harbor, malmö, Sweden. Urban Sci 4(1):11. https://doi.org/10.3390/urbansci4010011
Bollano S (2024) The study of "cities of the future" urban planning and development decision-making experience. Architectural Stud 10(1):47–57. https://doi.org/10.56318/as/1.2024.47
Bricker S, Jelének J, Keur P, Vigna F, O'Connor S, Ryżyński G, Venvik G (2024) Geoscience for cities: delivering Europe's sustainable urban future. Sustainability 16(6):2559. https://doi.org/10.3390/su16062559
Bulkeley H, Stripple J (2020) Climate smart city: new cultural political economies in the making in malmö, Sweden. New Polit Econ 26(6):937–950. https://doi.org/10.1080/13563467.2020.1810219
Choi H, Song S (2022) Direction for a transition toward smart sustainable cities based on the diagnosis of smart city plans. Smart Cities 6(1):156–178. https://doi.org/10.3390/smartcities6010009
Colding J, Barthel S, Sörqvist P (2019) Wicked problems of smart cities. Smart Cities 2(4):512–521. https://doi.org/10.3390/smartcities2040031
Dewasiri NJ, Singh R (2025) Advancing climate neutrality through sustainable production and consumption: pathways to a resilient future. In: Singh R, Filho WL (eds) Climate neutrality through smart eco-innovation and environmental sustainability. Climate Change Management. Springer, Cham. https://doi.org/10.1007/978-3-031-83250-5_1
Duan Z, Kim S (2023) Progress in research on net-zero-carbon cities: a literature review and knowledge framework. Energies 16(17):6279. https://doi.org/10.3390/en16176279
Esenarro D, Manosalva J, Sanchez L, Vargas C, Raymundo V, Ccalla J (2024) Resilient urban-design strategies for landslide risk mitigation in Huaraz, Peru. Urban Sci 8(3):90. https://doi.org/10.3390/urbansci8030090
Fan Y, Wei F (2022) Contributions of natural carbon sink capacity and carbon neutrality in the context of net-zero carbon cities: a case study of Hangzhou. Sustainability 14(5):2680. https://doi.org/10.3390/su14052680
Fankhauser S, Smith S, Allen M, Axelsson K, Hale T, Hepburn C, Wetzer T (2021) The meaning of net zero and how to get it right. Nat Clim Chang 12(1):15–21. https://doi.org/10.1038/s41558-021-01245-w
Fasanmi OO, Adeniyi BC, Oyenuga MO (2025) Exploring the role of innovation on economic performance of manufacturing firm in Nigeria. J Econom Finance Manage Stud 8(1):694–706. https://doi.org/10.47191/jefms/v8-i1-67
Fernández CG, Peek D (2020) Smart and sustainable? Positioning adaptation to climate change in the European smart city. Smart Cities 3(2):511–526. https://doi.org/10.3390/smartcities3020027
Huang-Lachmann J (2019) Systematic review of smart cities and climate change adaptation. Sustain Account Manag Policy J 10(4):745–772. https://doi.org/10.1108/sampj-03-2018-0052

Khan MS, Woo M, Nam K, Chathoth PK (2017) Smart city and smart tourism: a case of Dubai. Sustainability 9(12):2279. https://doi.org/10.3390/su9122279

Khanh D, Varquez A, Kanda M (2023) Impact of urbanization on exposure to extreme warming in megacities. Heliyon 9(4):e15511. https://doi.org/10.1016/j.heliyon.2023.e15511

Kirimtat A, Krejcar O, Kertész A, Taşgetiren M (2020) Future trends and current state of smart city concepts: a survey. IEEE Access 8:86448–86467. https://doi.org/10.1109/access.2020.2992441

Korol O, Shushunova N, Lopatkin D, Zanin A, Шушунова T (2018) Application of high-tech solutions in ecodevelopment. Matec Web Conf 251:06002. https://doi.org/10.1051/matecconf/201825106002

Kuntsman A (2020) Smart cities' environmental dreams and their dirty material politics. Aoir Sel Pap Internet Res. https://doi.org/10.5210/spir.v2020i0.11254

Landauer M, Juhola S, Klein J (2018) The role of scale in integrating climate change adaptation and mitigation in cities. J Environ Planning Manage 62(5):741–765. https://doi.org/10.1080/09640568.2018.1430022

Li J, Shirowzhan S, Pignatta G, Sepasgozar S (2024) Data-driven net-zero carbon monitoring: applications of geographic information systems, building information modelling, remote sensing, and artificial intelligence for sustainable and resilient cities. Sustainability 16(15):6285. https://doi.org/10.3390/su16156285

Maestosi P, Andreucci M, Civiero P (2021) Sustainable urban areas for 2030 in a post-covid-19 scenario: focus on innovative research and funding frameworks to boost transition towards 100 positive energy districts and 100 climate-neutral cities. Energies 14(1):216. https://doi.org/10.3390/en14010216

Musakwa W (2017) Perspectives on geospatial information science education: an example of urban planners in southern Africa. Geo-Spat Inf Sci 20(2):201–208. https://doi.org/10.1080/10095020.2017.1339403

Nik V, Perera A, Chen D (2020) Towards climate resilient urban energy systems: a review. Nat Sci Rev 8(3). https://doi.org/10.1093/nsr/nwaa134

Noy B (2022) The role of zoning policies on sustainable development and climate change. https://doi.org/10.33774/apsa-2022-hrccp

Obracht-Prondzyńska H, Duda E, Anacka H, Kowal J (2022) Greencoin as an AI-based solution shaping climate awareness. Int J Environ Res Public Health 19(18):11183. https://doi.org/10.3390/ijerph191811183

Oyedele OM (2025) The cobra effect: exploring the dark side of innovation and environmental sustainability in developing nations. In: Singh R, Joshi A, Filho WL, Khan S (eds) Zero carbon industry, eco-innovation and environmental sustainability. World Sustainability Series. Springer, Cham. https://doi.org/10.1007/978-3-031-80220-1_15

Oyenuga MO, Akwen PI (2025) Tourism and intercultural understanding in developing nations. Afr J Cult, Hist, Relig Tradit 8(1):1–14. https://doi.org/10.52589/ajchrt-okuxnmsq

Oyenuga MO, Jeresa S (2025) Adaptation and resilience: private, public, and individual change makers. In: Singh R, Filho WL (eds) Climate neutrality through smart eco-innovation and environmental sustainability. Climate Change Management. Springer, Cham. https://doi.org/10.1007/978-3-031-83250-5_2

Oyenuga MO, Omale SA (2024) Is Africa jinxed? Exploring the challenges of technology access and adoption in Africa. Afr J Econ Sustain Dev 7(4):142–161. https://doi.org/10.52589/AJESDULN1LRNF

Oyetunde TO, Oyenuga MO, Adoga GJ (2023) Green management practices and organisational efficiency in higher institutions. Bp Int Res CritS Linguist Educ (BirLE) J 6(3):276–290. https://bircu-journal.com/index.php/birle/article/view/7716

Parks D (2018) Energy efficiency left behind? policy assemblages in Sweden's most climate-smart city. Eur Plan Stud 27(2):318–335. https://doi.org/10.1080/09654313.2018.1455807

Pietrapertosa F, Salvia M, Hurtado S, D'Alonzo V, Church J, Geneletti D, Reckien D (2019) Urban climate change mitigation and adaptation planning: are Italian cities ready? Cities 91:93–105. https://doi.org/10.1016/j.cities.2018.11.009

Prestwood E, Longhurst J, Townsend I, Haines T, Tsiarapa E (2018). Facilitating stakeholder dialogues on a carbon neutral city: we need to talk about carbon (and Air Quality). https://doi.org/10.2495/air180461

Reckien D, Salvia M, Heidrich O, Church J, Pietrapertosa F, Gregorio-Hurtado S, Dawson R (2018) How are cities planning to respond to climate change? assessment of local climate plans from 885 cities in the EU-28. J Clean Prod 191:207–219. https://doi.org/10.1016/j.jclepro.2018.03.220

Romanelli M (2022) Smart cities and intellectual capital for urban innovation and the future. Eur Conf Knowl Manag 23(2):1010–1016. https://doi.org/10.34190/eckm.23.2.767

Sabry G (2022) Describing the relationship between smart cities and sustainability by focusing on the characteristics of smart cities to help progress that is more effective. 28(1):235–261. https://doi.org/10.21608/jsfc.2022.299829

Seto K, Churkina G, Hsu A, Keller M, Newman P, Qin B, Ramaswami A (2021) From low- to net-zero carbon cities: the next global agenda. Annu Rev Environ Resour 46(1):377–415. https://doi.org/10.1146/annurev-environ-050120-113117

Sharifi A, Khavarian-Garmsir A (2020) The COVID-19 pandemic: impacts on cities and major lessons for urban planning, design, and management. Sci Total Environ 749:142391. https://doi.org/10.1016/j.scitotenv.2020.142391

Singh R, Gupta A, Kandpal V (2024). ESG Excellence in the digital world. Fintech for ESG and the Circular Economy 193–204. https://doi.org/10.1002/9781394234035.ch10

Singh A, Singh R, Katragadda R (2025) Leveraging artificial intelligence for climate neutrality: human-centric mitigation strategies to combat global warming. In: Singh R, Filho WL (eds) Climate neutrality through smart eco-innovation and environmental sustainability. Climate Change Management. Springer, Cham. https://doi.org/10.1007/978-3-031-83250-5_7

Singh R, Joshi A, Rani S (2025) Transforming sustainable business models for manufacturing industry to build zero carbon industry. In: Singh R, Joshi A, Filho WL, Khan S (eds) Zero carbon industry, eco-innovation and environmental sustainability. World Sustainability Series. Springer, Cham. https://doi.org/10.1007/978-3-031-80220-1_1

Singh R, Sharma A, Kumar K (2025) Navigating carbon neutrality: strategies for sustainable future and environmental challenges. In: Singh R, Filho WL (eds) Climate neutrality through smart eco-innovation and environmental sustainability. Climate Change Management. Springer, Cham. https://doi.org/10.1007/978-3-031-83250-5_5

Singh R, Rani S, Filho WL (2025) Climate change and the path to climate neutrality: understanding causes, impacts, and action strategies. In: Singh R, Filho WL (eds) Climate neutrality through smart eco-innovation and environmental sustainability. Climate Change Management. Springer, Cham. https://doi.org/10.1007/978-3-031-83250-5_17

Singh R, Filho WL, Lohiya L (2025) Fostering human factors in smart eco-innovation: a comparative study on sustainable technology adoption in developing countries. In: Singh R, Filho WL (eds) Climate neutrality through smart eco-innovation and environmental sustainability. Climate Change Management. Springer, Cham. https://doi.org/10.1007/978-3-031-83250-5_10

Suwa A (2019) Local government and technological innovation: lessons from a case study of "Yokohama Smart City Project". Smart Environment Smart Cities 387–403. https://doi.org/10.1007/978-981-13-6822-6_10

Tarek S (2023) Smart eco-cities conceptual framework to achieve the UN SDGs: a case study application in Egypt. Civil Eng Architecture 11(3):1383–1406. https://doi.org/10.13189/cea.2023.110322

Yiğitcanlar T (2020) Smart city beyond efficiency: technology–policy–community at play for sustainable urban futures. Hous Policy Debate 31(1):88–92. https://doi.org/10.1080/10511482.2020.1846885

Yin C, Xiao J, Zhang T (2021) Effectiveness of Chinese regulatory planning in mitigating and adapting to climate change: comparative analysis based on q methodology. Sustainability 13(17):9701. https://doi.org/10.3390/su13179701

Yin Z (2024) Sustainable urban planning and climate change adaptation: disaster protection strategies for global urban development. Lecture Notes Education Psychol Public Media 46(1):60–65. https://doi.org/10.54254/2753-7048/46/20230587

Ziozias C, Kontogianni E, Ανθόπουλος Λ (2023) Carbon-neutral city transformation with digitization: guidelines from international standardization. Energies 16(15):5814. https://doi.org/10.3390/en16155814

Бакаева H, Suvorova M, Sheps R, Kormina A (2021) Adaptation techniques of urban planning due to climate change. E3s Web Conf 263:05013. https://doi.org/10.1051/e3sconf/202126305013

MIX
Papier aus verantwortungsvollen Quellen
Paper from responsible sources
FSC® C105338
FSC
www.fsc.org

If you have any concerns about our products,
you can contact us on
ProductSafety@springernature.com

In case Publisher is established outside the EU,
the EU authorized representative is:
Springer Nature Customer Service Center GmbH
Europaplatz 3, 69115 Heidelberg, Germany

Printed by Libri Plureos GmbH
in Hamburg, Germany